Erratum

Quantum Fields by N. N. Bogoliubov and D. V. Shirkov

On the back cover of this book, the bibliographical information concerning Professor D. V. Shirkov incorrectly states that he is deceased. He is alive and well. Our deepest apologies. Corrections will appear on our next printing of this book.

The Benjamin/Cummings Publishing Company, Inc.
Advanced Book Program
Reading, Massachusetts

January 1983

Quantum Fields

Quantum Fields

N. N. Bogoliubov
D. V. Shirkov
Joint Institute for Nuclear Research
Dubna, U.S.S.R.

Authorized translation from the Russian edition by
D. B. Pontecorvo

1983

The Benjamin/Cummings Publishing Company, Inc.
Advanced Book Program
Reading, Massachusetts

London · Amsterdam · Don Mills, Ontario · Sydney · Tokyo

N. N. Bogoliubov and D. V. Shirkov, Quantum Fields

First English Edition, 1982
Originally published in 1980 as Квантовые поля

by Izdatel'stov "Nauka," Moscow

Library of Congress Cataloging in Publication Data

Bogoliubov, N. N. (Nikolaĭ Nikolaevich), 1909–
Quantum fields.

Translation of: Kvantovye polia.
Bibliography: p.
Includes index.
1. Quantum field theory. I. Shirkov, D. V.
(Dmitriĭ Vasil'evich) II. Title.
QC174.45.B5813 1982 530.1'43 82-4366
ISBN 0-8053-0983-7

Printed in the United States of America

ABCDEFGHIJ-HA-898765432

CONTENTS

PREFACE

The main purpose of this book is to provide graduate students in physics with the necessary minimum of information on the fundamentals of modern quantum field theory.

It may turn out to be sufficient both for theoreticians, specializing in nuclear physics, quantum statistics and other fields, in which quantum field methods are utilized and which are based on quantum concepts, and also for experimental physicists in the fields of nuclear and high-energy physics. For the latter category of readers the present book should be supplemented by a course on particle physics and particle interactions. At the same time the book can be recommended as an introductury text for persons intending to work in the field of quantum field theory and of the theory of elementary interactions.

The material in this book corresponds to a course lasting one academic year. Our personal experience testifies that parallel practical studies at seminars are extremely desirable. For this purpose part of the technical material has been assembled at the end of the book in the form of Appendices. There, also, sets of exercises and problems, gathered together as assignments corresponding to chapters of the main text, are given.

The authors are grateful to the editor of this book D. A. Slavnov, to the reviewers M. A. Brown, L. V. Prokhorov, K. A. Ter-Martirosyan, and also to B. M. Barbashov, B. V. Medvediev, and N. M. Shumeiko for valuable comments on the typescript of the book.

N. N. Bogoliubov
D. V. Shirkov

PREFACE TO THE ENGLISH-LANGUAGE EDITION

This book is a text on the fundamentals of quantum field theory and renormalized perturbation theory (RPT). The traditional field of application of the latter for a long time was limited to quantum electrodynamics. During recent years, due to the creation of a unified theory of electroweak interactions and to the successes of quantum chromodynamics it has become clear that the physical scope of RPT is much wider. However, in the study of the quark–gluon interaction, as well as of possible mechanisms of the grand unification of interactions, a decisive part is played by the simultaneous use of results of RPT and of the apparatus of the renormalization-group method.

Therefore we have written a special Appendix IX, "The renormalization group," for the English-language edition of our book. Besides this, small editorial changes and corrections of noticed misprints have been made.

N. N. Bogoliubov
D. V. Shirkov

Quantum Fields

SOME BASIC CONCEPTS AND NOTATION

We shall take all components of four-vectors to be real. The four-vector $a = (a^0, \boldsymbol{a})$ consisting of the zero component a^0 and the space vector $\boldsymbol{a}$ will, according to tradition, be called a contravariant 4-vector, and we shall denote its components by upper indices

$$a^\nu = (a^0,\ a^1,\ a^2,\ a^3).$$

We define the product of two vectors a and b as

$$ab = a^0 b^0 - \boldsymbol{a}\boldsymbol{b} = a^0 b^0 - a^1 b^1 - a^2 b^2 - a^3 b^3.$$

It is conveniently written in the form

$$ab = \sum_{\mu,\ \nu} g_{\mu\nu} a^\mu b^\nu \qquad (\mu,\ \nu = 0,\ 1,\ 2,\ 3),$$

where $g_{\mu\nu}$ is a diagonal metric tensor

$$g_{\mu\nu} = 0 \ \text{ when } \mu \neq \nu, \quad g_{00} = 1; \quad g_{11} = g_{22} = g_{33} = -1,$$

differing in sign from the well-known Minkowski tensor. The transition from contravariant components a^ν to covariant ones a_μ (i.e. lowering of indices) is accomplished with the aid of the metric tensor:

$$a_\mu = g_{\mu\nu} a^\nu, \qquad a^\mu = g^{\mu\nu} a_\nu, \qquad g_{\mu\nu} = g^{\mu\nu},$$

that is,

$$a_0 = a^0, \qquad a_k = -a^k \quad (k = 1,\ 2,\ 3), \qquad a_\nu = (a^0,\ -a).$$

Here and below, summation over twice repeated indices is implied in all cases, and we will omit the summation sign. Indices representing summation over the three space components are denoted by Latin letters, taken from the middle of the alphabet:

$$ab = a_n b_n = a^k b^k = a_1 b_1 + a_2 b_2 + a_3 b_3,$$

while indices representing summation over all four components (0, 1, 2, 3) are denoted by Greek letters:

$$ab = a^\nu b_\nu = a_\mu b^\mu = \sum_{\mu,\,\nu} g^{\mu\nu} a_\mu b_\nu .$$

Occasionally, in order to simplify the form of a cumbersome expression, we shall lower or raise both Greek indices, that is

$$A_\nu B_\nu = A^\mu B^\mu \equiv AB = A_\nu B^\nu,$$
$$F_{\mu\nu} F_{\mu\nu} \equiv F_{\mu\nu} F^{\mu\nu} = \sum_{\mu,\,\nu} F_{\mu\nu} F^{\mu\nu}$$

Thus, the presence of two identical Greek indices in different factors *always* implies covariant summation independently of the positions of the indices.

Indices associated with internal symmetry groups (for example, isotopic indices) are denoted, as a rule, by latin letters taken from the beginning of the alphabet (a, b, . . .).

The symbol $\hat{a}$ stands for the contraction of the four-vector components a_ν with the Dirac matrices

$$\hat{a} = a^\nu \gamma_\nu = a_\nu \gamma^\nu .$$

For derivatives with respect to covariant and contravariant components the following abbreviated notation is often used:

$$\frac{\partial u}{\partial x_\nu} = \partial^\nu u = u^{;\nu}, \qquad \frac{\partial \varphi_a}{\partial x^\nu} = \partial_\nu \varphi_a = \varphi_{a;\,\nu} .$$

Naturally, we have

$$\varphi_a^{;\mu} = g^{\mu\nu} \varphi_{a;\,\nu} .$$

We represent the D'Alembert operator

$$\square = \Delta - \partial_0^2$$

in the form

$$\square = - \partial^\nu \partial_\nu .$$

Throughout the book the so-called *rational system of units* is used, in which the velocity of light and Planck's constant are taken equal to unity, i.e., $c = \hbar = 1$. In this system energy and momentum have the dimension of mass, or of a reciprocal length, and the time $x_0 = t$ has the dimension of length

$$[E] = [p] = m = l^{-1}, \qquad [x_0] = [x] = l = m^{-1}.$$

As a rule, the formulae for the four-dimensional Fourier transform are written in the form

$$f(x) \sim \int e^{-ipx} \tilde{f}(p)\, dp, \qquad \tilde{f}(p) \sim \int e^{ipx} f(x)\, dx.$$

The sign of the exponent is chosen to be such that the time component corresponds to the quantum-mechanical expression

$$f(x^0, \mathbf{x}) = f(t, \mathbf{x}) \sim \int e^{-iEt} \tilde{f}(E, \mathbf{x})\, dE.$$

The three-dimensional Fourier transform correspondingly is

$$\varphi(\mathbf{x}) \sim \int e^{i\mathbf{px}} \tilde{\varphi}(\mathbf{p})\, d\mathbf{p}, \qquad \tilde{\varphi}(\mathbf{p}) \sim \int e^{-i\mathbf{px}} \varphi(\mathbf{x}).\, d\mathbf{x}.$$

An exception will be made for the positive-frequency parts of field functions and the positive-frequency parts of the Green's functions. The normalizing factors of the Fourier transforms (powers of 2π) are chosen in different ways in various parts of this book.

Within each section single numbering of formulae is used: (1), (2), (3), . . . , which is used directly for references to formulae within the given section. Double numbering, like (8.12), (AII.6), indicates references to formulae of other sections. The first symbol points to the number of the section or appendix (Section 8, Appendix II), and the second one to the ordinal number of the formula in the section.

Bibiliographical references are denoted by the years printed after the author name. For example, "Medvediev (1977)" denotes a reference to the book by B. V. Medvediev, published in 1977, the full bibliographical title of which is presented in the list of references at the end of the present book. The only exception is made for the third edition of our book *Introduction to the Theory of Quantized Fields*, which will be referred to as the *Introduction.*

1. PARTICLES AND FIELDS

1.1. Particles and their main properties. Quantum field theory is a physical theory of elementary particles and their interactions. It is based on the connection arising between relativistic particles and quantum fields, when classical fields are quantized. The properties of the quantized fields closely correspond to the properties of the particles. Therefore, we shall first of all make a list of the main properties of particles.

An important attribute of a relativistic particle is its rest mass m. The theory of relativity relates the mass m to the particle energy E and its momentum p by the well known relation:

$$E^2 - c^2p^2 = m^2c^4.$$

In the rational system of units this relation assumes the form

$$E^2 - p^2 = m^2.$$

In this system mass can be measured in energy units. For the mass of the electron m_e we have* $m_e = 9.109534(47) \times 10^{-28}$ g $= 0.5110034(14)$ MeV. The proton mass $m_p = 938.2796(27)$ MeV $\simeq 0.94$ GeV, and so on.

The second significant characteristic of a particle is its spin (i.e., its intrinsic angular momentum). In accordance with general theorems of quantum mechanics the spins of particles turn out to be quantized—their absolute values are integer multiples of one-half of Planck's constant:

$$1.0545887(57) \times 10^{-27} \text{ erg} \cdot \text{s} = 6.582173(17) \times 10^{-22} \text{ MeV}.$$

---

*Here and below we present the latest data. The numbers in brackets represent the uncertainty, equal to one standard deviation, in terms of the last digit of the main number, i.e. $0.5110034(14) = 0.5110034 \pm 0.0000014$.

Therefore, in the rational system of units, which we use, the spins of the electron and the proton turn out to be equal to one-half, and the spin of the photon (γ-quantum) is one:

$$s_e = s_p = 1/2, \qquad s_\gamma = 1.$$

A third cardinal property of particles is the existence of electrical charge, the values of which are also quantized. The charges of all observed† particles are multiples of the so-called "elementary charge" $e = 4.803242(14) \times 10^{-10}$ CGS units $= 1.6021892(46) \times 10^{-19}$ C, equal to the charge of the electron. Unlike the spin, the quantum nature of which is quite clear, the discreteness of electrical charge represents an exciting mystery.

Finally, an important characteristic of a particle is its lifetime τ. The point is that only a few particles in the free state live for an indefinitely long time (i.e., are absoutely stable). Among these are the electron e^-, the proton p, the photon, and the neutrino ν,* and also their antiparticles—the positron e^+, the antiproton $\bar{p}$, and the antineutrino $\bar{\nu}$. All other particles are unstable and spontaneously decay, by the exponential law $\exp(-t/\tau)$, into other particles. The coefficient τ is called the lifetime. For example, the lifetime of the neutron, which decays according to the scheme (β-decay)

$$n \rightarrow p + e^- + \bar{\nu}_e,$$

turns out to be equal to $\tau_n = 918(14)$ s $= 15.3$ min. The charged pion π^+ decays into a muon and a neutrino:

$$\pi^+ \rightarrow \mu^+ + \nu_\mu$$

with a lifetime $\tau_{\pi^+} = 2.6030(23) \times 10^{-8}$ s, etc.

The instability of particles displays a most important property of the microcosm—the mutual transformation of particles into each other, which is a consequence of their interactions. The interactions of particles fall into four classes: strong, electromagnetic, weak, and gravitational. We shall not dwell in detail on the properties of the interactions (which are discussed below, in

†The electrical charges of quarks (hypothetical constituents of hadrons) are usually considered equal to simple fractions ($\pm\frac{1}{3}$, $\pm\frac{2}{3}$) of e. However, quarks in the free state have not been observed.

* Besides the two well-known types of neutrinos, the electronic ν_e and numonic ν_μ, there exists, probably, a third one—the ν_τ, a "partner" of the recently discovered heavy τ-lepton.

Section 10). We only point out that due to particle interactions, particles in some combination transform, as a rule, into other particles in some other combination, if such transitions are not forbidden by any conservation laws (of energy, momentum, angular momentum, electric charge, baryon number, strangeness, and some others).

The absence of any transition allowed by known conservation laws is considered to be an indication of the existence of some new, hitherto unknown conservation law.

1.2. Conservation laws. Generally speaking, conservation laws are the consequence of some symmetries, which reflect the property of nonobservability of certain characteristics of physical objects. It is well known that in conservative systems the law of energy conservation is the manifestation of symmetry with respect to the continuous operation of time translation. Invariance under time translation is, in turn, equivalent to the nonobservability of absolute time. Another sequence of the same type is made up by the parity conservation law, invariance under space inversion, and the conventionality of the concept of "right" and "left" (i.e., the nonobservability of "absolute right" and "absolute left").

Most of the conservation laws are connected with continuous symmetries and can be obtained from the latter with the aid of Noether's theorem (see below, Section 2.3). Among them the conservation laws of energy, momentum, and angular momentum may be singled out as following from symmetry of physical objects in the space-time. Such symmetry is a consequence of such deep and general properties as the nonobservability of absolute time and of absolute space coordinates (symmetry under space-time translations), the isotropy of space, and the equivalence of coordinate systems moving with respect to each other with constant velocities (symmetry under rotations in space and Lorentz rotations). The corresponding conservation laws are of a very general nature; they are characteristic of all particles and all interactions, and are obeyed in all transitions.

Within the accuracy of modern experiments the conservation laws of electric charge and baryon number are universal as well. These conservation laws can also be correlated with the already mentioned symmetries under phase transformations of the continuous type. The latter, however, have no explicit physical basis and are not connected with the space-time structure. Such symmetries are called *intrinsic symmetries.**

---

*Some recent theoretical speculations (models of grand unification of interactions) raise hope for the explanation of the quantum nature and conservation of the electric charge. At the same time, there seems to be a possibility of baryonic-charge nonconservation (in particular, in proton decay with lifetime $\geq 10^{31}$ years).

Table 1. *Interaction symmetry properties and conserved physical quantities*

Physical quantities	Interactions		
	Strong	Electromagnetic	Weak
Electric charge	+	+	+
Baryon number	+	+	+
Parity	+	+	−
Isospin	+	−	−
Strangeness	+	+	−

(+) = conserved; (−) = not conserved.

The class of intrinsic symmetries includes isotopic symmetry, and also some others (e.g., the so-called unitary symmetry). Most intrinsic symmetries and the related quantities conserved are of an approximate nature. With the two exceptions just mentioned (electric charge and baryon number) the respective conservation laws are obeyed in some interactions and violated in others. Isotopic invariance (the isotopic-spin conservation law) holds in strong interactions and is violated in electromagnetic and weak interactions. The law of parity (to be more precise, of space-parity) conservation, which is connected with the symmetry of the wave function under space inversion, holds in strong and electromagnetic interactions and is violated in weak interactions. Analogous behaviour is displayed by the conservation laws of strangeness (as well as of the new quantum numbers, charm and beauty). These properties are summarized in Table 1.

We shall not discuss herein the properties of the gravitational interaction, because of its very low intensity. For illustration we may point out that two protons are attracted with a gravitational force approximately 10^{37} times less than the force of their electric repulsion.

1.3. Particle–field correspondence. A field represents a physical system with an infinite number of degrees of freedom. The concept of field arises naturally when one attempts to reject the notion of instantaneous action of particles on each other at a distance (Newton's *actio in distans*)—a notion that contradicts the theory of special relativity (see Medvediev 1977, Part II, Section 6, and also Wigner 1971, p. 12). Considering the space between particles to be filled with a field, we charge the field with the task of transferring perturbations with finite speed from one particle to another. Thus, the introduction of classical fields into physics is dictated by reasons of relativistic invariance. In particle theory a central role is played by relativistic quantum fields.

The quantized wave field is a fundamental physical concept within the framework of which the dynamics of particles and their interactions is formulated. It allows one to describe various states of a system of many particles by means of a single physical object in ordinary space-time. Quantized fields appear when classical fields are quantized, as a result of which the field functions take on an operator meaning and are expressed in terms of particle creation and annihilation operators. Thereby there arises the possibility of describing a most important property of elementary particles—processes of mutual transformations.

An example, well known for a long time, of a classical field is the electromagnetic field, which describes light and at the same time the interaction of electrically charged particles. A classical description, based on Maxwell's equations, leads to a purely wave concept of electromagnetism. From a methodological viewpoint it sometimes turns out to be convenient to consider a continuous system—a field—as the limit of a discrete mechanical system with a number of degrees of freedom N, tending to infinity. One field oscillator corresponds to each degree of freedom.

A description of the corpuscular properties fo light is achieved as a result of a quantization procedure, in the course of which the field is associated with discrete energy quanta, corresponding to various possible energy states of the field oscillators. Quanta of the electromagnetic field—photons—have zero rest mass, have no electrical charge, and have a spin equal to one. The last fact corresponds to the polarization properties of classical fields and manifests itself in the fact that the electromagnetic field is multicomponent and is described by a set of field functions—the components of the electric or magnetic field strengths, or the components of the potential A_ν.

The properties of field functions, corresponding to other particles, also reflect the spin, charge, and other discrete characteristics of the respective particles. After quantization the quanta of the fields are usually identified with the particles.

To the real universe of interacting particles a system of coupled equations for the various fields is ascribed. After quantization the expressions representing the coupling of these equations (the interaction Lagrangian or Hamiltonian—see Section 10 below) describe elementary interactions between particles. These expressions acquire a clear interpretation in Feynman's rules of correspondence (see Sections 19, 20 below).

We shall postpone consideration of the interaction apparatus of particles and fields, and begin by examining properties of free fields and their quantization.

1.4. The representation of the Lorentz group. Let us consider the laws of transformation of field functions under relativistic transformations of coordinates. We recall some definitions.

The full (in other words, general orthochronous) *Lorentz group* consists of homogeneous linear transformations of the four coordiantes x^ν which leave invariant the quadratic form

$$x^2 = x_\nu x^\nu = (x^0)^2 - \boldsymbol{x}^2$$

and do not reverse the direction of time x^0. This group includes the usual spatial rotations, the Lorentz "rotations" in the x^0x^1, x^0x^2, and x^0x^3 planes (i.e., transformations under transitions to a coordinate system moving with respect to the initial one with constant velocity), the reflections of the three space axes, and all the products of these transformations. If time reversal is added, then we obtain the general Lorentz group.

In physics an important role is played by the group which consists of the full-Lorentz-group transformations together with translations along all four coordinate axes. This group is called the inhomogeneous Lorentz group (or the full *Poincaré group*). We shall call invariance under this group relativistic invariance, to be distinguished from Lorentz invariance, which corresponds to the Lorentz group.

Henceforth we shall be interested in the transformtion laws of field functions under transformations of coordinates belonging to the full Poincaré group

$$x \to x' = P\,(\omega;\ x), \tag{1}$$

where $\omega = (L,\ a)$ denotes the set of parameters describing translations (a) and rotations (L), and

$$x'_\nu = L_{\nu\mu} x^\mu + a_\nu, \quad L_{\nu\sigma} g^{\sigma\rho} L_{\mu\rho} = g_{\nu\mu}. \tag{2}$$

The field function $u(x)$ represents either one function (being a single-component function) or several functions (being a many-component function) of the four coordinates x^ν defined in each of the reference frames. A transition from one coordinate system x to another x' which is related to x by the transformation (2) corresponds to a linear homogeneous transformation of the components of the field functions

$$u\,(x) \to u'\,(x') = \Lambda\,(\omega)\, u\,(x), \tag{3}$$

where the transformation matrix Λ is completely determined by the matrix L of the Lorentz transformation (2), i.e., depends on the same parameters as L.

We emphasize that the transformation (3) can not be reduced to the replacement of the argument x by x', since it describes a transformation from one coordinate system to another, and not a displacement from one point of space to another.

To each Lorentz transformation L there corresponds a matrix Λ_L, to the unit element of the group L there corresponds the unit matrix $\Lambda = 1$, while to the product of two elements L_1 and L_2 of the Lorentz group there corresponds the product

$$\Lambda_{L_1L_2} = \Lambda_{L_1}\Lambda_{L_2}.$$

A system of matrices with such properties is called, in group theory, a linear representation of the group. Matrices Λ of finite rank form a finite-dimensional representation of the Lorentz group. The rank of a representation is determined by the dimension of the matrices, i.e., by the number of components of u.

Therefore, the possible kinds of wave functions and the laws of their transformation can be obtained by investigation of the finite-dimensional (irreducible) representations of the Lorentz group. Such an investigation represents a special part of the theory of group representations, the result of which comes to the following. The finite-dimensional representations of the Lorentz group may be single-valued or double-valued, i.e., the correspondence $L \to \Lambda_L$ may be single-valued or double-valued. The importance to physics of double-valued representations is due to the fact that the field functions, generally speaking, are not directly observable (specifically, fields which transform according to double-valued representations always enter into the observable quantities in the form of quadratic combinations). However, the lack of single-valuedness of the operator Λ_L must be such that the observables transform in a completely unique manner under any Lorentz transformation L. Besides, it is necessary that the operators Λ_L be continuous functions of the parameters of the transformation L, i.e., that to an infinitesimal transformation of the reference frame there correspond an infinitesimal transformation of the field functions. The combined requirements stated above lead to the representations of the Lorentz group being divided into two categories. The first category is characterized by the single-valuedness of the correspondence $L \to \Lambda_L$ and contains the single-valued so-called *tensor* and *pseudotensor** representations. The field functions which

*The difference between tensors and pseudotensors is related to the reflections of space axes and will be considered at the end of this section.

transform in accordance with the tensor representation are called tensors (Pseudotensors) and in some cases may be directly observable (e.g. the electromagnetic field). The second category is characterized by this correspondence being double-valued: $L \to \pm \Lambda_L$.

The transformation law for a (pseudo) tensor of the Nth rank, $T^{\nu_1,\nu_2,\ldots,\nu_N}$, under continuous transformations of coordinates has the form

$$T'_{\nu_1, \ldots \nu_N}(x') = \frac{\partial x'_{\nu_1}}{\partial x_{\mu_1}} \cdots \frac{\partial x'_{\nu_N}}{\partial x_{\mu_N}} T_{\mu_1, \ldots \mu_N}(x), \tag{4}$$

or, using the notation (2),

$$T'^{\nu_1, \ldots, \nu_N}(x') = L^{\nu_1}_{\mu_1} \ldots L^{\nu_N}_{\mu_N} T^{\mu_1, \ldots, \mu_N}(x).$$

The double-valued representations are called spinor representations, and the corresponding quantities are called *spinors*. The transformation law for spinor quantities has a more complex structure and is given for the simplest spinors in Appendix II. Here, we merely note that under the translation

$$x'^{\nu} = x^{\nu} + a^{\nu}$$

the transformation law for tensor quantities which follows from (4),

$$u(x) \to u'(x') = u(x) \tag{5}$$

remains valid also for spinors.

We shall now present the simplest tensor representations and the quantities which correspond to them. The tensor of zero rank which under any continuous transformation transforms in accordance with the law (5) is an invariant and is called a scalar (pseudoscalar).

The tensor of the first rank which transforms under rotations of coordinates in accordance with the law

$$u'^{\nu}(x') = L^{\nu}_{\mu} u^{\mu}(x) = L^{\nu\mu} u_{\mu}(x), \tag{6}$$

is called a contravariant vector (pseudovector). The covariant vector

$$u_{\nu}(x) = g_{\nu\mu} u^{\mu}(x)$$

connected with it transforms in accordance with the law

$$u'_\nu(x') = L^\mu_\nu u_\mu(x). \tag{7}$$

Corresponding formulas for covariant and contravariant tensors of the second and higher ranks may be written down without any difficulty.

Now consider the operation of space inversion P, i.e., of reflection of all three space axes:

$$x \rightarrow x' = Px, \quad x'_0 = x_0, \quad \mathbf{x}' = -\mathbf{x}. \tag{8}$$

The transformation laws of the field functions are not defined by the formulas (4) and must be formulated separately. Since the double application of the transformation ($P^2 = 1$) is equivalent to the identity operator due to the single-valuedness of the tensor representations, these laws for the components of the tensors $T^{\cdots}(x)$ can have only two forms:

$$PT(x) = T'(x') = T'(Px) = \pm T(x).$$

A tensor of zero rank which does not change sign under inversion,

$$Pu(x) = +u(x) \tag{9}$$

is called a scalar. In the other case, the corresponding quantity which satisfies the relation

$$Pu(x) = -u(x), \tag{10}$$

is called a pseudoscalar.

If a tensor of the first rank changes the signs of its space components under the P-transformation and does not change the sign of the time component:

$$PV^\mu(x) = V_\mu(x),$$

i.e.,

$$PV^0(x) = V^0(x), \quad P\mathbf{V}(x) = -\mathbf{V}(x), \tag{11}$$

it is then called a vector. If, on the other hand, only the time component changes sign and the space components do not:

$$PV^\mu(x) = -V_\mu(x),$$

i.e.,

$$PV^0(x) = -V^0(x), \quad P\mathbf{V}(x) = \mathbf{V}(x),$$

then it is a pseudovector (axial vector). Generally, it is possible to describe the transformation law of pseudotensors by the formula (4) multiplied by the determinant of the transformation of the coordinates. The relations (9), (10) and other similar relations define also the important property of *parity* of the field functions and of the respective particles. This characteristic plays an essential role in determining (see Section 10.2 below) the possible forms of interaction between various fields.

CHAPTER I

FREE CLASSICAL FIELDS

2. DYNAMICAL FIELD INVARIANTS

2.1. The Lagrangian. In this section we shall present the formalism which permits us to obtain, on a general basis, the equations of motion as well as quantities which are preserved in time and correspond to invariance properties with respect to one continuous transformation or another. Such a formalism applied to a mechanical system with a finite number of degrees of freedom is based on the Lagrangian function and is called the Lagrangian formalism.

The Lagrangian function L is a function of time, it depends on the dynamical variables of the system, and in mechanics it is expressed as a sum over all the material points of the system. For a continuous system of the type of a wave field this sum is replaced by a spatial integral of the density of the Lagrangian function,

$$L(x^0) = \int d\boldsymbol{x} \mathcal{L}(x^0, \boldsymbol{x}).$$

The latter function

$$\mathcal{L}(x^0, \boldsymbol{x}) = \mathcal{L}(x)$$

is called the Lagrangian.

The starting point of the Lagrangian formalism is represented by the action of the system, $\mathcal{A}$, which is the integral of L over time:

$$\mathcal{A} = \int dt\, L(t) = \int dx^0\, d\boldsymbol{x} \mathcal{L}(x^0, \boldsymbol{x}) = \int dx\, \mathcal{L}(x). \tag{1}$$

Hence it may be seen that the Lagrangian function $L(t)$ turns out to play an intermediate role in field theory, while the major role is ascribed to the Lagrangian $\mathcal{L}(x)$.

The equations of motion can be obtained with the aid of the principle of least action, according to which actual motion occurs in such a manner that the action $\mathscr{A}$ achieves an extremum, i.e. its variation vanishes. From the condition

$$\delta\mathscr{A} = 0,$$

together with the assumption that the variations of the field functions δu_i vanish at the surface of the four-volume over which the integral is taken, we obtain using integration by parts the Euler–Lagrange equations

$$\frac{\delta\mathscr{A}}{\delta u_a(x)} = \frac{\partial\mathscr{L}}{\partial u_a(x)} - \partial_\nu \frac{\partial\mathscr{L}}{\partial u_{a;\,\nu}} = 0. \tag{2}$$

Let us briefly list the main requirements imposed on the Lagrangian in quantum field theory. First, the Lagrangian is a function only of the dynamical variables, i.e., of the components of the field functions $u_a(x)$ and of their derivatives. The Lagrangian does not depend explicitly on the coordinates x, as such a dependence would violate relativistic invariance. That the theory is local is provided for by the fact that the value of the Lagrangian $\mathscr{L}$ at the point x is determined by the value of $u_a(x)$ and of a finite number of partial derivatives evaluated at the same point x. Such a Lagrangian is also referred to as local. The presence, for instance, of integrals in $\mathscr{L}$ would lead to the nonlocal case.

To obtain differential equations of not higher than the second order, the Lagrangian is considered to be a function of the field components u and of their *first* derivatives:

$$\mathscr{L}(x) = \Phi(u_a(x),\ u_{a;\,\nu}(x)). \tag{3}$$

Notice, also, that since the physical properties of a system are determined by the action $\mathscr{A}$, the expression for which involves the Lagrangian under the integral sign, the correspondence $\mathscr{A} \rightarrow \mathscr{L}$ is not one-to-one. Lagrangians differing from each other by a total four-divergence (of some four-vector)

$$\mathscr{L}'(x) = \mathscr{L}(x) + \partial_\nu F^\nu(x), \tag{4}$$

turn out to be physically equivalent. This is because the integral (1) of $F^\nu{}_{;\nu}$ is reduced, with the use of the Gauss-Ostrogradsky theorem, to the surface integral of F^ν over the three-dimensional boundary of the four-dimensional region of integration. Since the variations of the field functions δu vanish on this boundary, we find that the term $F^{\nu}{}_{;\nu}$ does not contribute to physical quantities.

Important requirements imposed on the properties of the Lagrangian are that it should be real (in the quantum case, Hermitian) and relativistically invariant. The Lagrangian being real leads to the dynamical invariants (energy, momentum, current, and so on) being real and, ultimately, to the unitarity of the S-matrix.

Relativistic invariance of the Lagrangian

$$\mathcal{L}'(x') = \Phi(u'_a(x'), u'_{a;\nu}(x')) = \Phi(u_a(x), u_{a;\nu}(x)) = \mathcal{L}(x) \quad (5)$$

means that $\mathcal{L}$ behaves like a (pseudo)scalar under the Poincaré transformations. As the infinitesimal element of the 4-integration region in (1), namely $dx = dx^0 d\boldsymbol{x}$, is also an invariant, we obtain that the value of the action does not change under transformations of the Poincaré group. Therefore, the Lagrangian being a scalar ensures invariance of the action.

2.2. Dynamical invariants. Energy-momentum. This latter property turns out to be essential for obtaining the so-called dynamical invariants, i.e., quantities which are preserved in time. The dynamical invariants include the energy, momentum, and angular momentum, as well as some other quantities which are conserved by virtue of the corresponding invariance of the action (or Lagrangian), such as e.g., the electric charge.

As an example we shall consider the energy-momentum four-vector P^{ν}. It can be represented in the form of the spatial integral

$$P^{\nu} = \int d\boldsymbol{x}\, T^{\nu 0}(x^0, \boldsymbol{x}) \quad (6)$$

of the corresponding components of the energy-momentum tensor, which in terms of the Lagrangian is given by the relation

$$T^{\nu\mu}(x) = \frac{\partial \mathcal{L}}{\partial u_{a;\mu}} u_{a;\nu}(x) - g^{\nu\mu}\mathcal{L}(x). \quad (7)$$

The independence of the integrals (6) of the time x^0 is a consequence of the tensor $T^{\nu\mu}$ satisfying the continuity equation

$$\frac{\partial T^{\nu\mu}}{\partial x^{\mu}} = 0. \quad (8)$$

To verify this, consider the integral

$$\int dx \frac{\partial T^{\nu\mu}}{\partial x^{\mu}}.$$

We transform it by the Gauss–Ostrogradsky theorem into a surface integral

$$\int dx \frac{\partial T^{\nu\mu}}{\partial x^{\mu}} = \int_{\sigma} d\sigma_{\mu}\, T^{\nu\mu}(x), \tag{9}$$

where σ is the boundary surface of the integration volume and $d\sigma_{\mu}$ is a surface element perpendicular to the x^{μ}-axis, and we assume the region of integration to expand to infinity in the spacelike directions and to be limited in time by two three-dimensional "surfaces" $\sigma_1(x^0 = t_1)$ and $\sigma_2(x^0 = t_2)$. Assuming also that at points at infinity in the spacelike directions the fields u_a, their derivatives, and the components of the energy-momentum tensor equal zero, we obtain from (8) and (9)

$$\int_{\sigma_1} d\boldsymbol{x}\, T^{\nu 0}(x) - \int_{\sigma_2} d\boldsymbol{x}\, T^{\nu 0}(x) = \int d\boldsymbol{x}\, T^{\nu 0}(t_1, \boldsymbol{x}) - \int d\boldsymbol{x}\, T^{\nu 0}(t_2, \boldsymbol{x}) = 0,$$

Q.E.D.

We shall now demonstrate that the continuity equation (8) is a consequence of the equations of motion. Consider the divergence of the right-hand side of (7):

$$\partial_{\mu} T^{\nu\mu} = \frac{\partial}{\partial x^{\mu}} \left(\frac{\partial \mathcal{L}}{\partial u_{a;\,\mu}} \right) u_{a}^{;\,\nu} + \frac{\partial \mathcal{L}}{\partial u_{a;\,\mu}} \partial_{\mu} (u_{a}^{;\,\nu}) - \partial^{\nu} \mathcal{L}.$$

Using the equations of motion (2) and changing the order of differentiation of the second derivative, we obtain

$$\partial_{\mu} T^{\nu\mu} = \frac{\partial \mathcal{L}}{\partial u_a} \frac{\partial u_a}{\partial x_{\nu}} + \frac{\partial \mathcal{L}}{\partial u_{a;\,\mu}} \frac{\partial u_{a;\,\mu}}{\partial x_{\nu}} - \frac{\partial \mathcal{L}}{\partial x_{\nu}} = 0.$$

Let us now draw some conclusions. We have established that each of the four quantities p^{ν} $(\nu = 0,1,2,3)$ is conserved in time by virtue of the corresponding spatial density $T^{\nu 0}$ being the zero component of the "four-vector" $\theta^{\mu}(\nu) \equiv T^{\nu\mu}$, satisfying the equation of continuity (8). The latter follows from Euler–Lagrange equations as well as from the fact that the Lagrangian $\mathscr{L}$ depends on the corrdinates x only through the field functions u_a and their first derivatives $u_{;\nu}$, as a result of which

$$\frac{\partial \mathscr{L}}{\partial x_\nu} = \frac{\partial \mathscr{L}}{\partial u_a} u_a^{;\nu} + \frac{\partial \mathscr{L}}{\partial u_{a;\mu}} \frac{\partial}{\partial x_\nu}\left(\frac{\partial u_a}{\partial x^\mu}\right).$$

It can be shown that the conservation of the projection of the "four-vector" p^ν on the ν-axis is a consequence of the covariance of the Lagrangian (and of the vanishing of the corresponding variation of the action) under translation along the x^ν-axis.

2.3 Noether's theorem. This statement represents a special case of the so-called Noether's theorem, which may be formulated as follows.

Let there be a continuous transformation of both coordinates and simultaneously of field functions, which depends on s parameters ω_k $(k = 1,2, \ldots, s)$:

$$x_\nu \to x'_\nu = f_\nu(x;\ \omega), \tag{10}$$

$$u_a(x) \to u'_a(x') = U_a(u_b(x);\ \omega) \tag{11}$$

and which makes the variation of the action $\mathscr{A} = \int \mathscr{L}(x)dx$ equal to zero:

$$\delta\mathscr{A} = 0.$$

There then exist s dynamical invariants C_k (i.e., functions, which are independent of time, of the field functions and their derivatives) which may be represented in the form of spatial integrals

$$C_k = \int d\boldsymbol{x}\, \theta^0_{(k)}(x) \tag{12}$$

of the zero components of certain "four-vectors"

$$\theta^\nu_{(k)} = \frac{\partial \mathscr{L}}{\partial u_{a;\nu}}\left(u_{a;\mu} X^\mu_k - \Psi_{a;k}\right) - X^\nu_k \mathscr{L}(x), \tag{13}$$

and

$$X_k^\mu = \left.\frac{\partial f^\mu(x;\ \omega)}{\partial \omega_k}\right|_{\omega=0}, \quad \Psi_{a;\,k} = \left.\frac{\partial U_a}{\partial \omega_k}\right|_{\omega=0}. \tag{14}$$

We shall not present the proof of this theorem here; we refer the interested reader to Section 2.1 of the *Introduction*.

We note, first of all, that the expressions (6), (7) presented above are special cases of the general expressions (12), (13) corresponding to

$$X_k^\nu = \delta_k^\nu, \quad \Psi_{a;\,k} = 0. \tag{15}$$

The values (15) correspond to the transformations

$$x'_\nu = x_\nu + \omega_\nu,$$
$$u'_a(x') = u_a(x),$$

i.e., to translations of all four coordinates.

The latter formula represents the transformation law of the field functions under translations. As is seen, it is identical for field functions having different tensor dimensionalities. Substituting the values (15) into the general formula (13), we obtain the expression (7) for the energy-momentum tensor. With the aid of (12) we arrive at formula (6) for the conserved energy-momentum four-vector P^ν.

2.4. Angular momentum and spin. We shall now obtain the dynamical invariants corresponding to the four-dimensional Lorentz rotations of the coordinate system. As follows from (14), it is sufficient to consider only infinitesimal four-rotations which are of the form

$$x'_\nu = x_\nu + x^\mu \delta L_{\nu\mu}, \tag{16}$$

where $\delta L_{\nu\mu}$ are the infinitesimal parameters of the rotations. Due to the antisymmetry of $\delta L_{\nu\mu}$, we may choose as independent parameters six of them:

$$\delta\omega_{(\mu\nu)} = \delta L_{\mu\nu} \qquad \text{at} \quad \mu < n, \tag{17}$$

which represent infinitesimal angles of rotation in the $x_\nu x_\nu$ plane. Thus the index k in formulas (12)–(15) stands for the double index $(\mu\nu)$.

Representing the variation δx in the form

$$\delta x_\nu = X_\nu^k \,\delta\omega_k = \sum_{\rho<\sigma} X_\nu^{(\rho\sigma)}\,\delta\omega_{(\rho\sigma)}, \tag{18}$$

and using (16), (17), we obtain

$$\delta x_\nu = x^\mu\,\delta L_{\nu\mu} = \sum_{\rho<\sigma} (x^\sigma\delta_\nu^\rho - x^\rho\,\delta_\nu^\sigma)\,\delta\omega_{(\rho\sigma)},$$

i.e.,

$$X_\nu^{(\rho\sigma)} = x^\sigma\delta_\nu^\rho - x^\rho\delta_\nu^\sigma. \tag{19}$$

We represent the infinitesimal variation of the field functions $u'_\alpha(x') = u_\alpha(x) + \delta u_a(x)$ in the form

$$\delta u_a\,(x) = \sum_{b,\,\rho<\sigma} A_a^{b\,(\rho\sigma)} u_b\,(x)\,\delta\omega_{(\rho\sigma)}. \tag{20}$$

In accordance with (1.5) and (1.6), for the scalar field

$$A^{\cdots}_{\cdots} = 0$$

and for the vector field

$$A_\mu^{\nu\,(\rho\sigma)} = \delta_\mu^\rho g^{\nu\sigma} - \delta_\mu^\sigma g^{\nu\rho}, \qquad \rho<\sigma.$$

Therefore, for the vector field

$$\Psi_\nu^{(\rho\sigma)} = A_\nu^{\mu\,(\rho\sigma)} u_\mu\,(x) = u^\sigma\,(x)\,\delta_\nu^\rho - u^\rho\,(x)\,\delta_\nu^\sigma. \tag{21}$$

Substituting (21) and (19) into (13), we obtain the angular-momentum tensor

$$M^{\tau\,(\rho\sigma)} = \frac{\partial \mathcal{L}}{\partial u_{\nu;\,\tau}}(u_{;\nu}^{\;\;\rho} x^\sigma - u_{;\nu}^{\;\;\sigma} x^\rho) + \mathcal{L}\,(x^\rho g^{\sigma\tau} - x^\sigma g^{\rho\tau}) - \frac{\partial \mathcal{L}}{\partial u_{\nu;\,\tau}} A_\nu^{\mu\,(\rho\sigma)} u_\mu\,(x). \tag{22}$$

By comparing the first two terms of the right-hand side with (7), we see that they can be represented in the form

$$x^\sigma T^{\rho\tau} - x^\rho T^{\sigma\tau} = M_0^{\tau\,(\rho\sigma)}. \tag{23}$$

This expression corresponds to the relation between the angular-momentum tensor and energy-momentum tensor in the mechanics of a point. Therefore, the quantity (23) should be identified with the orbital angular momentum of the wave field. In the case of a single-component (scalar, pseudoscalar) field the total angular momentum is reduced to the orbital angular momentum.

In the case of multicomponent fields the last term of the right-hand side of (22) does not equal zero. it characterizes the polarization properties of the fields and corresponds to the spin angular momentum of the field. Thus,

$$M^{\tau(\rho\sigma)} = M_0^{\tau(\rho\sigma)} + S^{\tau(\rho\sigma)},$$

where

$$S^{\tau(\rho\sigma)} = -\frac{\partial \mathcal{L}}{\partial u_{a;\,\tau}} A_a^{b\,(\rho\sigma)}\, u_b(x). \tag{24}$$

For the vector field, with the aid of (21), we obtain

$$S^{(\rho\sigma)} = \int d\boldsymbol{x} S^{0\,(\rho\sigma)} = \int d\boldsymbol{x} \left\{ u^{\rho}(x) \frac{\partial \mathcal{L}}{\partial \dot{u}_{\sigma}} - u^{\sigma}(x) \frac{\partial \mathcal{L}}{\partial \dot{u}_{\rho}} \right\}. \tag{25}$$

Here the symbol $\dot{u}$ stands for the partial derivative of u with respect to time. The three quantities $S^{(23)}$, $S^{(31)}$, $S^{(12)}$ represent the respective components of a spatial pseudovector—the spin vector

$$S^{n} = \varepsilon^{nmp} S^{(mp)} \qquad (n,\ m,\ p = 1,\ 2,\ 3). \tag{26}$$

2.5. The current vector and the charge. Finally, in the case of complex fields which correspond to charged particles (in the simplest case the charge is the electric charge) the Lagrangian turns out to be invariant with respect to the phase transformation of the field functions (the gauge transformation of the first kind), which does not operate on the coordinates.

We shall make use of the example of a system described by a single complex field in order to elucidate the point under discussion. By virtue of the property of reality, the Lagrangian as well as the dynamical quantities may depend on the complex field only through quadratic forms of the type $\overset{*}{u}_a u_b$, where u_a and $\overset{*}{u}_a$ are complex conjugate functions or their derivatives. Hence it immediately follows that the complex wave functions u may be multiplied by an arbitrary unitary phase factor of the form $\exp(i\alpha)$ which does not lead to any change in the quadratic form $\overset{*}{u}u$ and consequently does not lead to any observable effects.

Treating u and $\overset{*}{u}$ as linearly independent functions, we may write the gauge transformation of the first kind in the form

$$u_a \to u'_a = e^{i\alpha} u_a, \quad \overset{*}{u}_b \to \overset{*}{u}{}'_b = e^{-i\alpha} \overset{*}{u}_b. \tag{27}$$

Assuming α to be infinitesimal, we find that

$$u \to u + i\alpha u, \quad \overset{*}{u} \to \overset{*}{u} - i\alpha \overset{*}{u}.$$

Hence, in accordance with (14), it follows that

$$\Psi_a = iu_a \quad \text{for all} \quad u_a,$$
$$\Psi_b = -i\overset{*}{u}_b \quad \text{for all} \quad \overset{*}{u}_b,$$

and also that $X = 0$. Substituting these expressions into (13), we arrive at an expression that has the tensor dimensionality of a vector:

$$J^\nu(x) = i\left(\frac{\partial \mathscr{L}}{\partial \overset{*}{u}_{a;\,\nu}} \overset{*}{u}_a - \frac{\partial \mathscr{L}}{\partial u_{b;\,\nu}} u_b\right). \tag{28}$$

This four-vector satisfies the equation of continuity

$$\partial_\nu J^\nu = 0 \tag{29}$$

and therefore is usually identified with the 4-vector of the *current*. The spatial integral of its zero component,

$$Q = \int J^0(x)\, dx = i \int dx \left(\frac{\partial \mathscr{L}}{\partial \overset{*}{u}_{a;\,0}} \overset{*}{u}_a - \frac{\partial \mathscr{L}}{\partial u_{b;\,0}} u_b\right) \tag{30}$$

is time-independent and represents a conserved characteristic of the field. It is identified with the *charge*. The integral of motion Q can describe not only the electric charge, but also other conserved characteristics, like baryonic charge, strangeness, charm, colour, SU(3) symmetry of the gauge quark-gluon model of strong interactions, etc.

Clearly, systems with similar integrals of motion are also described by complex (in the general case, multicomponent) fields. We may point out, also, that these transformations ($\Psi \neq 0$, $X = 0$) which do not act on the coordinates are called transformations of the *internal symmetries*. A physically important example of *internal* symmetries is represented by isotopic symmetry. The transformations corresponding to them can also be investigated with the aid of Noether's theorem, which leads to the concept of the conserved isotopic-spin vectors.

3. The simplest fields

3.1. The scalar field. The simplest field is the real scalar field $\varphi(x)$ which describes spinless particles of the same sort. The free Lagrangian of this field is chosen to be as follows:

$$\mathcal{L} = \frac{1}{2} \varphi_{;\nu}(x)\, \varphi^{;\nu}(x) - \frac{m^2}{2} \varphi^2(x), \tag{1}$$

so that the equation of motion (2.2), derived from it, turns out to be the Klein–Gordon equation

$$\frac{\partial \mathcal{L}}{\partial \varphi} - \frac{\partial}{\partial x^\nu} \frac{\partial \mathcal{L}}{\partial \varphi_{;\nu}} = -m^2 \varphi - \frac{\partial^2 \varphi}{\partial x^\nu\, \partial x_\nu} = (\Box - m^2)\, \varphi(x) = 0. \tag{2}$$

Here

$$\Box = -\partial_\nu \partial^\nu = \Delta - \frac{\partial^2}{\partial t^2}$$

is the d'Alembert operator.

With the aid of (2.7) we obtain from (1) the energy-momentum tensor

$$T^{\mu\nu}(x) = \varphi^{;\nu}(x)\, \varphi^{;\mu}(x) - g^{\mu\nu} \mathcal{L}. \tag{3}$$

The angular-momentum tensor can hence be obtained now by using the relation (2.23). The spin angular momentum is equal to zero.

Substituting into (3) the explicit form of the Lagrangian (1), we obtain the energy density

$$T^{00} = (1/2)\,[\dot{\varphi}^2 + (\nabla \varphi)^2 + m^2 \varphi^2] \tag{4}$$

and the density of the momentum vector

$$T^{0k} = -\varphi^{;\,0}\varphi^{;\,k} \qquad (k=1,\ 2,\ 3)$$

3.2. Momentum representation. In the theory of particles the momentum representation is used very frequently. First, it is more convenient to describe physical problems when particles are characterized by their energies and momenta (but not by space-time coordinates). Second, the dynamical variables acquire a more compact and clear structure in this representation.

So we write the field function $\varphi(x)$ in the form of a four-dimensional Fourier integral:

$$\varphi(x) = (2\pi)^{-2} \int dk\, e^{ikx} \tilde{\varphi}(k), \quad dk = dk_0\, d\boldsymbol{k}.$$

The condition that $\varphi(x)$ should be real, $\overset{*}{\varphi}(x) = \varphi(x)$, gives

$$\overset{*}{\tilde{\varphi}}(k) = \tilde{\varphi}(-k).$$

In accordance with (2) the Fourier amplitude $\tilde{\varphi}$ satisfies the equation

$$(k^2 - m^2)\, \tilde{\varphi}(k) = 0$$

and may therefore be represented in the form

$$\tilde{\varphi}(k) = \sqrt{2\pi}\, \delta(k^2 - m^2)\, \varphi(k).$$

The delta function establishes the relation

$$(k_0)^2 - \boldsymbol{k}^2 = m^2 \tag{5}$$

between the frequency k_0, the wave vector $\boldsymbol{k}$, and the parameter m. As will be shown later, the frequency k_0 is to be identified with the energy, and the wave vector with the momentum vector, as a result of which the relation (5) turns out to be the well-known relation from the relativistic mechanics of a particle in which m represents the mass.

The Fourier transform takes on the form

$$\varphi(x) = (2\pi)^{-3/2} \int dk\, \delta(k^2 - m^2)\, e^{ikx} \tilde{\varphi}(k).$$

Because of the delta function under the integral sign, the integration is in fact carried out not over the whole four-dimensional k-space but only over the two three-dimensional hyperboloids

$$k^0 = \pm\sqrt{\mathbf{k}^2 + m^2},$$

one of which lies entirely within the upper light cone and the other within the lower one. Noting further that the two hyperboloids are separately Lorentz-invariant, we arrive at the following Lorentz-invariant decomposition of the field function into two terms:

$$\varphi(x) = \varphi^+(x) + \varphi^-(x), \tag{6}$$

where

dk δ(k²−m²)θ(k₀) Lorentz invariant element with θ(k₀>0)

$$\varphi^\pm(x) = (2\pi)^{-3/2} \int dk\, e^{\pm ikx} \delta(k^2 - m^2)\, \tilde{\varphi}^\pm(k). \tag{7}$$

Here

$$\tilde{\varphi}^\pm(k) = \theta(k_0)\, \varphi(\pm k),$$

and θ is the well-known step function

$$\theta(x) = \begin{cases} 1, & x > 0, \\ 0, & x < 0. \end{cases}$$

The functions φ^+ and φ^- will in the future be referred to as the positive-frequency and negative-frequency parts respectively of the function $\varphi(x)$. As may be seen, the sign of the frequency is determined by the sign of the product kx (more precisely, by the sign of the "time" term $k^0x^0 = x^0\sqrt{\mathbf{k}^2 + m^2}$ in the exponent of the integrand). In this connection we note that in the contemporary literature the converse notation is sometimes used, i.e. the sign of the frequency being taken the same as that of the form $k_\mu x_\mu = \mathbf{kx} + k_4x_4 = -kx$. Such a notation is associated with the customary way of writing down the quantum-mechanical wave function, which is proportional to $\exp(-iEx_0)$, where E is the energy. Our choice of notation is connected with the fact that expressions of the type of φ^+ correspond in quantum theory to the creation of the field particles, while expressions of the type of φ^- to their annihilation. Therefore, in the system of notation adopted by us the signs + and − symbolize the physical meaning of the corresponding quantum operators.

As will be shown later, the above decomposition also turns out to be very convenient for writing down the dynamical variables in the momentum representation, since they may be expressed as quadratic forms in $\varphi^+(k)$ and $\varphi^-(k)$.

We also note that in accordance with (7) the rules of complex conjugation for $\varphi^\pm(k)$ have the form

$$(\varphi^\pm(k))^* = \varphi^\mp(k).$$

Carrying out in (7) the integration over k_0, we obtain

$$\varphi^\pm(x) = \frac{1}{(2\pi)^{3/2}} \int \frac{d\boldsymbol{k}}{\sqrt{2k_0}} e^{\pm ikx} \varphi^\pm(\boldsymbol{k}). \tag{8}$$

Here the following notation has been introduced:

$$\varphi^\pm(\boldsymbol{k}) = \frac{\varphi^\pm(k_0, \boldsymbol{k})}{\sqrt{2k_0}}, \quad k_0 = +\sqrt{\boldsymbol{k}^2 + m^2}.$$

Substituting (6) and (8) into (4), we obtain

$$\begin{aligned} P^0 &= \int T^{00}\, d\boldsymbol{x} = (1/2) \int [(\partial_\nu \varphi(x))^2 + m^2 \varphi^2(x)]\, d\boldsymbol{x} = \\ &= (1/2) \int d\boldsymbol{x}\, \{\partial_\nu \varphi^+ \partial_\nu \varphi^+ + 2\, \partial_\nu \varphi^+ \partial_\nu \varphi^- + \partial_\nu \varphi^- \partial_\nu \varphi^- + \\ &\qquad + m^2 [\varphi^+(x)\varphi^+(x) + 2\varphi^+(x)\varphi^-(x) + \varphi^-(x)\varphi^-(x)]\}. \end{aligned}$$

Note that in this expression *all summations* over ν are *noncovariant.* It is not difficult to demonstrate that the terms in which the product of the functions $\varphi^\pm$ of identical frequency are present do not contribute to the dynamical invariant P^0. Thus, for example,

$$\begin{aligned} &\int d\boldsymbol{x}\, [(\partial_\nu \varphi^+(x))^2 + m^2 (\varphi^+(x))^2] = \\ &= \frac{1}{(2\pi)^3} \int\int \frac{d\boldsymbol{k}\, d\boldsymbol{k}'}{2\sqrt{k_0 k_0'}} \varphi^+(\boldsymbol{k}) \varphi^+(\boldsymbol{k}') e^{i(k_0 + k_0')x_0} (m^2 - k_\nu k_\nu') \int d\boldsymbol{x}\, e^{-i(\boldsymbol{k}'+\boldsymbol{k})\boldsymbol{x}} = \\ &\qquad = \int \frac{d\boldsymbol{k}}{2k_0} \varphi^+(\boldsymbol{k}) \varphi^+(-\boldsymbol{k}) e^{2ik_0x_0} (m^2 - k_0^2 - \boldsymbol{k}^2), \end{aligned}$$

and, since $m^2 - k_0^2 + \boldsymbol{k}^2 = 0$, we obtain

$$\int d\boldsymbol{x}\, [(\partial_\nu \varphi^+(x))^2 + m^2 (\varphi^+(x))^2] = 0.$$

An analogous relation holds also for the quadratic form in φ^-. Therefore

$$P^0 = \int dx\, [\partial_\nu \varphi^+(x)\, \partial_\nu \varphi^-(x) + m^2 \varphi^+(x)\, \varphi^-(x)].$$

Inserting (8) into this expression, with the aid of a similar calculation we now find

$$P^0 = \int dk\, k^0 \varphi^+(k)\, \varphi^-(k).$$

The corresponding expression for the momentum vector has the form

$$P^n = \int T^{0n}\, dx = \int dk\, k^n \varphi^+(k) \varphi^-(k), \qquad n = 1,\ 2,\ 3.$$

Combining these expressions, we write them in the form which would also hold if throughout the whole calculation the functions φ^+ and φ^- were treated as noncommuting (i.e., their order were not changed):

$$P^\nu = \frac{1}{2} \int dk\, k^\nu\, [\varphi^+(k)\, \varphi^-(k) + \varphi^-(k)\, \varphi^+(k)]. \tag{9}$$

Now the point of introducing the particular normalizing factors in (8) has become clear. The products

$$n(k) = \varphi^+(k)\, \varphi^-(k) = |\varphi^+(k)|^2$$

represent the average densities of uncharged, spinless particles of mass m, momentum k, and energy $\sqrt{m^2 + k^2}$. Under quantization these products transform into operators with integer eigenvalues.

An important field, from a practical point of view, is the field which corresponds to a pi-mesons (pions). This field is a pseudoscalar. The three pions π^+, π^0, π^-, which differ from each other only by their electric charge (equal to 0, $\pm$ 1) form an isotopic triplet. They are thus assoicated with a three-component field function $\varphi = \{\varphi_1, \varphi_2, \varphi_3\}$, the constituents of which form a vector in the three-dimensional isotopic space.

The free Lagrangian is written down in the form

$$\mathscr{L} = \frac{1}{2} (\varphi_{;\nu} \varphi^{;\nu}) - \frac{m^2}{2} (\varphi\varphi) \tag{10}$$

and is the sum of the Lagrangians of each of the isotopic-vector components. Therefore the dynamical quantities associated with the transformations of the

Poincaré group are represented by sums of the expressions obtained in Sections 3.1 and 3.2 for each of the isotopic componnents.

The novel conserved quantity corresponds to transformations of rotations in the isotopic space and is called the *isotopic spin vector.*

3.3. The vector field. As a next example, we shall consider the vector field, which is used for describing particles with spin 1. The field U_ν consists of four ($\nu = 0,1,2,3$) components forming a covariant four-vector, i.e., transforming under Lorentz rotations (2.16) in accordance with the law

$$U_\mu(x) \to U'_\mu(x') = U_\mu(x) + \delta U_\mu(x); \quad \delta U_\mu = \delta\Omega_{\mu\nu} U^\nu(x).$$

Generally, the Lagrangian of the vector field may be constructed by various means. The point is that the simplest generalization of the Lagrangian (1) in the form of the covariant sum over the components

$$\mathcal{L} = -\frac{1}{2} U_{\mu;\nu} U^{\mu;\nu} + \frac{m^2}{2} U_\mu U^\mu \tag{11}$$

is not the only possible Lorentz-invariant expression. To it one may add a term proportional to

$$U_{\mu;\nu} U^{\nu;\mu}, \tag{12}$$

which by virtue of (2.4) is equivalent to the expression

$$U^{;\mu}_\mu U^\nu_{;\nu} = (\partial U/\partial x)^2.$$

To verify this, it is sufficient to establish that the difference between the written expressions can be represented as the four-divergence $\partial_\nu F^\nu$.

The Lagrangian (11) is a "good" Lagrangian, as all dynamical variables, from its point of view, are covariant sums (over μ) of the corresponding expressions from (1) for the single-component field. However, since the term connected with the component U_0 enters into such sums with the negative sign, its contribution to the energy turns out to be negative. The way out of this difficulty consists of imposing on the components U_μ the invariant subsidiary condition

$$\partial^\mu U_\mu = \frac{\partial}{\partial x_\mu} U_\mu(x) \equiv (\partial U) = 0. \tag{13}$$

This condition reduces the number of linearly independent components from four to three and, as will be shown later, guarantees that the energy of the vector field is positive definite. The three remaining components correspond to the three possible values of the component of the spin along a given axis, which are respectively 1, 0, −1, i.e., which describe particles with spin 1. The imposition of the subsidiary condition is equivalent to the exclusion of a particle of spin zero, which in this formulation would lead to a negative energy.

The condition (13) is compatible with the equations of motion. Moreover, by appropriate modification of the Lagrangian (11) (addition of a term of the form of (12)),

$$\mathscr{L}=-\frac{1}{4}H_{\mu\nu}H^{\mu\nu}+\frac{m^2}{2}U_\nu U^\nu,$$

where

$$H_{\mu\nu}=\partial_\mu U_\nu-\partial_\nu U_\mu, \tag{14}$$

we can obtain as a consequence of the equations of motion

$$\partial_\nu H^{\nu\mu}-m^2U^\mu=(\Box-m^2)\,U^\mu+\partial^\mu\,\partial_\nu U^\nu=0. \tag{15}$$

These equations are called *Proca's equations*. Differentiating them with respect to x^μ and performing simple transformations, we get the subsidiary condition (13). Thus, Proca's equations are equivalent to the set of the Klein–Gordon equations together with the condition (13).

The dynamical quantities corresponding to the analogous Lagrangian of the complex vector field

$$\mathscr{L}=-\frac{1}{2}\overset{*}{H}_{\mu\nu}H^{\mu\nu}+m^2\overset{*}{U}_\nu U^\nu, \tag{16}$$

have the following form: the energy-momentum tensor

$$T_{\mu\nu}=\overset{*}{H}_{\mu\sigma}U^\sigma_{;\nu}+\overset{*}{U}{}^\sigma_{;\nu}H_{\mu\sigma}-g_{\mu\nu}\mathscr{L}, \tag{17a}$$

the current vector

$$J_\nu=i\left(\overset{*}{U}{}^\sigma H_{\sigma\nu}-\overset{*}{H}_{\sigma\nu}U^\sigma\right), \tag{17b}$$

and the spin angular-momentum tensor

$$S^{\nu\,(\mu\sigma)} = \overset{*}{U}{}^{\mu} H^{\sigma\nu} - \overset{*}{H}{}^{\mu\nu} U^{\sigma} + \overset{*}{H}{}^{\sigma\nu} U^{\mu} - \overset{*}{U}{}^{\sigma} H^{\mu\nu}. \qquad (17c)$$

For further calculations it is necessary to decompose the potentials into positive- and negative-frequency parts and go over to the three-dimensional representation:

$$U^{\pm}_{\nu}(x) = \frac{1}{(2\pi)^{3/2}} \int \frac{d\boldsymbol{k}}{\sqrt{2k_0}}\, e^{\pm ikx}\, U^{\pm}_{\nu}(\boldsymbol{k}) \qquad (18)$$

(and analogously for $\overset{*}{U}{}^{\pm}$).

Substituting the expansions (18) into the zero ($\nu = 0$) components of the expressions (17) and integrating over $d\boldsymbol{x}$, we obtain the dynamical quantities:

$$P_{\nu} = -\int d\boldsymbol{k}\, k_{\nu} \left[\overset{*}{U}{}^{-}_{\mu}(\boldsymbol{k})\, U^{+,\,\mu}(\boldsymbol{k}) + \overset{*}{U}{}^{+}_{\mu}(\boldsymbol{k})\, U^{-,\,\mu}(\boldsymbol{k})\right], \qquad (19)$$

$$Q = \int d\boldsymbol{k} \left[\overset{*}{U}{}^{-}_{\mu}(\boldsymbol{k})\, U^{+,\,\mu}(\boldsymbol{k}) - \overset{*}{U}{}^{+}_{\mu}(\boldsymbol{k})\, U^{-,\,\mu}(\boldsymbol{k})\right], \qquad (20)$$

$$\boldsymbol{S} = i \int d\boldsymbol{k} \left\{\left[\overset{*}{\boldsymbol{U}}{}^{+}(\boldsymbol{k}) \times \boldsymbol{U}^{-}(\boldsymbol{k})\right] - \left[\overset{*}{\boldsymbol{U}}{}^{-}(\boldsymbol{k}) \times \boldsymbol{U}^{+}(\boldsymbol{k})\right]\right\}. \qquad (21)$$

Taking into account tlie conditions, following from the definitions (18), for complex conjugation,

$$(U^{\pm}_{\nu}(\boldsymbol{k}))^{*} = \overset{*}{U}{}^{\mp}_{\nu}(\boldsymbol{k}),$$

we see that the contribution, corresponding to $\mu = 0$, to the right-hand side of (19) turns out to be negative, while the energy P_0 is of undetermined sign.

As was pointed out above, this difficulty is removed by the imposition of additional conditions, which in the momentum representation (18) take the form

$$k^{\nu} U^{\pm}_{\nu}(\boldsymbol{k}) = 0, \quad k^{\nu} \overset{*}{U}{}^{\pm}_{\nu}(\boldsymbol{k}) = 0. \qquad (22)$$

By virtue of these conditions the components U_{ν} are no longer independent. Expressing with their aid the components $U^{\pm}_{0}$ through the remaining ones:

$$U^{\pm}_{0}(\boldsymbol{k}) = \frac{1}{k_0} k_n U^{\pm}_{n}(\boldsymbol{k}), \quad \overset{*}{U}{}^{\pm}_{0}(\boldsymbol{k}) = \frac{1}{k_0} k_n \overset{*}{U}{}^{\pm}_{n}(\boldsymbol{k}) \qquad (n = 1,\ 2,\ 3),$$

we obtain for the quadratic form under the integral sign in (19) the following expression which depends only on the "spatial" components of the potential:

$$-\overset{*}{U}{}^{\pm}_{\nu} U^{\pm,\nu} = \overset{*}{U}{}^{\pm}(\boldsymbol{k})\, U^{\mp}(\boldsymbol{k}) - \frac{1}{k_0^2}\left(\boldsymbol{k}\overset{*}{U}{}^{\pm}(\boldsymbol{k})\right)(\boldsymbol{k}U^{\mp}(\boldsymbol{k})). \quad (23)$$

3.4. The local reference frame. This expression is diagonalized by the linear substitution

$$U(\boldsymbol{k}) = \boldsymbol{e}_1 a_1(\boldsymbol{k}) + \boldsymbol{e}_2 a_2(\boldsymbol{k}) + \frac{\boldsymbol{k}}{|\boldsymbol{k}|} \cdot \frac{k_0}{m} a_3(\boldsymbol{k}). \quad (24)$$

Here $\boldsymbol{e}_1$ and $\boldsymbol{e}_2$ are unit vectors, orthogonal to the wave vector $\boldsymbol{k}$ and to each other:

$$(\boldsymbol{e}_i \cdot \boldsymbol{e}_j) = \delta_{ij}, \quad \boldsymbol{e}_3 = \boldsymbol{k}/|\boldsymbol{k}| \quad (i,\ j = 1,\ 2,\ 3),$$

i.e., they represent transverse-polarization vectors. The substitution (24) represents the decomposition of the vector potential into its longitudinal and transverse components with respect to the momentum, i.e. the transition to the local reference frame in momentum space. With the aid of (24) we obtain

$$-\overset{*}{U}{}^{\pm}_{\nu}(\boldsymbol{k})\, U^{\mp,\nu}(\boldsymbol{k}) = \overset{*}{a}{}^{\pm}_{n}(\boldsymbol{k})\, a^{\mp}_{n}(\boldsymbol{k}). \quad (25)$$

Inserting this expression into (19) and (20), we arrive at the diagonal expressions for the energy-momentum and charge, the energy being in terms of the new variables manifestly positive definite:

$$P_\nu = \int d\boldsymbol{k}\, k_\nu \left[\overset{*}{a}{}^{+}_{n}(\boldsymbol{k})\, a^{-}_{n}(\boldsymbol{k}) + \overset{*}{a}{}^{-}_{n}(\boldsymbol{k})\, a^{+}_{n}(\boldsymbol{k})\right],$$
$$Q = \int d\boldsymbol{k} \left[\overset{*}{a}{}^{+}_{n}(\boldsymbol{k})\, a^{-}_{n}(\boldsymbol{k}) - \overset{*}{a}{}^{-}_{n}(\boldsymbol{k})\, a^{+}_{n}(\boldsymbol{k})\right].$$

However, the expression for the contribution to the component of the spin along the direction of the momentum vector,

$$\Delta S_3 \sim i\left(\overset{*}{a}{}^{+}_{1}(\boldsymbol{k})\, a^{-}_{2}(\boldsymbol{k}) - \overset{*}{a}{}^{+}_{2} a^{-}_{1} + \overset{*}{a}{}^{-}_{2} a^{+}_{1} - \overset{*}{a}{}^{-}_{1} a^{+}_{2}\right) \quad (26)$$

turns out to be nondiagonal.

The linear substitution

$$a^{\pm}_{1} = \frac{b^{\pm}_{1} + b^{\pm}_{2}}{\sqrt{2}}, \quad a^{\pm}_{2} = \frac{b^{\pm}_{1} - b^{\pm}_{2}}{i\sqrt{2}}, \quad a^{\pm}_{3} = b^{\pm}_{3} \quad (27)$$

(and similarly for the conjugate amplitudes $\overset{*}{a}_n{}^{\pm}$) diagonalizes this expression, leaving unchanged the diagonal form of P_ν and Q:

$$P_\nu = \int d\boldsymbol{k}\, k_\nu \left[\overset{*}{b}{}_n^{+}(\boldsymbol{k})\, b_n^{-}(\boldsymbol{k}) + \overset{*}{b}{}_n^{-}(\boldsymbol{k})\, b_n^{+}(\boldsymbol{k})\right], \tag{28}$$

$$Q = \int d\boldsymbol{k} \left[\overset{*}{b}{}_n^{+}(\boldsymbol{k})\, b_n^{-}(\boldsymbol{k}) - \overset{*}{b}{}_n^{-}(\boldsymbol{k})\, b_n^{+}(\boldsymbol{k})\right], \tag{29}$$

$$\Delta S_3 \sim \left[\overset{*}{b}{}_1^{+}(\boldsymbol{k})\, b_1^{-}(\boldsymbol{k}) - \overset{*}{b}{}_1^{-} b_1^{+} + \overset{*}{b}{}_2^{-} b_2^{+} - \overset{*}{b}{}_2^{+} b_2^{-}\right]. \tag{30}$$

Hence it may be seen that the quadratic combinations of the amplitudes $\overset{*}{b}{}^{\pm}$ and $b^{\mp}$ may be considered as the densities of the average number of particles which have definite values of energy, momentum, charge, and component of spin along the direction of motion. Thus, for example, the quantity $\overset{*}{b}{}_2^{-}(\boldsymbol{k})b_2^{+}(\boldsymbol{k})$ represents the density of particles with momentum $\boldsymbol{k}$, energy k_0, charge -1, and component of spin along the direction of motion equal to $+1$; similarly, $\overset{*}{b}{}_3^{-}(\boldsymbol{k})b_3^{+}(k)$ is the density of particles with momentum $\boldsymbol{k}$, energy k_0, charge -1, zero component spin, etc. As we shall see later, after quantization the amplitude $\overset{*}{b}{}_2^{+}(\boldsymbol{k})$ describes the creation of a particle of energy k_0, momentum $\boldsymbol{k}$, charge -1, and spin component $+1$, while the amplitude $b_2^{-}(\boldsymbol{k})$ describes its annihilation, etc.

Therefore, in accordance with (24) and (27) the amplitudes $a_{1,2}$ correspond to linearly polarized oscillations, while $b_{1,2}$ correspond to circularly polarized oscillations.

Thus, the complex vector field describes positively and negatively charged particles of mass $m \neq 0$ and with three possible values of the component of the spin vector along the direction of motion, equal respectively to 1, 0, -1.

4. The electromagnetic field

4.1. Potential of the electromagnetic field. The free electromagnetic field satisfies the Maxwell equations

$$\operatorname{curl} \boldsymbol{E} = -\dot{\boldsymbol{H}}, \quad \operatorname{curl} \boldsymbol{H} = \dot{\boldsymbol{E}},$$
$$\operatorname{div} \boldsymbol{E} = 0, \qquad \operatorname{div} \boldsymbol{H} = 0.$$

For a more symmetric description of the electromgnetic field, a real convariant four-vector of the electromagnetic potential $A_\nu = (A_0, -\boldsymbol{A})$ is introduced in such a way that

$$\boldsymbol{E} = -\operatorname{grad} A_0 - \dot{\boldsymbol{A}}, \quad \boldsymbol{H} = \operatorname{curl} \boldsymbol{A}.$$

The components of the "four-dimensional curl" of the potential form the well-known antisymmetric tensor of the electromagnetic field

$$F_{\mu\nu} = \partial_\mu A_\nu - \partial_\nu A_\mu, \tag{1}$$

whose components are related to the components of the vectors representing the electric and magnetic field strengths by the relations

$$E^n = -F_{n0}, \quad H^m = \frac{1}{2}\varepsilon_{nmp}F_{np} \qquad (n,\ m,\ p = 1,\ 2,\ 3) \tag{2}$$

It is convenient to exhibit this relation in the following explicit form:

$$F^{\mu\nu} = \begin{pmatrix} 0 & -E^1 & -E^2 & -E^3 \\ E^1 & 0 & H^3 & -H^2 \\ E^2 & -H^3 & 0 & H^1 \\ E^3 & H^2 & -H^1 & 0 \end{pmatrix}.$$

The Maxwell equations are equivalent to the following equations for the components of the tensor F:

$$\left.\begin{aligned} \partial^\nu F_{\nu\mu} = 0, \\ \partial_\sigma F_{\mu\nu} + \partial_\nu F_{\sigma\mu} + \partial_\mu F_{\nu\sigma} = 0. \end{aligned}\right\} \tag{3}$$

In making the further transition in these equations from F to the potential A, we see that the first four equations of (3) are a consequence of the definition (1) and do not lead to any equation for A, while the next four equations yield

$$\Box\, A_\nu + \partial_\nu \partial^\mu A_\mu = 0 \tag{4}$$

which represents an analog of Proca's equations (3.15).

From this it follows that the formulation of the equations of the electromagnetic field in terms of the potential A_ν is really simple, symmetric, and manifestly covariant.

4.2. Gauge invariance and the Lorentz condition. The method of introducing the four-potential A_ν is nonunique. It is not difficult to verify that the observable quantities $\boldsymbol{E}$, $\boldsymbol{H}$, $F_{\mu\nu}$, and also the Maxwell equations are invariant under a so-called gradient transformation of the second kind of the potential A_ν:

$$A_\nu(x) \to A'_\nu(x) = A_\nu(x) + \partial_\nu f(x), \tag{5}$$

which is often called a *gauge transformation*.

Therefore the potential A, which is not an observable quantity, turns out to be nonunique. The function b on the right-hand side of (5) is an arbitrary function. Usually a smoothness condition is imposed on it.

Such a nonuniqueness signifies that the dynamical description of the electromagnetic field in terms of A involves a nonphysical degree of freedom which is usually associated with the value of the four-divergence

$$\partial^\nu A_\nu(x) \equiv \partial A(x) = \chi(x),$$

for which we have introduced a special notation. The requirement that this quantity be equal to zero,

$$\partial A(x) = 0 \tag{6}$$

is known in the electromagnetic field theory as the *Lorentz condition*. Since, if manifestly relativistic covariance is observed, after quantization (see Section 8.5 below) it is not possible to satisfy the Lorentz condition as an operator relation, it will be of use to us to investigate the degree of freedom which is represented by the arbitrary scalar function $\chi(x)$, and to discuss its physical meaning.

For this we shall first digress for a while from the free electromagnetic field and assume that the potential A does not satisfy equation (4). It may be considered to describe a field which is generated by some external sources and, for example, satisfies the inhomogeneous equation

$$\Box A_\nu(x) + \partial_\nu(\partial A(x)) = J_\nu(x),$$

where J_ν is the current of the sources.

Second, we shall pass over to the momentum representation

$$A_\nu(x) = \int dk\, e^{ikx} A_\nu(k),$$

and we shall use the same symbol A to denote both functions and their Fourier transforms. In this representation the operation of differentiation with respect to the configurational variables (and also the converse operations) are algebraic. The gauge transformation (5) acquires the form

$$A'_\nu(k) = A_\nu(k) + k_\nu \tilde{f}(k), \tag{7}$$

where the function $\tilde{f}(k)$ differs by the imaginary unit from the Fourier transform of the function $f(x)$.

We introduce the projection operators

$$P^{\text{tr}}_{\mu\nu}(k) = g_{\mu\nu} - \frac{k_\mu k_\nu}{k^2}, \quad P^{\text{long}}_{\mu\nu}(k) = \frac{k_\mu k_\nu}{k^2}. \tag{8}$$

Note that the operation $(k^2)^{-1}$ is an integral operator in the x-representation and can be written in the form

$$\frac{1}{k^2} A_\nu \sim \int G(x-y)\, A_\nu(y)\, dy,$$

where G is the Green's function for the d'Alembert equation defined by the relation

$$\Box G(x) = -\delta(x). \tag{9}$$

The Green's functions for the equations for free fields are considered below in Section 18. It is shown therein that to define uniquely the solutions of equation (9), it is necessary to take into account the boundary conditions. Here this fact is of no significance, and it will be quite sufficient for us that the function G satisfies equation (9). Let us now decompose the four-potential A into its transverse and longitudinal components, using the projection operators (8):

$$A_\nu = A^{\text{tr}}_\nu + A^{\text{l}}_\nu, \quad A^{\text{tr}}_\nu = P^{\text{tr}}_{\nu\mu} A^\mu, \quad A^{\text{l}}_\nu = P^{\text{long}}_{\nu\mu} A^\mu. \tag{10}$$

These components satisfy the relations

$$k^\nu A^{\text{tr}}_\nu(k) = 0, \quad k^\nu A^{\text{l}}_\nu(k) = kA(k). \tag{11}$$

Here it is convenient to introduce the special notation

$$kA(k) = \xi(k)$$

for the quantity which coincides, up to a factor of the imaginary unit, with the Fourier transform of the function $\chi(x)$.

From the formulae (10) and (11) it follows that the longitudinal component of the electromagnetic field is determined by the function ξ. Actually,

$$A^{l}_{\nu}(k)=\frac{k_{\nu}}{k^2}\,\xi(k).$$

Under the gauge transformation (7) the latter transforms according to the formula

$$\xi'(k)=\xi(k)+k^2\tilde{f}(k).$$

At the same time the transverse component A^{tr} is not changed under the transformation (7). It presents no difficulty to verify also that the gauge-invariant tensor F does not depend on A^{l} (i.e. on ξ). It is therefore clear that the degree of freedom described by the scalar function $\xi(k)$ involves all the arbitrariness connected with the nonuniqueness of the four-potential A and is nonphysical.

4.3. The generalized Lagrangian. The equations of motion (4) may be obtained with the use of the variational principle from the Lagrangian

$$\mathscr{L}_{\mathrm{tr}}=-\frac{1}{4}F_{\mu\nu}F^{\mu\nu}, \tag{12}$$

which represents the well-known expression $\frac{1}{2}(\boldsymbol{E}^2-\boldsymbol{H}^2)$ written in terms of F. The subscript "tr" signifies that the "transverse" Lagrangian (12) is gauge-invariant and contains no dependence on the longitudinal components A^{l}.

Let us point out that the differential operator of the equation of motion (4),

$$K^{\mathrm{tr}}_{\mu\nu}=\partial_{\mu}\partial_{\nu}-g_{\mu\nu}\partial^2,$$

which is proportional to the transverse projection operator P^{tr}, has no inverse operator, and this complicates the solution of the corresponding inhomogeneous equation of motion. The transverse Lagrangian $\mathscr{L}^{\mathrm{tr}}$ is specified from the point of view of the Hamiltonian formalism. We recall that in the Hamiltonian formalism an important part is played by the canonical momenta, which are defined as the derivatives of the Lagrangian with respect to the velocities $v_i = \dot{q}_i$, i.e., with respect to the time derivatives of the system coordinates q_i. In the case of a system with an infinite number of degrees of freedom—a field—the field functions $\varphi(\boldsymbol{x}, t) = \varphi(x)$ are chosen to be the generalized coordinates. Therefore, the generalized canonical momenta of the classical field are defined in the following manner:

$$\pi_i(x) = \frac{\partial \mathscr{L}}{\partial \dot{\varphi}_i(x)}.$$

The Hamiltonian formalism represents the basis of the so-called canonical quantization, which is presented in Section 6. We refer the reader to this section for a more detailed presentation of the Hamiltonian formalism for fields, and now we return to the Lagrangian (12).

Choosing as generalized coordinates the components of the four-potential A_ν, we arrive, by elementary calculation, at the conclusion that the generalized momenta corresponding to the zero component A_0 are equal to zero:

$$\pi_0(x) = F_{00} = 0.$$

In other words, the Lagrangian $\boldsymbol{L}_{\text{tr}}$ does not contain any dependence on the "velocity" $\dot{A}_0$. Lagrangians with such properties are sometimes called degenerate or *singular*. They correspond to systems with *constraints*. For the description of systrems with constraints the generalized Hamiltonian dynamics introduced by Dirac (1968) is used. In our case the transition to the generalized dynamics is equivalent to the following modification of the Lagrangian:

$$\mathscr{L}_{\text{tr}} \rightarrow \mathscr{L} = -\frac{1}{4} F_{\mu\nu}F^{\mu\nu} + \frac{a}{2}(\partial A)^2, \tag{13}$$

where a is a numerical coefficient.

This Lagrangian leads to the following equations of motion:

$$\partial^\nu F_{\mu\nu} + a\partial_\mu(\partial A) = \Box A_\mu + (1+a)\,\partial_\mu(\partial A) = 0, \tag{14}$$

from which, in particular, it follows that the longitudinal component of the electromagnetic field satisfies the d'Alembert equation

$$\Box\, \partial^\nu A_\nu(x) = \Box\, \chi(x) = 0. \tag{15}$$

Note that this property of the longitudinal component is preserved also in the case when the electromagnetic field interacts with other fields, i.e., when the right-hand side of the equations of motion (14) contains the components of an external current J_ν, satisfying the equation of continuity $\partial^\nu J_\nu = 0$, Thus, the longitudinal degree of freedom $\chi(x)$ interacts neither with the transverse components A^{tr} nor with any other fields.

As is seen, the Lagrangian (13) and equations (14) are no longer, generally speaking, invariant under the transformation (5). Thus, we conclude that the insertion of the term $\sim(\partial A)^2$ into the Lagrangian fixes the gauge. At the same time the expressions (13), (14) do not change under the specialized gauge transformation

$$A'_\nu(x) = A_\nu(x) + \partial_\nu f_0(x), \tag{16}$$

the gauge function of which obeys the d'Alembert equation

$$\Box f_0(x) = 0. \tag{17}$$

The insertion of the gauge-fixing term into the Lagrangian leads also to an operator in the equation of motion

$$K_{\mu\nu} = g_{\mu\nu}\Box + (1+a)\partial_\mu\partial_\nu \to (1+a)k_\mu k_\nu = k^2 P^{\text{tr}}_{\mu\nu}(k) - ak^2 P^{1}_{\mu\nu}(\boldsymbol{k})$$

which is no longer singular, i.e., to the existence of its inverse

$$K^{-1}_{\mu\nu} = \frac{1}{k^2} P^{\text{tr}}_{\mu\nu}(k) + \frac{d_1}{k^2} P^{1}_{\mu\nu}(k), \quad d_1 = \frac{-1}{a}. \tag{18}$$

4.4. The diagonal gauge. If the factor d_1 fixing the gauge is equal to one, we then come to diagonal expressions for K and K^{-1}:

$$d_1 = 1: \; K_{\mu\nu} = g_{\mu\nu}k^2, \quad K^{-1}_{\mu\nu} = g_{\mu\nu}(k^2)^{-1}.$$

Such a gauge is called diagonal. In this gauge the Lagrangian

$$\mathscr{L}_{\text{diag}} = -\frac{1}{4} F_{\mu\nu}F^{\mu\nu} - \frac{1}{2}(\partial A)^2$$

may, by integration by parts, be transformed into the form

$$\mathscr{L}_{\text{diag}} = -\frac{1}{2}\partial^\nu A_\mu(x)\,\partial_\nu A^\mu(x), \tag{19}$$

which represents the covariant sum of the Lagrangians for each of the components A_μ separately.

The Lagrangian (19) leads to equations of motion

$$\Box A_\mu(x) = 0 \tag{20}$$

in the form of d'Alembert equations. This Lagrangian will turn out to be the most convenient one for quantization. According to its physical meaning (the observed fields $\boldsymbol{E}$, $\boldsymbol{H}$), the Lagrangian (19), as well as the more general expression (13), is equivalent to the initial "transverse" Lagrangian (12).

We may point out, furthermore, that the equations of motion (20) and the Lagrangian (19), while representing a specific case of the general relations (13) and (14), are invariant under the transformation (16). In any fixed reference frame the function f_0 of this transformation can be chosen so that the zero component of the potential A_0, which satisfies the free equation (20), vanishes. The Lorentz condition then takes the form

$$\operatorname{div} \boldsymbol{A} = 0. \tag{21}$$

The corresponding gauge is called the *Coulomb gauge*.

In the momentum representation equation (21) looks as follows:

$$\boldsymbol{k}\boldsymbol{A}(k) = 0.$$

Evidently, in the Coulomb gauge the electromagnetic field satisfies the condition of transversality in the usual (three-dimensional) sense. Thus, notwithstanding the fact that the electromagnetic field is described by a four-component potential, only the two linearly independent components, orthogonal to the wave vector $\boldsymbol{k}$, have physical meaning. The actual reduction of the four-component field to a two-component one which occurs due to gauge invariance is closely connected with the vanishing of the rest mass of the field particles—the photons. Namely, by virtue of this important property, only derivatives of A_ν are present in the field equations and there arises the property of gauge invariance of the electromagnetic field.

From the Lagrangian (19) by standard methods we obtain the energy-momentum tensor

$$T^{\mu\nu} = - A_\sigma^{;\mu} A^{\sigma;\nu} - g^{\mu\nu}\mathcal{L};$$

the spatial densities of the four-vector P^{ν} components

$$T^{00} = -\frac{1}{2}(\dot{A}^{\nu}\dot{A}_{\nu} + \partial_n A^{\nu}\partial_n A_{\nu}), \quad T^{0k} = \dot{A}_{\nu}\partial^k A^{\nu};$$

the spin angular-momentum tensor

$$S^{\tau(\nu\mu)} = A^{\mu}A^{\nu;\tau} - A^{\nu}A^{\mu;\tau};$$

and the spatial density of the spin vector

$$\boldsymbol{S} = [\boldsymbol{A}(x) \times \dot{\boldsymbol{A}}(x)].$$

The Lorentz condition (6) must be imposed to complement the Lagrangian formalism.

For calculating the dynamical invariants we perform decomposition into frequency terms and introduce the momentum representation for the free field:

$$A_{\nu}^{\pm}(x) = \frac{1}{(2\pi)^{3/2}} \int \frac{d\boldsymbol{k}}{\sqrt{2k_0}} e^{\pm ikx} A_{\nu}^{\pm}(\boldsymbol{k}). \qquad (22)$$

4.5. Transition to the local reference frame. To perform diagonalization, it is necessary in the representation (22) to decompose the field into components related to the local reference frame. As the rest mass is equal to zero ($m = 0$), the transformation (3.24) must be modified. We represent $A(\boldsymbol{k})$ as a sum of the transverse, longitudinal, and time components:

$$A_{\nu}^{\pm}(\boldsymbol{k}) = \overset{1}{e}_{\nu} a_1^{\pm}(\boldsymbol{k}) + \overset{2}{e}_{\nu} a_2^{\pm}(\boldsymbol{k}) + \overset{3}{e}_{\nu} a_3^{\pm}(\boldsymbol{k}) + \overset{0}{e}_{\nu} a_0^{\pm}(\boldsymbol{k}). \qquad (23)$$

Here $e^1(\boldsymbol{k})$ and $e^2(\boldsymbol{k})$ are purely spatial transverse unit vecotrs orthogonal to each other and similar to the ones introduced in (3.24), $e^3 = \boldsymbol{k}/|\boldsymbol{k}|$ is the longitudinal unit orthogonal vector,

$$e_0^i = 0, \quad \boldsymbol{e}^i\boldsymbol{e}^j = \delta_{ij}, \quad [\boldsymbol{e}^i \times \boldsymbol{e}^j] = \boldsymbol{e}^k \qquad (i, j, k = 1, 2, 3),$$

and e^0 is the unit time component of the 4-vector: $e_{\nu}^0 = \delta_{\nu 0}$. It is easy to verify that under the transformation (23) the basic quadratic form remains diagonal:

$$A_\nu^+(\boldsymbol{k})\,A_\nu^-(\boldsymbol{k}) = g^{\sigma\rho} a_\sigma^+(\boldsymbol{k})\,a_\rho^-(\boldsymbol{k}). \tag{24}$$

Inserting the expansion (23) into the Lorentz condition written separately for the positive- and negative-frequency components, we obtain two relations

$$|\boldsymbol{k}|\,a_3^{\pm}(\boldsymbol{k}) - k_0 a_0^{\pm}(\boldsymbol{k}) = 0.$$

Noting that since the mass m is zero, $|\boldsymbol{k}| = k_0$, we hence obtain

$$a_3^+(\boldsymbol{k})\,a_3^-(\boldsymbol{k}) = a_0^+(\boldsymbol{k})\,a_0^-(\boldsymbol{k}). \tag{25}$$

The obtained relation (25) means that because of the Lorentz condition the densities of the average numbers of the "longitudinal" photons $a_3^+a_3^-$ and of the "timelike" photons $a_0^+a_0^-$ are equal to each other. Since their contributions to the right-hand side of (24) are of opposite sign, one may therefore speak of the "longitudinal" and "timelike" photons as "compensating" each other. Substituting (25) into (24), we obtain

$$-A_\nu^+(\boldsymbol{k})\,A_\nu^-(\boldsymbol{k}) = a_1^+(\boldsymbol{k})\,a_1^-(\boldsymbol{k}) + a_2^+(\boldsymbol{k})\,a_2^-(\boldsymbol{k}).$$

Further, on computing the energy-momentum four-vector

$$P^\nu = \sum_{s=1,\,2} \int d\boldsymbol{k}\, k^\nu a_s^+(\boldsymbol{k})\,a_s^-(\boldsymbol{k}),$$

we find that in the case under consideration, as in the case of the vector field, the energy turns out to be positive definite only due to the Lorentz condition.

Defining, further, the spin vector

$$\boldsymbol{S} = i\int d\boldsymbol{k}\,[\boldsymbol{A}^+(\boldsymbol{k})\times\boldsymbol{A}^-(\boldsymbol{k})] = i\int d\boldsymbol{k}\,\varepsilon_{abc} e^a a_b^+(\boldsymbol{k})\,a_c^-(\boldsymbol{k}),$$

we find its component along the direction of the propagation vector in the form

$$\Delta S_3 \sim i\,[a_1^+(\boldsymbol{k})\,a_2^-(\boldsymbol{k}) - a_2^+(\boldsymbol{k})\,a_1^-(\boldsymbol{k})].$$

Introducing the new amplitudes b_1 and b_2 as in Section 3, we obtain for P_ν and S_3 the "diagonal" expressions

$$P_\nu = \int dk\, k_\nu [b_1^+(\boldsymbol{k})\, b_1^-(\boldsymbol{k}) + b_2^+(\boldsymbol{k})\, b_2^-(\boldsymbol{k})], \tag{26}$$

$$\Delta S_3 \sim [b_1^+(\boldsymbol{k})\, b_1^-(\boldsymbol{k}) - b_2^+(\boldsymbol{k})\, b_2^-(\boldsymbol{k})], \tag{27}$$

from which it follows directly that the products

$$n_s(\boldsymbol{k}) = b_s^+(\boldsymbol{k})\, b_s^-(\boldsymbol{k}) \qquad (s = 1, 2) \tag{28}$$

may be regarded as densities of the average numbers of particles of zero mass, momentum $\boldsymbol{k}$, and energy $|\boldsymbol{k}|$, having a component of the spin angular momentum along the direction of the propagation vector $\boldsymbol{k}$ equal to $+1$ $(S = 1)$ and -1 $(S = 2)$. We are in fact dealing with *photons*.

Thus, the transition to the momentum representation allows one to see directly that the electromagnetic field describes transverse photons with two possible values of the component of spin along the direction of motion.

5. The Dirac field

5.1. The Dirac equation and Dirac matrices. Fields which transform in accordance with the spinor representations of the Poincaré group are usually called spinor fields. These fields after quantization describe particles of half-integer spin ($\frac{1}{2}, \frac{3}{2}, \frac{5}{2}, \ldots$). The simplest of spinor fields is the field of spin $\frac{1}{2}$. Such a field for particles of nonzero mass was introduced by Dirac for the description of electrons. It also describes muons, nucleons (protons and neutrons), and certain hyperons. It is this field that we shall henceforth call the *Dirac field*. Following Dirac, we shall obtain the corresponding equations by means of the "factorization" of the Klein-Gordon operator

$$\Box - m^2 = P_\nu P^\nu - m^2,$$

where for convenience the usual quantum-mechanical notation

$$P_\nu = i\frac{\partial}{\partial x^\nu} = i\partial_\nu.$$

has been adopted. The operator $\Box - m^2$ is quadratic in the derivatives ∂_ν and, as may easily be seen, cannot be represented in the form of two factors linear in P_ν with numerical coefficients. Indeed, if by analogy with the formula for factoring the difference of two squares, one attempts to write it in the form

$$\Box - m^2 = (P + m)(P - m),$$

where P is some linear combination of the operators P_ν with coefficients γ^ν:

$$P = P_\nu \gamma^\nu,$$

then one must demand that the relation

$$P_\nu P^\nu = (P_\mu \gamma^\mu)^2,$$

should hold, on expanding the right-hand side of which one finds the condition determining the coefficients γ:

$$\gamma^\nu \gamma^\mu + \gamma^\mu \gamma^\nu = 2g^{\mu\nu}. \tag{1}$$

Since, in accordance with this condition, the quantities γ^ν with different indices anticommute, they are not ordinary numbers, but they may be chosen in the form of *matrices*.

With the aid of these quantities the Klein-Gordon operator may be represented as a product of two commuting matrix operators

$$\Box - m^2 = (i\gamma^\nu \partial_\nu + m)(i\gamma^\mu \partial_\mu - m),$$

and in order that the field function may satisfy the Klein-Gordon equation

$$(\Box - m^2)\psi = 0,$$

we can demand that it should satisfy one of the following first-order equations:

$$(i\gamma^\nu \partial_\nu - m)\psi(x) = 0 \quad \text{or} \quad (i\gamma^\nu \partial_\nu + m)\psi(x) = 0.$$

By tradition, we shall choose as the equation of motion for the spinor field the first of these equations:

$$(i\hat{\partial} - m)\psi(x) = 0, \quad \hat{\partial} \equiv \gamma^\nu \partial_\nu. \tag{2}$$

Equation (2) is, certainly, less general than the Klein–Gordon equation. Therefore one may expect it to contain more detailed information. As is known, this actually is the case, since with the aid of equation (2) Dirac for the first time succeeded in describing the spin of the electron equal to $\frac{1}{2}$.

Equation (2) is called the *Dirac equation*, and the matrices γ^ν defined by the relations (1) are called the *Dirac matrices*.

Due to the matrix character of the differential operator of the Dirac equation, the field function ψ turn out to be multicomponent. As is shown in Appendix II, in the simplest case the Dirac field ψ is a four-component spinor; it is often represented as a column

$$\psi = \begin{pmatrix} \psi_1 \\ \psi_2 \\ \psi_3 \\ \psi_4 \end{pmatrix},$$

which, for reasons of economy of space, we shall sometimes write in the following manner: $\psi = (\psi_1, \psi_2, \psi_3, \psi_4)^T$.

Thus, the single matrix equation (2) contains four equations

$$(i\gamma^\nu \partial_\nu - mI)_{\alpha\beta} \psi_\beta (x) = 0 \qquad (\alpha, \ \beta = 1, \ 2, \ 3, \ 4).$$

The four quantities γ^ν are square matrices of rank four with matrix elements $\gamma^\nu_{\alpha\beta}$, and I is the unit diagonal matrix ($I_{\alpha\beta} = \delta_{\alpha\beta}$, the symbol for which we shall often omit. The main properties of the Dirac matrices are determined by the (anti)commutation relation (1) and are expounded in detail in Appendix II. From (1)it specifically follows that the Dirac matrices may be chosen to be unitary:

$$\gamma^\nu \overset{+}{\gamma}{}^\nu = \overset{+}{\gamma}{}^\nu \gamma^\nu = I, \quad \text{(no summation over } \nu\text{)},$$

if the conditions of Hermitian conjugation are imposed in the form of

$$\overset{+}{\gamma}{}^\nu = g_{\nu\mu} \gamma^\mu \equiv \gamma_\nu. \tag{3}$$

As is seen, the operation of lowering (raising) the Lorentz indices of the Dirac matrices is performed according to the usual rules. The Hermitian-conjugate spinor $\overset{*}{\psi}$ represents a 4-component row $\overset{*}{\psi} = (\overset{*}{\psi}_1, \overset{*}{\psi}_2, \overset{*}{\psi}_3, \overset{*}{\psi}_4)$, the elements of which are derived from the components ψ_α by complex conjugation. Multiplying the spinor on the right by the matrix γ^0, we obtain the so-called Dirac-conjugate (also referred to as simply the adjoint) spinor

$$\bar{\psi}(x) = \overset{*}{\psi}(x) \gamma^0.$$

Taking into account (13), it is possible to demonstrate (see Appendix II) that the adjoint spinor $\bar{\psi}$ satisfies the equation

$$i\partial_\nu \bar{\psi}(x)\,\gamma^\nu + m\bar{\psi}(x) = 0. \tag{4}$$

We recall that each of the components of the spinor $\psi(x)$ satisfies the Klein–Gordon equation. To verify this, it is sufficient to operate on the Dirac equation (2) on the left with the operator $i\hat{\partial} + m$ and apply the commutation relation (1).

Note that custom implies that the product $\bar{\psi}O\psi$, where O is a matrix, is meant to contain summation over the spinor indices

$$\bar{\psi}O\psi = \bar{\psi}_\alpha O_{\alpha\beta}\psi_\beta$$

and represents an ordinary number, while an expression of the form $\psi\bar{\psi}$ is a matrix

$$\psi\bar{\psi} = A, \quad A_{\alpha\beta} = \psi_\alpha\bar{\psi}_\beta.$$

5.2. Lagrangian formalism. The Dirac equation (2) and its conjugate equation (4) may be obtained from the Lagrangian

$$\mathscr{L} = (i/2)\left[\bar{\psi}(x)\,\gamma^\nu\partial_\nu\psi - \partial_\nu\bar{\psi}(x)\,\gamma^\nu\psi(x)\right] - m\bar{\psi}(x)\,\psi(x) \tag{5}$$

with the use of the variational principle. A peculiarity of this Lagrangian is that it vanishes if $\bar{\psi}$ and ψ in it satisfy the equations of motion.

Using Noether's theorem, we hence obtain the energy-momentum tensor

$$T^{\mu\nu} = (i/2)\left[\bar{\psi}(x)\,\gamma^\mu\partial^\nu\psi(x) - \partial^\nu\bar{\psi}(x)\,\gamma^\mu\psi(x)\right] \tag{6}$$

and the current vector

$$J^\mu(x) = \bar{\psi}(x)\,\gamma^\mu\psi(x). \tag{7}$$

In deriving (6) the equations of motion were used.

In order to calculate the spin tensor, we note that after summing over the spin indices, the formula (2.24) may be written in the form

$$S^{\tau(\mu\nu)} = -\frac{\partial\mathcal{L}}{\partial\psi_{;\tau}} A^{\psi(\mu\nu)}\psi(x) - \bar{\psi}(x) A^{\bar{\psi}(\mu\nu)} \frac{\partial\mathcal{L}}{\partial\bar{\psi}_{;\tau}},$$

The infinitesimal generators A^{ψ}, $A^{\bar{\psi}}$ occurring in it are defined, with the aid of the formulae of the infinitesimal Lorentz transformations

$$\psi'(x') = \left(1 - \frac{i}{2}\sigma^{\mu\nu}\varphi\right)\psi(x), \quad \bar{\psi}'(x') = \bar{\psi}(x)\left(1 + \frac{i}{2}\sigma^{\mu\nu}\varphi\right),$$

which follow from the formulae (AII.16, 18, 20), the form (see Exercise S29)

$$A^{\psi(\mu\nu)} = -\frac{i}{2}\sigma^{\mu\nu}, \quad A^{\bar{\psi}(\mu\nu)} = \frac{i}{2}\sigma^{\mu\nu},$$

where $\sigma^{\mu\nu}$ is the so-called matrix spin tensor (see (AII.2)). Therefore

$$S^{\tau(\mu\nu)} = (1/4)\,\bar{\psi}(x)\,(\gamma^{\tau}\sigma^{\mu\nu} + \sigma^{\mu\nu}\gamma^{\tau})\,\psi(x). \tag{8}$$

5.3 Momentum representation. Taking into account that the components of the Dirac spinor $\psi_{\alpha}(x)$ satisfy the Klein–Gordon equation, we write the momentum representation in the form

$$\psi(x) = (2\pi)^{-3/2}\int dp\, e^{ipx}\delta(p^2 - m^2)\,\tilde{\psi}(p),$$

where the amplitude $\tilde{\psi}$ satisfies the Dirac equation in the momentum representation,

$$(\hat{p} + m)\,\tilde{\psi}(p)\,|_{p^2 = m^2} = 0.$$

(Recall that in our notation $\hat{p} = \gamma^{\nu}p_{\nu}$.)

Decomposing, as usual, the field function into the frequency components $\psi = \psi^{+} + \psi^{-}$, and explicitly integrating over p_0, we define the formulae of the three-dimensional momentum representation in the form

$$\psi^{\pm}(x) = (2\pi)^{-3/2}\int d\boldsymbol{p}\,\psi^{\pm}(\boldsymbol{p})\exp(\pm ipx). \tag{9}$$

Here in the integrand (and throughout the whole section)

$$p_0 = +\sqrt{\boldsymbol{p}^2 + m^2}. \tag{10}$$

The spinors $\psi^{\pm}(p)$ obey the matrix equations

$$(m \pm \hat{p})\,\psi^{\pm}(\boldsymbol{p}) = 0.$$

The corresponding equations for the adjoint spinors are

$$\bar{\psi}^{\pm}(\boldsymbol{p})\,(m \mp \hat{p}) = 0. \tag{11}$$

The explicit form of the components $\psi^{\pm}{}_{\alpha}$ and $\psi^{\pm}{}_{\beta}$ depends on the representation of the Dirac matrices γ and may be defined in the following manner. By virtue of its covariance, equation (10) may be considered in some fixed reference frame, keeping in mind that transition to any other frame can always be accomplished through the transformations given in Appendix II. Choosing as such a system the one in which $\boldsymbol{p} = 0$, we find from (10)

$$(\gamma^0 \pm I)\,\psi^{\pm}(0) = 0.$$

From this, in the standard representation (AII.5), we obtain

$$\psi^-_{\alpha}(0) = c_1\delta_{1\alpha} + c_2\delta_{2\alpha}, \quad \psi^+_{\beta}(0) = c_3\delta_{3\beta} + c_4\delta_{4\beta}. \tag{12}$$

Here γ, β are spinor indices, and $\delta_{\alpha\beta}$ are the Kronecker symbols. The solution for an arbitrary nonzero $\boldsymbol{p}$ may be obtained from (12) by an appropriate Lorentz transformation.

The equations satisfied by ψ^+ and ψ^- may also be written in the form

$$(\pm\gamma^0 p_0 - \boldsymbol{\gamma p} + m)\,\psi^{\pm}(\pm p) = 0,$$

in which they differ from each other only by the sign of p_0. As has just been established, each of them has two linearly independent solutions. Hence it follows that the Dirac equation for each given value of the four-vector p (the sign of the component p_0 being fixed) has only two linearly independent solutions.

Corresponding to this, the adjoint spinor $\bar{\psi}$ also has two linearly independent solutions for a fixed sign of p_0. Since the considered solutions of the Dirac equation are complex, it follows that they describe particles differing by their charges (to be more precise, particles and antiparticles). The existence of two linearly independent solutions means that these particles may exist in two different states which, as will be shown below, differ in the sign of the component of spin along the direction of motion.

5.4. Decomposition into spin states. Denote by $v^{s,+}(\boldsymbol{p})$ the normalized (see the normalization relations (16) below) linearly independent solutions for $p_0 > 0$, i.e., of the first of the equations (10), and by $v^{s,-}(\boldsymbol{p})$ those for $p_0 < 0$, i.e., of the second of the equations (10) ($s = 1,2$). We then write the expansions of the functions $\psi^{\pm}(\boldsymbol{p})$ into spin states in the form

$$\psi_\alpha^{\pm}(\boldsymbol{p}) = \sum_{s=1,\,2} a_s^{\pm}(\boldsymbol{p})\, v_\alpha^{s,\,\pm}(\boldsymbol{p}) = a_s^{\pm}(\boldsymbol{p})\, v_\alpha^{s,\,\pm}(\boldsymbol{p}). \tag{13}$$

Similarly, for the adjoint spinor

$$\overline{\psi}^{\pm}(\boldsymbol{p}) = \overset{*}{a}{}_s^{\pm}(\boldsymbol{p})\, \overline{v}{}^{s,\,\pm}(\boldsymbol{p}) \equiv \overset{*}{a}{}_s^{\pm}(p)\, \overset{*}{v}{}^{s,\,\pm}(p)\, \gamma^0. \tag{14}$$

Since $\overline{\psi}^{+}$ and $\overline{\psi}^{-}$ are the positive- and negative-frequency parts of the function $\overline{\psi}$, the conditions of Hermitian conjugation for the normalized spinors v have the form

$$(v^{s,\,\pm}(\boldsymbol{p}))^* = \overset{*}{v}{}^{s,\,\mp}(\boldsymbol{p}). \tag{15}$$

Therefore the conditions for the orthonormality of the spinors may be written in the form

$$\overset{*}{v}{}^{s,\,\pm}(\boldsymbol{p})\, v^{r,\,\mp}(\boldsymbol{p}) \equiv \sum_{\alpha=1,\,2,\,3,\,4} \overset{*}{v}{}_\alpha^{s,\,\pm}(\boldsymbol{p})\, v_\alpha^{r,\,\mp}(\boldsymbol{p}) = \delta^{sr}. \tag{16}$$

By means of purely algebraic transformations, one may obtain from (15) and (16) and the Dirac equations a number of relations for quadratic forms in the spinors v and $\overline{v} = \overset{*}{v}\gamma^0$, the most important of which are the following: the condition of orthonormality for the adjoint spinors,

$$\overline{v}{}^{s,\,\pm}(\boldsymbol{p})\, v^{r,\,\mp}(\boldsymbol{p}) = \pm\frac{m}{p_0}\,\delta^{sr}; \tag{17}$$

the condition of mutual orthogonality for the spinors v with arguments differing in sign,

$$\overset{*}{v}^{s,\pm}(\boldsymbol{p})\, v^{r,\pm}(-\boldsymbol{p}) = 0; \tag{18}$$

the relations ($l, n = 1,2,3$)

$$\left.\begin{aligned} &\overset{*}{v}^{s,\pm}(\boldsymbol{p})\,(p^n\gamma^l - p^l\gamma^n - m\gamma^n\gamma^m)\, v^{r,\pm}(-\boldsymbol{p}) = 0, \\ &\sum_n p_n \overset{*}{v}^{s,\pm}(\boldsymbol{p})\,(\gamma^n\gamma^l - \gamma^l\gamma^n)\, v^{r,\mp}(-\boldsymbol{p}) = 0 \end{aligned}\right\} \tag{19}$$

and finally, the formulas for summation over the spin index,

$$v_\alpha^{s,\pm}(\boldsymbol{p})\, \bar{v}_\beta^{s,\mp}(\boldsymbol{p}) = \frac{(\hat{p} \mp m)_{\alpha\beta}}{2p_0}. \tag{20}$$

5.5. Dynamical invariants. Substituting (9), (13), (14) into (6) and integrating $T^{0\nu}$ over $d\boldsymbol{x}$, we obtain, taking into account the condition of orthonormality (16), the energy-momentum four-vector

$$P_\nu = \int d\boldsymbol{k}\, k_\nu [\overset{*}{a}{}_s^+(\boldsymbol{k})\, a_s^-(\boldsymbol{k}) - \overset{*}{a}{}_s^-(\boldsymbol{k})\, a_s^+(\boldsymbol{k})]. \tag{21}$$

Considering that, in accordance with the definitions used, the rules of complex conjugation of the amplitudes in the momentum representation have the form $(a^\pm)^* = \overset{*}{a}{}^\pm$, we come to the conclusion that the expression

$$P_0 = \int d\boldsymbol{k}\, k_0 [\overset{*}{a}{}_s^+(\boldsymbol{k})\, a_s^-(\boldsymbol{k}) - \overset{*}{a}{}_s^-(\boldsymbol{k})\, a_s^+(\boldsymbol{k})] \tag{22}$$

for the energy which follows from (21) is not positive definite. Positive definiteness of the energy of the spinor field is achieved only in quantum theory by Fermi–Dirac quantization.

Proceeding to the calculation of the spin vector, we note that in the standard representation (AII.5) the components of the spin tensor $\sigma^{\mu\nu}$ may be expressed in terms of the matrices α_i and σ_j in the following way:

$$\sigma^{\mu\nu} = \begin{pmatrix} \sigma^{00} & \sigma^{01} & \sigma^{02} & \sigma^{03} \\ \sigma^{10} & \sigma^{11} & \sigma^{12} & \sigma^{13} \\ \sigma^{20} & \sigma^{21} & \sigma^{22} & \sigma^{23} \\ \sigma^{30} & \sigma^{31} & \sigma^{32} & \sigma^{33} \end{pmatrix} = \begin{pmatrix} 0 & i\alpha_1 & i\alpha_2 & i\alpha_3 \\ -i\alpha_1 & 0 & \sigma_3 & -\sigma_2 \\ -i\alpha_2 & -\sigma_3 & 0 & \sigma_1 \\ -i\alpha_3 & \sigma_2 & -\sigma_1 & 0 \end{pmatrix}. \tag{23}$$

Therefore, setting $\tau = 0$ and $\mu, \nu = 1, 2, 3$ in (8), we see that the spatial density of the spin vector is given by the matrix vector σ:

$$S = (^1/_2) \int \overset{*}{\psi}(x)\, \sigma\psi(x)\, dx. \tag{24}$$

In contrast to the spins of the vector and electromagnetic fields, the spin vector (24) of the Dirac field is not conserved in time (a consequence of the lack of symmetry in the energy-momentum tensor). However, in the case that the field functions ψ, $\bar{\psi}$ do not depend on some of the coordinates $x^0, x^1, \ldots,$ it is possible to make the equation of continuity hold for some of the components of the tensor S, so that the corresponding integrals remain conserved in time. Thus, setting $\partial/\partial x_1 = \partial/\partial x_2 = 0$, we obtain

$$\partial_\nu S^{\nu(12)} = 0,$$

from which it follows that the component of the spin vector along the x^3-axis,

$$S_3 = \int d\boldsymbol{x}\, S^{0(12)}$$

is conserved in time. In the momentum representation this statement is equivalent to the conservation of the component of the spin vecor along the direction of motion.

By passing in (24) to the three-dimensional momentum representation and carrying out the integration over the three-dimensional configuration space, we obtain

$$\Delta S \sim (^1/_2)\{\overset{*}{\psi}{}^+(\boldsymbol{k})\, \sigma\psi^-(\boldsymbol{k}) + \overset{*}{\psi}{}^-(\boldsymbol{k})\, \sigma\psi^+(\boldsymbol{k}) + \\ + e^{2ik_0x_0}\overset{*}{\psi}{}^+(\boldsymbol{k})\, \sigma\psi^+(\boldsymbol{k}) + e^{-2ik_0x_0}\overset{*}{\psi}{}^-(\boldsymbol{k})\, \sigma\psi^-(\boldsymbol{k})\}.$$

Restricting ourselves to the consideration of only the S_3 component, we make use of the first of the relations (19), which we write in the form

$$\overset{*}{v}{}^{\pm}(\boldsymbol{k})\, \sigma v^{\pm}(-\boldsymbol{k}) = \frac{i}{m}\overset{*}{v}{}^{\pm}(\boldsymbol{k})\, [\boldsymbol{k} \times \gamma]\, v^{\pm}(-\boldsymbol{k}).$$

By virtue of this relation the time-dependent terms in S_3 disappear in the reference frame in which $k_1 = k_2 = 0$, and we arrive at an expression of the following form:

$$\Delta S_3 \sim (^1/_2)\, [\overset{*}{a}{}^+_s(\boldsymbol{k})\, a^-_r(\boldsymbol{k})\, \overset{*}{v}{}^{s,+}(\boldsymbol{k})\, \sigma_3 v^{r,-}(\boldsymbol{k}) + \\ + \overset{*}{a}{}^-_s(\boldsymbol{k})\, a^+_r(\boldsymbol{k})\, \overset{*}{v}{}^{s,-}(\boldsymbol{k})\, \sigma_3 v^{r,+}(\boldsymbol{k})], \tag{25}$$

In order to proceed it is convenient to refer to a definite representation of the Dirac matrices. In the standard representation the matrix σ_3 is diagonal:

$$\sigma_3 = \begin{pmatrix} 1 & \cdot & \cdot & \cdot \\ \cdot & -1 & \cdot & \cdot \\ \cdot & \cdot & 1 & \cdot \\ \cdot & \cdot & \cdot & -1 \end{pmatrix}.$$

Choosing the normalized spinors in the reference frame $k_1 = k_2 = 0$ in the form

$$\left.\begin{aligned} v^{1,-} &= N^{-1}\left(1, \quad 0, \quad \frac{k_3}{k_0+m}, \quad 0 \right)^{\mathrm{T}}, \\ v^{2,-} &= N^{-1}\left(0, \quad 1, \quad 0, \quad \frac{-k_3}{k_0+m} \right)^{\mathrm{T}}, \\ v^{1,+} &= N^{-1}\left(\frac{k_3}{k_0+m}, \quad 0, \quad 1, \quad 0 \right)^{\mathrm{T}}, \\ v^{2,+} &= N^{-1}\left(0, \quad \frac{-k_3}{k_0+m}, \quad 0, \quad 1 \right)^{\mathrm{T}}, \end{aligned}\right\} \tag{26}$$

where N is the normalization constant equal to

$$N = \left[1 + \left(\frac{k_3}{k_0+m}\right)^2\right]^{1/2} = \left(\frac{2k_0}{k_0+m}\right)^{1/2},$$

we find that the expression (25) assumes the form

$$\Delta S_3 \sim (1/2)\,[\overset{*}{a}{}_1^{+} a_1^{-} - \overset{*}{a}{}_2^{+} a_2^{-} + \overset{*}{a}{}_1^{-} a_1^{+} - \overset{*}{a}{}_2^{-} a_2^{+}]. \tag{27}$$

Comparing the expressions (21) for energy-momentum, (27) for the component of the spin vector, and the expression for the charge

$$Q = \int \overset{*}{\psi}(x)\,\psi(x)\,dx = \int d\boldsymbol{k}\,[\overset{*}{a}{}_s^{+}(\boldsymbol{k})\,a_s^{-}(\boldsymbol{k}) + \overset{*}{a}{}_s^{-}(\boldsymbol{k})\,a_s^{+}(\boldsymbol{k})], \tag{28}$$

which follows directly from (7), we find that the Dirac field corresponds to charged particles, with the possible values of the component of the spin along any given axis being equal to $\pm\frac{1}{2}$. A more detailed classification of the possible values of energy-momentum, charge, and spin component will be given after quantization (Section 9), where it will receive a complete and unambiguous foundation.

QUANTIZATION OF FREE FIELDS

6. QUANTIZATION OF FIELDS.

6.1. The essence of the field quantization procedure. The quantization of wave fields is a procedure which is, essentially, similar to quantization in nonrelativistic quantum mechanics. Its specific feature consists in that unlike the physical systems with finite numbers of degrees of freedom, usually considered in quantum mechanics, wave fields correspond to systems with an infinite number of degrees of freedom. Such a system may be regarded as the limit of a system with a very large number of degrees of freedom, N, as $N \to \infty$. In conventional quantum mechanics an example of this would be a system consisting of a large number of particles. A particular case of this kind is represented by a system of N oscillators.

An important technical aspect of the procedure of quantization of fields consists of the *occupation-number representation* (or, in other words, the second-quantization representation*). This representation is introduced in the simplest manner by using as an example the quantum-mechanical oscillator, and is widely utilized for examining systems consisting of a large number of identical particles. We refer the reader to standard books on quantum mechanics for technical details (see, for example, Davydov (1973), Section 33; Messiah (1978), Chapter XII; Elyutin and Krivchenkov (1976), Chapter III, Section 12; Medvediev (1977), Part III, Section 10); here we shall only recall the essentials of the method of reasoning.

We write the Hamiltonian of a one-dimensional oscillator of unit mass in the form

$$H = \frac{1}{2}(p^2 + \omega^2 q^2). \tag{1}$$

*The term "second" seems to imply the existence of a first quantization. Actually, quantization is carried out only once, and this term turns out to be misleading. Whenever possible, we shall avoid using it.

The wave functions Ψ_n of the various states of the oscillator satisfy the equation

$$H\psi_n = E_n\psi_n, \quad E_n = \hbar\omega\,(n + 1/2)$$

and can be obtained from each other with the aid of certain operators a and $\overset{+}{a}$:

$$\overset{+}{a}\psi_n = \sqrt{n+1}\;\psi_{n+1}, \quad a\psi_n = \sqrt{n}\;\psi_{n-1}, \tag{2}$$

which are usually represented through combinations of the operators of differentiation and of multiplication by the coordinate expressed in terms of the variables p and/or q (here we set $\hbar = 1$), i.e.,

$$a = \sqrt{\frac{\omega}{2}}\left(\check{q} + \frac{i\check{p}}{\omega}\right), \quad \overset{+}{a} = \sqrt{\frac{\omega}{2}}\left(\check{q} - \frac{i\check{p}}{\omega}\right),$$

with $\check{p} = -i\partial/\partial q$, $\check{q} = q$ in the coordinate representation and $\check{p} = p$, $\check{q} = i\partial/\partial p$ in the momentum representation. (Here the sign ˇ over a letter denotes an operator.) The operator $\overset{+}{a}$ raises the state index by one (and the energy by $\hbar\,\omega$) and is called the *raising* operator. Correspondingly, a is the *lowering* operator. The ground state satisfies the equation

$$a\psi_0 = 0. \tag{3}$$

The numerical factors $\sqrt{n+1}$, $\sqrt{n}$ are introduced in (2) for the normalization of the states (see (6) below).

From the formulas (2) it follows that the states ψ_n are eigenfunctions of the products $a\overset{+}{a}$ and $\overset{+}{a}a$ with eigenvalues $n + 1$ and n. Therefore one may write

$$a\overset{+}{a} = \check{n} + 1, \quad \overset{+}{a}a = \check{n}, \quad \check{n}\psi_n = n\psi_n,$$

and also

$$\left.\begin{aligned} \left[a,\ \overset{+}{a}\right] &= a\overset{+}{a} - \overset{+}{a}a = 1, \\ [a,\ a] &= \left[\overset{+}{a},\ \overset{+}{a}\right] = 0. \end{aligned}\right\} \tag{4}$$

The operator $\check{n}$ is appropriately called the state index operator. The Hamiltonian may now be represented in the form

$$H = \omega\left(\check{n} + {}^1/_2\right) = \frac{\omega}{2}\left(a\overset{+}{a} + \overset{+}{a}a\right). \tag{5}$$

6.2. Interpretation of the occupation-number representation from a corpuscular point of view. The transition to the occupation-number representation consists in that, instead of the ψ-function being defined as a function of the coordinates (the coordinate representation) or as a function of the momenta (the momentum representation), the state of a system is characterized by the occupation number n, i.e., the index of the excited state. The linear dependence of the Hamiltonian (5) upon the operator of the occupation number $\check{n}$ represents the basis of the corpuscular interpretation of the oscillator problem.

We shall for this purpose agree to regard an oscillator which is in the excited state ψ_n $(n \geq 1)$ as the collection of n excitation quanta each of which has the energy ω (i.e., $\hbar\,\omega$). The fact that the levels are equidistant, i.e., that there is a linear dependence of the energy on the number n, leads to an important property of the excitation quanta, namely, to their being indistinguishable. In accordance with this interpretation the operators raising and lowering the numbers of the excited states $\overset{+}{a}$ and a acquire the meaning of *creation* and *annihilation operators* of the excitation quanta. The index n is now equal to the number of excitation quanta in the state ψ_n, and the operator $\check{n}$ becomes the *operator of the number of quanta.*

The state ψ_n in which there are n quanta (particles) may be obtained from the ground state ψ_0 in which there are no particles (and which, therefore, is appropriately called the vacuum state) by applying the creation operator n times in succession:

$$\psi_n = \frac{(\overset{+}{a})^n}{\sqrt{n!}}\,\psi_0. \tag{6}$$

The numerical factor $(n!)^{-1/2}$ occurs in this formula in accordance with the first of the formulas (2) and provides equality between the norms of the states ψ_n with different n.

The relations (2)–(6) represent the formulation of the quantum-mechanical oscillator problem in the framework of the occupation-number representation (or, in other words, the Fock representation). In particular we emphasize that particles are associated not with the oscillator itself but with its excitation quanta. The particles are "generated" as a result of ordinary quantization.

For the transition to the case of many degrees of freedom ($N > 1$), a set of N oscillators with various frequencies ω_k is introduced. The index k assumes the N values 1, 2, . . . ,N. The Hamiltonian is represented by the sum

$$H = \sum_k H_k = \sum_k \omega_k \left(\overset{+}{a}_k a_k + {}^1/_2\right).$$

The operators $\overset{+}{a}_k$, a_l satisfy the commutation relations (4) for coinciding indices $k = l$ and commute with each other if $k \neq l$, i.e.,

$$\left[a_k, \overset{+}{a}_l\right] = \delta_{kl}, \quad [a_k, a_l] = \left[\overset{+}{a}_k, \overset{+}{a}_l\right] = 0.$$

An arbitrary state of the system is characterized here by a set of occupation numbers $n_1, n_2, \ldots, n_N$. The state

$$\psi(\ldots n_k \ldots) = \prod_{1 \leq k \leq N} \left\{\frac{(a_k^+)^{n_k}}{\sqrt{n_k!}}\right\} \psi_0 \tag{7}$$

contains n_1 particles of the first kind with energy ω_1, n_2 particles of the second kind with energy ω_2, etc.

We shall now show that a real scalar field function $u(x)$ which satisfies the Klein–Gordon equation

$$(\Box - m^2)\, u(x) = 0$$

may be represented in the form of a set of oscillators. For this we shall first pass from the infinite configurational space over to a "large cube" of volume $V = L^3$, and then impose on the solutions of the equation the periodicity conditions:

$$\begin{aligned} u(x) = u(t, x_1, x_2, x_3) &= u(t, x_1 + L, x_2, x_3) = \\ = u(t, x_1, x_2 + L, x_3) &= u(t, x_1, x_2, x_3 + L). \end{aligned}$$

Representing u in the form of a Fourier sum

$$u(t, \boldsymbol{x}) = \sum_{k_1, k_2, k_3} \left(\frac{2\pi}{V\omega_k}\right)^{1/2} \left\{a(t, \boldsymbol{k})\, e^{i\boldsymbol{k}\boldsymbol{x}} + \overset{*}{a}(t, \boldsymbol{k})\, e^{-i\boldsymbol{k}\boldsymbol{x}}\right\},$$

we obtain the equation of motion for the amplitude $a(t, k)$:

$$\ddot{a}(t, \boldsymbol{k}) + \omega_k^2 \, a(t, \boldsymbol{k}) = 0,$$

where $\omega_k^2 = \boldsymbol{k}^2 + m^2$. This equation is actually the equation for an oscillator. Thus, the amplitude $a(t, k)$ is the amplitude of an oscillator with the frequency $\omega_k = (\boldsymbol{k}^2 + m^2)^{1/2}$.

The condition of periodicity in x-space leads to the discreteness of the allowed values of momentum:

$$\boldsymbol{k}(n_1, n_2, n_3) = \left\{\frac{2\pi}{L} n_1, \frac{2\pi}{L} n_2, \frac{2\pi}{L} n_3\right\}.$$

To each $\boldsymbol{k}(n)$ there corresponds an oscillator with the energy

$$\omega(n) = \left\{m^2 + \frac{4\pi^2}{L^2}(n_1^2 + n_2^2 + n_3^2)\right\}^{1/2}.$$

Thus, the relativistic field satisfying the Klein–Gordon equation may be put into correspondence with a Hamiltonian of the oscillator type in which three integers n_1, n_2, n_3 are associated with the oscillators.

Transition to the real infinite configuration space ($L \to \infty$) leads to a continuous momentum space. All sums are then replaced by integrals according to the rule

$$\left(\frac{2\pi}{L}\right)^3 \sum_{\boldsymbol{k}} \to \int d\boldsymbol{k},$$

and the Kronecker symbol is replaced by the delta function:

$$\left(\frac{L}{2\pi}\right)^3 \delta_{\boldsymbol{k}\boldsymbol{k}'} \to \delta(\boldsymbol{k} - \boldsymbol{k}').$$

Taking into account the change of the normalization of the operators,

$$a_{\boldsymbol{k}} \to \left(\frac{2\pi}{L}\right)^{3/2} a(\boldsymbol{k})$$

we obtain the commutation relations

$$\left[a(\boldsymbol{k}),\ \overset{+}{a}(\boldsymbol{q})\right]=\delta(\boldsymbol{k}-\boldsymbol{q}),$$
$$[a(\boldsymbol{k}),\ a(\boldsymbol{q})]=\left[\overset{+}{a}(\boldsymbol{k}),\ \overset{+}{a}(\boldsymbol{q})\right]=0. \tag{8}$$

Equation (7) retains its form in the continuous case. The corresponding states contain quanta with strictly fixed momenta. The wave functions of these states in the configuration representation correspond to plane waves. Therefore, such states have no finite norm (see Assignment O3).

We may point out, furthermore, that as a result of the quantization of (8) the field function $u(x) = u(t, \boldsymbol{x})$, which satisfies the Klein–Gordon equation, turns out to be represented in the form of a three-dimensional momentum integral of a linear form in the operators $a(\boldsymbol{k})$, $\overset{+}{a}(\boldsymbol{k})$ and has, thus, itself become an operator.

6.3. Canonical quantization. The operator formulation, discussed above, of the oscillator quantum-mechanical problem may also be quite easily derived from the classical oscillator problem presented with the aid of the canonical formalism of classical mechanics.

In the canonical formalism the fundamental variables are the coordinates q and the momenta conjugate to them:

$$p=\frac{\partial\mathscr{L}}{\partial\dot{q}}\qquad\left(\dot{q}\equiv\frac{\partial q}{\partial t}\right).$$

The Hamiltonian of an oscillator in terms of the variables q, p has the form (1). The equations of motion for the dynamical variable A, being a function of q, p and hence not depending explicitly on the time, are written down in the form

$$\frac{dA(q,p)}{dt}=\{A,\ H\}, \tag{9}$$

where the braces denote the classical Poisson brackets

$$\{a,\ b\}\equiv\frac{\partial a}{\partial q}\frac{\partial b}{\partial p}-\frac{\partial a}{\partial p}\frac{\partial b}{\partial q}. \tag{10}$$

Specifically, therefore,

$$\{q, p\} = 1 \tag{11}$$

and

$$\dot{q} = \{q, H\} = p, \quad \dot{p} = \{p, H\} = -\omega^2 q.$$

From the last formulas, which represent the canonical equations of motion, it follows that the solutions may be decomposed into positive- and negative-frequency parts:

$$q(t) = \frac{a^{(+)}(t) + a^{(-)}(t)}{\sqrt{2\omega}}, \quad p(t) = i\sqrt{\frac{\omega}{2}}\,[a^{(+)}(t) - a^{(-)}(t)], \tag{12}$$

while

$$\dot{a}^{(\pm)}(t) = \pm i\omega a^{(\pm)}(t), \quad a^{(\pm)}(t) = a^{(\pm)}(0)\exp(\pm i\omega t).$$

Using relations inverse to (12), as well as (11), we obtain

$$\{a^{(-)}(t), a^{(+)}(t)\} = -i. \tag{13}$$

Substituting (12) into the Hamiltonian (1), we find

$$H = (1/2)\,[a^{(+)}(t)\,a^{(-)}(t) + a^{(-)}(t)\,a^{(+)}(t)] = (1/2)\,(\overset{+}{a}a + a\overset{+}{a}).$$

We have here left unchanged the order of factors which follows from (1) (such a precaution will prove to be useful for the transition to the quantum case), as well as adopted the more compact notation.

$$a^{(-)} \equiv a, \quad a^{(+)} \equiv \overset{+}{a}.$$

The quantization procedure may now be carried out with the aid of the following *canonical quantization postulate*: We assume that variables such as q, p, a, $\overset{+}{a}$ as well as functions of them (for instance, the Hamiltonian H) are operators acting on the state ψ-function.

The laws governing the mutual transpositions of these operators are established according to the following "correspondence principle": in the canonical-formalism formulas such as (9), (11), (13) the classical Poisson brackets (10) are replaced by the quantum Poisson brackets defined by the relation

$$\{a,\ b\}_{\text{quant}} = \frac{1}{i}[a,\ b] = \frac{1}{i}(ab - ba),$$

i.e.,

$$\{a,\ b\}_{\text{class}} \to \frac{1}{i}[a,\ b]. \tag{14}$$

Therefore in the quantum case the equation of motion (9) for the operator quantity A takes the form

$$i\frac{dA}{dt} = [A,\ H], \tag{15}$$

while the formulas (11), (13) transform into the commutation relations

$$[q,\ p] = i, \quad \left[a,\ \overset{+}{a}\right] = 1. \tag{16}$$

Taking account of the latter commutation relation, the Hamiltonian H may be written in the form (5). Thus, we arrive at the conclusion that the above-mentioned canonical quantization postulate for an oscillator is completely equivalent to the usual quantization formulation presented in Section 6.1. If applied to a physical field it may be regarded as the basis for the quantization of fields. We then immediately obtain the quantum formulation in the representation of occupation numbers.

For further discussion of the interrelations between various formulations of the quantization of systems with large numbers of degrees of freedom, it is necessary to recall some elementary information from the representation theory of quantum mechanics.

6.4. The Schrödinger and Heisenberg representations. The most widespread representation in quantum mechanics is the Schrödinger representation. In this representation the evolution of a system in time is described by means of a time-dependent wave function $\psi(t)$ which satisfies the Shrödinger

$$i\frac{\partial\psi(t)}{\partial t}=H\psi(t).$$

equation Here H is the Hamiltonian operator which corresponds to the total energy of the system and which does not depend on time for closed systems. In the Schrödinger representation the dynamical variables are characterized by operators B that do not explicitly depend on time. Their expectation values

$$\overline{B_t}=\overset{*}{\psi}(t)\,B\psi(t)\,, \tag{17}$$

however, may depend on the time through the wave functions $\psi(t)$.

In (17) the evolution in time of the expectation value of the dynamic variable B is due to $\psi(t)$. This dependence may be transferred directly into the operator B. For this purpose we formally integrate the Schrödinger equation. We obtain

$$\psi(t)=U(t)\,\psi, \tag{18}$$

where

$$U(t)=e^{-iHt}. \tag{19}$$

With the aid of (18) we represent (17) in the form

$$\overset{*}{\psi}(t)\,B\,\psi(t)=\overset{*}{\psi}\overset{*}{U}(t)\,BU(t)\,\psi=\overset{*}{\psi}B_H(t)\,\psi,$$

where

$$B_H(t)=\overset{*}{U}(t)\,BU(t)=e^{iHt}\,Be^{-iHt}. \tag{20}$$

Thus, the expectation value $\overline{B}_t$ is represented as the quantum average over the time-independent functions ψ of the operator (20) which does depend on time explicitly. We have here arrived at the Heisenberg representation, in which the dependence on time has been transferred from the wave function to the operator. Differentiating (20) with respect to t, we obtain the equation of motion

$$i\frac{dB_H(t)}{dt}=[B_H(t),\ H]. \tag{21}$$

Note that we have obtained two different methods of describing the evolution in time of a system. For transition from the instant of time 0 to instant t it is necessary *either* to transform the wave function ψ according to (18) *or* to transform operator B according to (20).

Comparing (21) and (15), we conclude that the scheme for canonical quantization presented in Section 6.2 leads to the Heisenberg representation for quantized wave fields.

In the canonical formalism the time plays a specific role. Therefore the impression may arise that use of the above method in the theory of wave fields may endanger the relativistic invariance of the latter. Actually this does not happen, and as a result of canonical quantization of the fields we obtain expressions which possess the required invariance. There exists, however, one more scheme for quantization, in which relativistic invariance is preserved at all intermediate stages. This scheme is based on the theory of representations or, to be more precise, on the requirements of correspondence between the transformation laws of quantized and classical fields under transformations of the fields and coordinate reference systems.

We shall call this scheme, which arose from the work of Schwinger during the late forties, the *relativistic scheme*, and describe its main points. (A more complete presentation is to be found, for instance, in Section 9 of the *Introduction.*)

6.5. The relativistic scheme of quantized fields. As a result of quantization, the field functions acquire an operator meaning and are expressed linearly in terms of the particle *creation* and *annihilation operators* between which required commutation relations are set up. These operators act on a wave function Φ common to all the fields.

In accordance with ordinary quantum mechanics, the wave function Φ characterizes completely the physical state of the system described by the quantized wave fields. Similarly to the usual ψ-function, the quantity Φ may be thought of as a vector in some linear space. In view of this the function Φ is called the *state amplitude* (or *state vector*). As in quantum mechanics, not all vectors have finite norms (for instance, those corresponding to plane waves). However, the norms of real physical states can always be chosen so as to be finite. The expectation values of dynamical quantities and transition probabilities are given by quadratic forms in Φ.

Let us now consider the transformation properties of the vectors Φ under transformations of coordinates and of the field functions introduced in Sections 1, 2 (see (1.1), (1.3), (2.10)):

$$x \to x' = P(\omega;\ x) \quad u(x) \to u'(x') = \Lambda(\omega)\, u(x).$$

To such transformations there corresponds a certain transformation of the state vector which, owing to the superposition principle, should be linear:

$$\Phi \rightarrow \Phi' = U(\omega)\,\Phi \tag{22}$$

and, as the norm is conserved, unitary:

$$\overset{*}{U}(\omega)\,U(\omega) = 1. \tag{23}$$

In the simplest case of the translation $u'(x) = u(x - \omega)$, the operator U has the form

$$U(\omega) = \exp(iP_\mu \omega^\mu). \tag{24}$$

The exponential behavior of the dependence of ω on the parameters follows from the group character of the transformation. By virtue of the unitarity condition (23), the translation operators must be Hermitian: $\overset{+}{P}_\mu = P_\mu$. Note that (24) in the particular case of $\omega_\nu = \delta_{\nu 0} t$ corresponds to (19). The operator P_μ is thus the four-momentum operator. The analog of formula (20) for the operator field function will be

$$u'(x) = u(x - \omega) = U^{-1}(\omega)\,u(x)\,U(\omega). \tag{25}$$

For the case of infinitesimal ω the apparent form of the transformation of the state amplitude

$$\Phi' = (1 + iP\omega)\,\Phi$$

coincides with the infinitesimal transformation of the field function

$$u'(x) = (1 + ip\omega)\,u(x). \tag{26}$$

If we consider u to be an ordinary wave function, then the coefficients p_ν may be regarded as quantum-mechanical operators of the four-momentum:

$$p_\nu = i\partial_\nu = i\partial/\partial x^\nu.$$

We emphasize that p_ν is the generator of displacements (the momentum operator) in the space of the field functions $u(x)$, and P_ν is the generator of displacements in the space of state vectors Φ, i.e., in the representation of occupation numbers or the representation of second quantization. (It was just because of this correspondence that the word "second" was used.)

The formulae written down may be generalized to the case of of the general transformations belonging to the Poincaré group:

$$U = \exp i\left(P_\nu a^\nu + \frac{1}{2} M_{\mu\nu}\omega^{\mu\nu}\right), \quad \Phi' = \left(1 + iPa + \frac{i}{2} M\omega\right)\Phi, \tag{27}$$

$$u'(x) = \left(1 + ipa + \frac{i}{2} m\omega\right) u(x), \qquad m_{\mu\nu} = i(x_\mu\partial_\nu - x_\nu\partial_\mu). \tag{28}$$

Here $m_{\mu\nu}$ are generators of four-dimensional rotations. By virtue of the correspondence principle we shall therefore interpret P and M in (27) as the operators for the energy-momentum four-vector and the angular-momentum tensor, respectively.

Similarly, under phase transformations of the field functions,

$$\varphi \to \varphi' = e^{i\alpha}\varphi, \quad \overset{*}{\varphi} \to \overset{*}{\varphi}{}' = e^{-i\alpha}\overset{*}{\varphi} \tag{29}$$

the unitary operator for the transformation

$$\Phi' = U_\alpha \Phi$$

has an exponential structure

$$U_\alpha = \exp(i\alpha Q), \tag{30}$$

in which the Hermitian operator Q should be interpreted as the charge operator. As the fundamental *postulate for the quantization* of wave fields we shall assume that the Hermitian operators for the energy-momentum four-vector P, the angular-momentum tensor M, the charge Q, and so on, which are generators of infinitesimal transformations of the state vectors (see (27) and (30)), are expressed in terms of the operator field functions by the same relations—of the type (2.6), (2.22), (2.30)—as in classical field theory, with the operator factors, of course, being arranged in this case in appropriate order. This postulate represents a further application of the correspondence principle and defines the transformation law of second-quantized state amplitudes.

7. COMMUTATION RELATIONS.

7.1. The physical meaning of the frequency components. Let us examine the relation (6.25) in the case of an infinitesimal translation. Substituting into it the expansions in powers of ω_μ and retaining the linear terms, we obtain the four equations

$$i\frac{\partial u(x)}{\partial x^\mu} = [u(x),\ P_\mu] \tag{1}$$

which generalize the equation of motion (6.21). These equations turn out to be very convenient for analyzing the frequency components of the field functions in the momentum representation. Substituting into (1) three-dimensional momentum Fourier expansions of the type of (3.8), (4.13), (5.9), we obtain the algebraic operator equations

$$k_\mu u^\pm(\boldsymbol{k}) = \mp [u^\pm(\boldsymbol{k}),\ P_\mu]. \tag{2}$$

Let us now introduce a state with a definite value p_ν of the energy-momentum 4-vector and described by the amplitude

$$P_\nu \Phi_p = p_\nu \Phi_p. \tag{3}$$

Multiplying the first of the equations (2) on the right by Φ_p and taking account of (3), we obtain

$$P_\nu u^+(\boldsymbol{k})\,\Phi_p = (p_\nu + k_\nu)\,u^+(\boldsymbol{k})\,\Phi_p.$$

In the same manner, from the second equation we find

$$P_\nu u^-(\boldsymbol{k})\,\Phi_p = (p_\nu - k_\nu)\,u^-(\boldsymbol{k})\,\Phi_p,$$

and in both cases $k_0 = \sqrt{\boldsymbol{k}^2 + m^2} > 0$.

From the relations obtained it follows that the expression $u^+(\boldsymbol{k})\Phi_p$ either is equal to zero or represents the amplitude of the state with energy-momentum $p + k$, and $u^-(\boldsymbol{k})\Phi_p$ is the amplitude of the state with energy-momentum $p - k$ (or zero). As at the same time the relation $k^2 = m^2$ holds, one may consider that the operator $u^+(\boldsymbol{k})$ describes the creation of a particle of mass m and four-momentum k, while the operator $u^-(\boldsymbol{k})$ corresponds to the annihilation of the same particle.

We emphasize that this property of the frequency parts of the field operators is quite general and valid for fields of arbitrary tensor dimensionality, both real and complex, and does not depend on the specific form of the commutation relations (i.e., it is valid under both Bose–Einstein quantization and Fermi–Dirac quantization; see Section 7.4 below).

Note, also, that if we start from a relation like (6.25), then for the phase transformations (6.29), (6.30)

$$e^{i\alpha}u = U^{-1}(\alpha)\, uU(\alpha), \quad e^{-i\alpha}\overset{*}{u} = U^{-1}(\alpha)\overset{*}{u}U(\alpha)$$

by equating the coefficients of the terms linear in α we obtain the following:

$$u(x) = [u(x),\ Q] \quad \overset{*}{u}(x) = -\left[\overset{*}{u}(x),\ Q\right]. \tag{4}$$

The equations (4) complement the equations (2). From them, specifically, it follows that the operators u (as well as the frequency components $u^{\pm}$) are "lowering" with respect to the charge of the system (i.e., to the eigenvalue of the charge operator Q). The adjoint operators $\overset{*}{u}$, $\overset{*}{u}^{\pm}$ raise the eigenvalue by one. We leave it to the reader to perform the corresponding calculations on his own.

Thus, in the case of the complex field the operator $\overset{*}{u}^{+}$ creates a particle with charge $+1$, while the operator $\overset{*}{u}^{-}$ annihilates a particle with charge -1, etc.

7.2. The vacuum amplitude and the Fock representation. Now it is possible to define the vacuum state in a natural way and establish rules for constructing the amplitudes corresponding to states with a definite number of various particles.

We shall consider a dynamical system consisting of several noninteracting quantized fields characterized by the operator functions $u_1(x)$, $\ldots, u_n(x)$. For convenience of notation, we shall also include in this sequence the corresponding adjoint functions in those cases in which $\overset{*}{u}_i$ differs from u_i.

Let us define the vacuum state amplitude Φ_0 for the given dynamical system. Since there are no particles in the vacuum, the momentum of the vacuum equals zero, while the energy of the vacuum is minimal (it also may be set equal to zero). As the negative-frequency operators u^{-} diminish the energy and the energy of the vacuum state Φ_0 is minimal, the following relations must hold for all x:

$$u_i^{-}(x)\,\Phi_0 = 0 \qquad (i = 1,\ \ldots,\ n) \tag{5}$$

Analogous reasoning leads to the relation

$$\overset{*}{\Phi}_0 u_i^{+}(x) = 0.$$

Going over to the momentum representation, we correspondingly obtain

$$u_i^{-}(\boldsymbol{k})\,\Phi_0 = 0. \tag{6}$$

Equation (6) and its adjoint relation $\overset{*}{\Phi}_0 u_j^+(\boldsymbol{k}) = 0$, together with the normalization condition

$$\overset{*}{\Phi}_0 \Phi_0 = 1 \tag{7}$$

may be considered to be the definition of the vacuum state of free fields.

The amplitude of any arbitrary state of the dynamical system under consideration may now be represented through the vacuum amplitude introduced above and the creation operators for the appropriate particles. Thus, the amplitude of the state which contains s particles of the kinds $j_1, \ldots, j_s$, respectively (some of the indices may coincide) will be given by an expression of the form

$$\int F_s^{(j_1, \ldots, j_s)}(\boldsymbol{k}_1, \ldots, \boldsymbol{k}_s)\, u_{j_1}^+(\boldsymbol{k}_1) \ldots u_{j_s}^+(\boldsymbol{k}_s)\, d\boldsymbol{k}_1 \ldots d\boldsymbol{k}_s \Phi_0.$$

The functions $F_s^{(\ldots j \ldots)}$ entering into this expression have the meaning of ordinary quantum-mechanical wave functions of a system of s particles in the momentum representation. The amplitude of the state which contains exactly s particles with the given characteristics $j_1, \ldots, j_s$ is completely described by one such function.

In the general case, when the number of particles is not fixed, the state amplitude is characterized by a sequence of functions F_s with various values of the index s. Summing over the discrete indices $j_1, \ldots, j_s$ and over s, we write

$$\Phi = \sum_{(j,\, s \geqslant 0)} \int F_s^{(\ldots j \ldots)}(\ldots \boldsymbol{k} \ldots)\, u_{j_1}^+(\boldsymbol{k}_1) \ldots u_{j_s}^+(\boldsymbol{k}_s)\, d\boldsymbol{k}_1 \ldots d\boldsymbol{k}_s \Phi_0. \tag{8}$$

We have here obtained the *Fock representation* for the state amplitude. Passing over to the configuration representation through the relations

$$F_s(\ldots \boldsymbol{k} \ldots) = \frac{1}{(2\pi)^{3s/2}} \int e^{i \sum_j \boldsymbol{k}_j \boldsymbol{x}_j} \varphi_s(\ldots \boldsymbol{x} \ldots)\, d\boldsymbol{x}_1 \ldots d\boldsymbol{x}_s,$$

$$u^+(\boldsymbol{k}) = \frac{1}{(2\pi)^{3/2}} \int e^{i\boldsymbol{k}\boldsymbol{x}} u^+(0, \boldsymbol{x})\, d\boldsymbol{x}, \quad u(0, \boldsymbol{x}) = u(x)|_{x^0=0},$$

we shall obtain for the state amplitude the following expression:

$$\Phi = \sum_{(j,\, s \geqslant 0)} \int \varphi_s(\boldsymbol{x}_1, \ldots, \boldsymbol{x}_s) \prod_{1 \leqslant \nu \leqslant s} \{u_{j_\nu}^+(0, \boldsymbol{x}_\nu)\, d\boldsymbol{x}_\nu\} \Phi_0. \tag{9}$$

in place of (8). The functions φ_s occurring here are to be interpreted in the nonrelativistic case as ordinary wave functions in configuration space.

As is seen, the dependence of Φ on the time has dropped out; this is quite natural, since in the representation chosen by us the state amplitude in the absence of interaction turns out to be constant.

7.3. Types of commutation relations. We now proceed to set up the commutation relations for the operator wave functions. In the classical theory of free fields with a quadratic Lagrangian the canonical formalism is introduced by expressing the field functions linearly in terms of mutually conjugate coordinates and momenta. The classical Poisson brackets for the field functions $\{u(x), u(y)\}$ then turn out to be certain functions of x and y (to be more precise, of the difference $x-y$) independent of u. Therefore, starting with the correspondence principle, it is customary to assume in the quantum theory of a free field that the commutation rule for the operator field functions has the form

$$\{u_a(x),\ u_b(y)\}_- \equiv [u_a(x),\ u_b(y)] = \\ = u_a(x)\, u_b(y) - u_b(y)\, u_a(x) = \Delta_{ab}(x-y). \tag{10B}$$

This commutation relation, however, turns out to be too stringent and does not embrace a number of physically important cases. Therefore, as an alternative to (10B) the following can be assumed:

$$\{u_a(x),\ u_b(y)\}_+ \equiv u_a(x)\, u_b(y) + u_b(y)\, u_a(x) = \Delta_{ab}(x-y), \tag{10F}$$

i.e., it is assumed that, instead of the commutator, the anticommutator of two field operators is a c-number.

The fact that field operators anticommute is expressed by commutation relation (10F), and it leads to the commutators of dynamical quantities, which are quadratic forms in the field operators, being expressed in terms of the commutation functions Δ. It is in this sense that the correspondence principle should be interpreted for the second case.

The quanta of fields satisfying the relations (10B) obey the Bose–Einstein statistics. the corresponding particles are called *bosons*. The quanta of fields which satisfy (10F) obey the Fermi–Dirac statistics, and the corresponding particles are called *fermions*.

The exact form of the commutation functions Δ for any arbitrary field is determined by equations (1), (4) and by the structure of the energy operator of the given field. However, independently of the specific form of the commutation relations, it may be demonstrated that the commutation function of free fields depends only on the difference $x - y$, i.e.,

$$\{u_a(x), \ u_b(y)\} = \Delta_{ab}(x - y). \tag{11}$$

To establish this fact, we shall consider the commutation relations (10) in the momentum representation. Since the formulas for the Fourier transformation are linear, the commutators or anticommutators of the frequency components in the momentum representation, $u^{\pm}(\boldsymbol{k})$, must also be c-numbers. We shall first show that operators of the same frequency must strictly commute or anticommute:

$$\{u_a^{\pm}(\boldsymbol{k}), \quad u_b^{\pm}(\boldsymbol{q})\} = 0. \tag{12}$$

Here, as in (11), the symbol $\{\ldots, \ldots\}$ denotes either a commutator or an anticommutator.

Consider the state amplitude Φ_p which corresponds to a definite value of the four-momentum and which satisfies equation (3). Operating on it with the operators $u_a^+(\mathbf{k})$ and $u_b^+(\mathbf{q})$ successively in different orders, we obtain new amplitudes

$$\Phi_1 = u_a^+(\boldsymbol{k})\, u_b^+(\boldsymbol{q})\, \Phi_p, \quad \Phi_2 = u_b^+(\boldsymbol{q})\, u_a^+(\boldsymbol{k})\, \Phi_p,$$

which in accordance with what was established above satisfy the equations

$$P_0 \Phi_{1,2} = (p_0 + k_0 + q_0)\, \Phi_{1,2}.$$

Adding and subtracting these equations, we find

$$P_0(\Phi_1 \pm \Phi_2) = P_0 \{u_a^+(\boldsymbol{k}), \ u_b^+(\boldsymbol{q})\} \Phi_p = (p_0 + k_0 + q_0) \{u_a^+, \ u_b^+\} \Phi_p.$$

If we now assume that $\{u_a^+, \ u_b^+\}$ is a c-number differing from zero, then canceling it out, we obtain

$$P_0 \Phi_p = (p_0 + k_0 + q_0)\, \Phi_p,$$

i.e., we come to a contradiction with the initial equation (3). Thus, (12) has been proven. In the same manner it may be proven that

$$\{u_a^{\pm}(\boldsymbol{k}), \ u_b^{\mp}(\boldsymbol{q})\} = 0 \text{ when } \boldsymbol{k} \neq \boldsymbol{q}. \tag{13}$$

The relations (12) and (13) have a simple physical meaning. They mean that acts of creation of particles of any arbitrary field do not interfere with each other and do not affect other acts of particle annihilation or the creation and annihilation of particles with different momenta.

If we now assume that in the general case, instead of zero on the right-hand side of (13), there occurs a $\delta(\boldsymbol{k}-\boldsymbol{q})$, then in the configuration representation we obtain the relations (11). Note that this formula reflects the translational invariance of the commutation relations. It is quite natural that for its proof we used the properties of the energy-momentum four-vector operator, which is the generator of translations.

It is not difficult to demonstrate that in the case of complex fields, besides (12), the commutators (anticommutators) of operators associated with particles having different charges are also always equal to zero, i.e.,

$$\{u^+,\ u^-\}=\{\overset{*}{u}{}^+,\ \overset{*}{u}{}^-\}=0. \tag{14}$$

For this it is necessary to consider the state amplitude Φ_q with a definite value of the charge and to carry out reasoning similar to the above for the proof of (12). We leave it to the reader to do this. The physical meaning of the relations (14) consists in the mutual independence of the acts of creation and annihilation of particles with different charges.

Thus, it is established that for any arbitrary (complex) field, only the c-numbers $\{u_\alpha^-(\boldsymbol{p}), \overset{*}{u}{}_\alpha^+(p)\}$ and $\{\overset{*}{u}{}_\alpha^-(\boldsymbol{p}), u_\alpha^+(\boldsymbol{p})\}$, (i.e., the (anti)commutators of the creation and annihilation operators corresponding to particles with coinciding energy-momentum four-vectors and charges) may differ from zero. The corresponding expressions in the configuration representation are invariant under translations:

$$\{u^\pm(x), \overset{*}{u}{}^\mp\ (y)\}=\Delta^\pm\,(x-y),$$

and their sum gives the total commutation function

$$\{u\,(x),\ \overset{*}{u}\,(y)\}=\Delta^+\,(x-y)+\Delta^-\,(x-y)=\Delta\,(x-y).$$

To determine the commutation functions, it is necessary to refer to the explicit form of the operator of the energy-momentum 4-vector and use the equations (1) and (4).

7.4 Fermi–Dirac and Bose–Einstein quantization. To establish the concrete form of the commutation relations, we shall make use of equations (1), (2), with the energy-momentum operator written in the form

$$P_\nu = \sum_s \int dq\, q_\nu [\overset{*}{a}{}_s^+(q)\, a_s^-(q) \pm \overset{*}{a}{}_s^-(q)\, a_s^+(q)]. \tag{15}$$

In accordance with the classical expressions (3.9), (3.28), (4.26), and (5.21) we have expressed it in terms of the independent amplitudes $a^\pm$, $\overset{*}{a}{}^\pm$ connected with $u^\pm$, $\overset{*}{u}{}^\pm$ by linear relations of the form

$$u_a^\pm(k) = \sum_s v_a^{s,\pm}(k)\, a_s^\pm(k),$$

where the coefficients $v^\pm$ are c-numbers.

As the operators $a^\pm$, $\overset{*}{a}{}^\pm$ do not commute, their order in (15) corresponds to to the order of the functions u and $\overset{*}{u}$ in the Lagrangians. We recall that the operators $a^\pm$ and $\overset{*}{a}{}^\pm$ are related by the conditions of complex (now Hermitian) conjugation,

$$(a^\pm(k))^* = \overset{*}{a}{}^\mp(k).$$

The upper sign (+) of the second term on the right-hand side of (15) stands for fields of integral spin (scalar, vector, electromagnetic fields), and the lower one (−) corresponds to fields of half-integral spin (the spinor field).

To derive the commutation relations, we first note that the commutators of the operators $a^\pm$ and the quadratic combinations occurring on the right-hand side of (15) may be represented in the form

$$[a^\pm(k),\ \overset{*}{a}{}^+(q)\, a^-(q)] = \\ = \{a^\pm(k),\ \overset{*}{a}{}^+(q)\}\, a^-(q) - \overset{*}{a}{}^+(q)\, \{a^-(q),\ a^\pm(k)\}. \tag{16}$$

Here, as was pointed out, the symbol $\{a, b\}$ denotes *either* the commutator *or* the anticommutator of the operators a and b, and it is necessary to use simultaneously the same definition in the first and second terms on the right-hand side.

Analogously,

$$[a^\pm(k),\ \overset{*}{a}{}^-(q)\, a^+(q)] = \\ = \{a^\pm(k),\ \overset{*}{a}{}^-(q)\}\, a^+(q) - \overset{*}{a}{}^-(q)\, \{a^+(q),\ a^\pm(k)\}. \tag{17}$$

On the basis of (12) and (14) we have

$$[a^+(\boldsymbol{k}),\ \overset{*}{a}{}^+(\boldsymbol{q})\,a^-(\boldsymbol{q})]=[a^-(\boldsymbol{k}),\ \overset{*}{a}{}^-(\boldsymbol{q})\,a^+(\boldsymbol{q})]=0,$$
$$[a^-(\boldsymbol{k}),\ \overset{*}{a}{}^+(\boldsymbol{q})\,a^-(\boldsymbol{q})]=\{a^-(\boldsymbol{k}),\ \overset{*}{a}{}^+(\boldsymbol{q})\}\,a^-(\boldsymbol{q}),$$
$$[a^+(\boldsymbol{k}),\ \overset{*}{a}{}^-(\boldsymbol{q})\,a^+(\boldsymbol{q})]=\{a^+(\boldsymbol{k}),\ \overset{*}{a}{}^-(\boldsymbol{q})\}\,a^+(\boldsymbol{q}).$$

Substituting (15) into (2), we now obtain

$$k_\nu a_a^-(\boldsymbol{k})=\int d\boldsymbol{q}\,q_\nu\sum_b\{a_a^-(\boldsymbol{k}),\ \overset{*}{a}{}_b^+(\boldsymbol{q})\}\,a_b^-(\boldsymbol{q}),$$

$$\mp k_\nu a_a^+(\boldsymbol{k})=\int d\boldsymbol{q}\,q_\nu\sum_b\{a_a^+(\boldsymbol{k}),\ \overset{*}{a}{}_b^-(\boldsymbol{q})\}\,a_b^+(\boldsymbol{q}),$$

from which it follows that

$$\{a_a^-(\boldsymbol{k}),\ \overset{*}{a}{}_b^+(\boldsymbol{q})\}=\delta_{ab}\delta(\boldsymbol{k}-\boldsymbol{q}), \tag{18}$$
$$\{a_a^+(\boldsymbol{k}),\ \overset{*}{a}{}_b^-(\boldsymbol{q})\}=\mp\,\delta_{ab}\delta(\boldsymbol{k}-\boldsymbol{q}), \tag{19}$$

where the explicitly shown indefiniteness in the sign ($\pm$) follows from Equation (15). Thus for fields of each type we have obtained two kinds of commutation relations. The requirement that these relations be symmetric with respect to the change of charge (or, to be more precise, to the substitution of antiparticles for particles),

$$a_a^\pm(\boldsymbol{k})\leftrightarrow\overset{*}{a}{}_a^\pm(\boldsymbol{k}) \tag{20}$$

uniquely defines the recipe for quantization in each case. The symmetry of (20) reflects the fact that the choice between the "fundamental" field function u and its complex conjugate $\overset{*}{u}$ is conventional, and the opposite choice $w=\overset{*}{u}$, $\overset{*}{w}=u$ leads only to the substitution of antiparticles for the "fundamental" particles. such a substitution affects the charge operator, which is not invariant under (20), but should not influence the equation of motion (1) or the expression for the energy-momentum four-vector. The symmetry conditions with respect to (20) also provide the correct transition from the complex to the real field: $\overset{*}{u}(x)=u(x)$, $\overset{*}{a}{}^\pm(\boldsymbol{k})=a^\pm(\boldsymbol{k})$.

Note that a transformation of the form (20) is called *charge conjugation*, and the corresponding symmetry *charge symmetry* (see Section 9.3 below). If we now take the upper sign on the right-hand side of (19) (which corresponds to integral-spin fields), then the symmetry of (18) and (19) under the transformation (20) implies that the symbol $\{a, b\}$ must be understood to be a commutator. Otherwise (i.e., for the lower sign on the right-hand side of (19)—a spinor field) the symmetry is provided by the condition

$$\{a, b\} = \{a, b\}_+.$$

Thus, we have arrived at the conclusion that fields of integral spin are subject to Bose–Einstein quantization:

$$[a_a^-(\boldsymbol{k}), \overset{*}{a}{}_b^+(\boldsymbol{q})] = [\overset{*}{a}{}_a^-(\boldsymbol{k}), a_b^+(\boldsymbol{q})] = \delta_{ab}\delta(\boldsymbol{k} - \boldsymbol{q}), \tag{21}$$

while fields of half-integral spin are subject to Fermi–Dirac quantization:

$$\{a_a^-(\boldsymbol{k}), \overset{*}{a}{}_b^+(\boldsymbol{q})\}_+ = \{\overset{*}{a}{}_a^-(\boldsymbol{k}), a_b^+(\boldsymbol{q})\}_+ = \delta_{ab}\delta(\boldsymbol{k} - \boldsymbol{q}). \tag{22}$$

We point out here that the commutation relations fix the norms of the operator field functions.

7.5. Connection between spin and statistics. The results obtained represent a particular case of the fundamental *Pauli theorem* that establishes the relationship between the transformation properties of the field and the way of quantizing it (the connection between spin and statistics):

> Fields describing particles with integral spin are subject to Bose–Einstein quantization; fields describing particles of half-integral spin are subject to Fermi–Dirac quantization.

The Pauli theorem is applicable to fields of arbitrary (no matter how high) spin.

For our proof we made use of symmetry under charge conjugation. However, other reasoning is possible. Violation of the connection between spin and statistics, established by the Pauli theorem, leads to a number of deep contradictions.

Thus, if instead of (20) the positiveness of a metric in a Hilbert space,

$$\overset{*}{\Phi}\overset{*}{A}A\Phi = \overset{*}{\Phi} |A|^2 \Phi > 0,$$

is used, then one may arrive at the formulas (18), (19) only via the condition (17) (see Assignment O8).

At the same time, the Fermi–Dirac quantization of the integral-spin field leads to a contradiction with the properties of commutators in space-time (see Assignment O9).

The appearance of the negative sign of the second term in the operator expression (5.22) for the energy of the spinor field leads to the impossibility of Bose–Einstein quantization of this field.

8. FIELDS OF INTEGRAL SPIN

8.1. The scalar field. In this section we shall examine the quantum theory of the scalar, vector, and electromagnetic fields.

In accordance with the results of Section 7.4, the commutation relations for the scalar field in the momentum representation have the form

(a) for the real field,

$$[\varphi^-(\boldsymbol{k}),\ \varphi^+(\boldsymbol{q})]=\delta(\boldsymbol{k}-\boldsymbol{q}),\quad [\varphi^\pm(\boldsymbol{k}),\ \varphi^\pm(\boldsymbol{q})]=0; \tag{1}$$

(b) for the complex field,

$$[\overset{*}{\varphi}{}^-(\boldsymbol{k}),\ \varphi^+(\boldsymbol{q})]=[\varphi^-(\boldsymbol{k}),\ \overset{*}{\varphi}{}^+(\boldsymbol{q})]=\delta(\boldsymbol{k}-\boldsymbol{q}) \tag{2}$$

(all the other commutators are equal to zero).

Passing over to the configuration representation, we obtain instead of the delta function on the right-hand sides the frequency components

$$\frac{1}{(2\pi)^3}\int\frac{dk}{\sqrt{2k_0}}\frac{dq}{\sqrt{2q_0}}e^{i(qy-kx)}\delta(\boldsymbol{k}-\boldsymbol{q})=\frac{1}{(2\pi)^3}\int\frac{dk}{2k_0}e^{ik(y-x)}=\frac{1}{i}D^-(x-y),$$

$$\frac{1}{(2\pi)^3}\int\frac{dk}{2k_0}e^{ik(x-y)}=\frac{1}{i}D^-(y-x)=iD^+(x-y)$$

of the so-called Pauli–Jordan commutation function

$$D(x)=D^+(x)+D^-(x)=i(2\pi)^{-3}\int e^{-ikx}\varepsilon(k^0)\,\delta(k^2-m^2)\,dk. \tag{3}$$

Limiting ourselves to the real field, we write down the commutation relations for the field functions in the configuration representation:

$$\begin{aligned}[\varphi^-(x),\ \varphi^+(y)]&=-iD^-(x-y),\\ [\varphi^+(x),\ \varphi^-(y)]&=-iD^+(x-y),\\ [\varphi(x),\ \varphi(y)]&=-iD(x-y).\end{aligned} \tag{4}$$

The explicit form of the functions $D^\pm$ and D is presented in Appendix V.

We here point out an important property of the function $D(x)$: it vanishes outside the light cone—see (18.17) below. Therefore

$$[\varphi(x), \varphi(y)] = 0 \quad \text{if} \quad (x-y)^2 < 0. \tag{5}$$

This property is called the property of *local commutativity*, and it reflects the causal independence of events separated by spacelike intervals.

Before writing down the Lagrangian and the dynamical variables for the quantum case, we shall give some useful definitions.

8.2 The form of dynamical variables in terms of normal products. We introduce the concept of an operator written in normal form and the concept of the normal product of operators.

The *normal form* of an operator is the form in which in each term all the creation operators u^+ are written to the left of all the annihilation operators u^-.

It is not difficult to see that the normal form of operators is the most convenient one from the point of view of carrying out calculations. Indeed, in order to calculate the matrix element $\overset{*}{\Phi} A \Phi$ of any arbitrary operator A in its normal form, it is necessary merely to commute the operators u^- occurring in A with all the a^+ occurring in the state amplitude Φ, and the u^+ occurring in A with all the a^- occurring in $\overset{*}{\Phi}$, until one of the u^- operates on Φ_0, or one of the u^+ operates on $\overset{*}{\Phi}_0$, which gives zero.

Consider an example. We write down in normal form the product of two Bose operators $\overset{*}{u}(x)$ and $u(y)$. We then obtain in turn

$$\overset{*}{u}(x)\, u(y) = \overset{*}{u}{}^+(x)\, u^+(y) + \overset{*}{u}{}^+(x)\, u^-(y) + \overset{*}{u}{}^-(x)\, u^+(y) + \overset{*}{u}{}^-(x)\, u^-(y) =$$

$$= \overset{*}{u}{}^+(x)\, u^+(y) + \overset{*}{u}{}^+(x)\, u^-(y) + u^+(y)\, \overset{*}{u}{}^-(x) + \overset{*}{u}{}^-(x)\, u^-(y) - i\Delta^-(x-y).$$

It is evident that in the more general case, by reducing to the normal form the product of a certain number of operator field functions u, we shall obtain a sum of products of the components u^+, u^- and of commutation Δ^--functions. The general prescription for such a product will be considered by us later (Section 17) and comprises the first Wick's theorem. The whole expression may be conventionally regarded as a "polynominal" in powers of Δ^--functions. The zero-power term of this polynominal, i.e., the sum of terms not involving any Δ^--functions, is called the normal product of the original operator field functions. The normal product may also be defined as the original product reduced to its normal form with all the commutation functions being taken equal to zero in the process of reduction.

The normal product of the operators $u_1, u_2, \ldots, u_n$ is denoted by the symbol $:u_1 u_2 \ldots u_n:$. The result of the last computation may be written in the form

$$\overset{*}{u}(x)\, u(y) = :\overset{*}{u}(x)\, u(y): - i\Delta^-(x-y).$$

We now agree by definition to express all the dynamical variables which depend quadratically on operators of the same arguments (like Lagrangian, energy-momentum, current, etc.) in the form of normal products. For example, we shall write the Lagrangian for a complex scalar field in the form

$$\mathscr{L} = :\overset{*}{\varphi}_{;\,\nu}\varphi^{;\,\nu}: - m^2 :\overset{*}{\varphi}\varphi:.$$

It is not difficult to see that by virtue of the definition of the vacuum amplitude Φ_0,

$$\varphi^-(x)\,\Phi_0 = \overset{*}{\varphi}{}^-(x)\,\Phi_0 = 0$$

and of the conjugate relation

$$\overset{*}{\Phi}_0\varphi^+(x) = \overset{*}{\Phi}_0\overset{*}{\varphi}{}^+(x) = 0$$

it follows from the preceding that the expectation values of all the dynamical variables vanish for the vacuum state. Thus, we automatically eliminate all pseudophysical quantities of the type of of vacuum energy, vacuum charge, etc.

It is clear, also, that all the conservation laws established in the classical theory remain valid also for the quantum expressions written in normal form, since the algebraic identities which we have used in the course of proving Noether's theorem are not violated.

We now write the fundamental quantities of the complex scalar field in the form

$$T_{\mu\nu} = :(\overset{*}{\varphi}_{;\,\mu}\varphi_{;\,\nu} + \overset{*}{\varphi}_{;\,\nu}\varphi_{;\,\mu}): - g_{\mu\nu}\mathscr{L}, \tag{7}$$

$$P^\nu = \int dk\, k^\nu\,[\overset{*}{\varphi}{}^+(k)\,\varphi^-(k) + \varphi^+(k)\,\overset{*}{\varphi}{}^-(k)], \tag{8}$$

$$Q = \int dk\,[\overset{*}{\varphi}{}^+(k)\,\varphi^-(k) - \varphi^+(k)\,\overset{*}{\varphi}{}^-(k)]. \tag{9}$$

From the structure of the operators P^ν, Q it follows $\overset{*}{\varphi}{}^+(\boldsymbol{k})$ is the creation operator of a particle of energy-momentum k and charge $+1$, $\varphi^-(\boldsymbol{k})$ is the annihilation operator of the same particle, $\varphi^+(\boldsymbol{k})$ is the creation operator of a particle of energy-momentum k and charge -1, and $\overset{*}{\varphi}{}^{-}(\boldsymbol{k})$ is the annihiliation operator of the same particle.

Note that the reasoning which established the sign of the frequency components of the field functions φ^{+} and φ^{-} has not led to the definition of their norm. This norm may now be established with the aid of the expression (8) for the energy-momentum four-vector. We consider the amplitude of the state containing a single scalar particle with an unnormalized momentum distribution function $c(\boldsymbol{k})$:

$$\Phi_1 = \int c(\boldsymbol{k})\, \varphi^{+}(\boldsymbol{k})\, d\boldsymbol{k}\, \Phi_0, \tag{10}$$

and calculate the expectation value of the operator P^{ν} in (8) over this state. After carrying out the commutations of the operators, we obtain

$$\langle P^{\nu}\rangle_1 = \overset{*}{\Phi}_1 P^{\nu}\Phi_1 / \overset{*}{\Phi}_1\Phi_1 = \int \overset{*}{c}(\boldsymbol{k})\, c(\boldsymbol{k})\, k^{\nu}\, d\boldsymbol{k} \Big/ \int |c(\boldsymbol{k})|^2\, d\boldsymbol{k}.$$

The transition to a state with a given value of the 4-momentum may be performed by localizing the function $c(\boldsymbol{k})$ in a small region around the value $k = K$ (for example, by the limiting process $c(\boldsymbol{k}) \to \delta(\boldsymbol{k} - \boldsymbol{K})$). In the above, $k^{\nu} \to K^{\nu}$ and we obtain

$$\langle P^{\nu}\rangle_1 \to K^{\nu}.$$

Thus, the expectation value of the operator P^{ν} over a state with the given value of the four-momentum K^{ν} is exactly equal to K^{ν}, and consequently the normalization of the operators $\varphi^{\pm}(\boldsymbol{k})$, $\overset{*}{\varphi}{}^{\pm}(\boldsymbol{k})$ which was established by the commutation relations (1), (2) corresponds to the normalization of expressions of the type (8), (9) for dynamical quantities.

8.3. The complex vector field. In order to set up the quantization rules for the four-potential of the vector field, we note that a straightforward generalization to this case of the quantization rules of the scalar field (i.e., the independent quantization of each component of the potential U_{ν} following the example of the scalar field) turns out to be impossible, since such a procedure does not guarantee that the expectation value of the energy will be positive and turns out to be incompatible with the subsidiary condition (3.13). The quantization procedure must therefore take cognizance of the subsidiary condition, which, as we have seen in Section 3, automatically guarantees that P^0 will be positive. In particular, it was shown that as a result of going over from U_{ν} to the longitudinal and transverse components by means of the formulas (3.24), (3.27), the energy-momentum four-vector may be expressed in terms of three linearly independent amplitudes $b_{\alpha}(k)$ in the following manner:

$$P^{\nu} = \int d\boldsymbol{k}\, k^{\nu} [\overset{*}{b}{}^{+}_{a}(\boldsymbol{k})\, b^{-}_{a}(\boldsymbol{k}) + \overset{*}{b}{}^{-}_{a}(\boldsymbol{k})\, b^{+}_{a}(\boldsymbol{k})] \tag{3.28}$$

This way of writing the classical four-momentum takes into account the subsidiary condition (3.13) and guarantees that the unquantized P^0 will be positive. It is also clear that the quantization of the three independent amplitudes b_a ($a = 1, 2, 3$) directly in accordance with Bose–Einstein rules will guarantee that the expectation value of the energy operator will be positive definite.

As a result of this, the operators b_a must be made to satisfy the following commutation relations:

$$[b^{-}_{a}(\boldsymbol{k}),\ \overset{*}{b}{}^{+}_{b}(\boldsymbol{q})]_{-} = [\overset{*}{b}{}^{-}_{a}(\boldsymbol{k}),\ b^{+}_{b}(\boldsymbol{q})] = \delta_{ab}\delta(\boldsymbol{k}-\boldsymbol{q}) \tag{11}$$

(all the other commutators are equal to zero).

Similar commutation relations exist for the operators $a^{\pm}_{a}$, $\overset{*}{a}{}^{\pm}_{b}$.

The expressions for the charge and for the component of the spin vector along the direction of motion in accordance with the formulas (3.29), (3.30) are

$$Q = \int d\boldsymbol{k}\, [\overset{*}{b}{}^{+}_{a}(\boldsymbol{k})\, b^{-}_{a}(\boldsymbol{k}) - b^{+}_{a}(\boldsymbol{k})\, \overset{*}{b}{}^{-}_{a}(\boldsymbol{k})], \tag{12}$$

$$\Delta S_3 \sim [\overset{*}{b}{}^{+}_{1}(\boldsymbol{k})\, b^{-}_{1}(\boldsymbol{k}) - \overset{*}{b}{}^{+}_{2}(\boldsymbol{k})\, b^{-}_{2}(\boldsymbol{k}) - b^{+}_{1}(\boldsymbol{k})\, \overset{*}{b}{}^{-}_{1}(\boldsymbol{k}) + b^{+}_{2}(\boldsymbol{k})\, \overset{*}{b}{}^{-}_{2}(\boldsymbol{k})]. \tag{13}$$

By evaluating the corresponding expectation values we convince ourselves that $\overset{*}{b}{}^{+}_{1}(\boldsymbol{k})$ and $b^{-}_{1}(\boldsymbol{k})$ are respectively the creation and annihilation operators for a particle of momentum $\boldsymbol{k}$, charge $+1$, and spin component $+1$ along the direction of motion; $\overset{*}{b}{}^{+}_{2}(\boldsymbol{k})$, $b^{-}_{2}(\boldsymbol{k})$ are the creation and annihilation operators for a particle of momentum $\boldsymbol{k}$, charge $+1$, and spin component -1; $\overset{*}{b}{}^{+}_{3}(\boldsymbol{k})$ and $b^{-}_{3}(\boldsymbol{k})$ are the creation and annihilation operators for a particle of momentum $\boldsymbol{k}$, charge $+1$, and spin component 0. The meaning of the operators b^{+}_{a} and $\overset{*}{b}{}^{-}_{a}$ may be obtained from the foregoing by means of the following rule: the transition from $b^{\pm}_{a}$ to $\overset{*}{b}{}^{\pm}_{a}$ corresponds to a change in the sign of the charge and in the sign of the spin component.

We see from the above that the complex vector field describes charged particles of mass m and with three possible values of the spin component (1, 0, -1) along a given direction. In other words, quanta of this field are charged mesons of spin 1 (they are also called vector mesons).

The amplitudes a_c also have a simple meaning. Thus, the amplitudes a_3 correspond to paricles of zero spin component along the direction of motion, while a_1 and a_2 describe combinations of states with spin components 1 and -1, which correspond to linear polarizations.

We also write down the commutation relations for the four dependent amplitudes U_ν. With the aid of (3.24) we find without difficulty the commutation relations for the three amplitudes U_n ($n = 1, 2, 3$):

$$[\overset{*}{U}{}_n^-(\boldsymbol{k}),\ U_m^+(\boldsymbol{q})] = [U_n^-(\boldsymbol{k}),\ \overset{*}{U}{}_m^+(\boldsymbol{q})] = \delta(\boldsymbol{k}-\boldsymbol{q})\left[\delta_{nm} + \frac{k_n k_m}{m^2}\right].$$

Utilizing the relation $k_0 U_0 - k_n U_n = 0$, we then determine the commutation relations which involve U_0:

$$[\overset{*}{U}{}_0^-(\boldsymbol{k}),\ U_n^+(\boldsymbol{q})] = [U_0^-(\boldsymbol{k}),\ \overset{*}{U}{}_n^+(\boldsymbol{q})] = -\frac{k_0 k_n}{m^2}\,\delta(\boldsymbol{k}-\boldsymbol{q}),$$

$$[\overset{*}{U}{}_0^-(\boldsymbol{k}),\ U_0^+(\boldsymbol{q})] = [U_0^-(\boldsymbol{k}),\ \overset{*}{U}{}_0^+(\boldsymbol{q})] = \left(\frac{k_0^2}{m^2} - 1\right)\delta(\boldsymbol{k}-\boldsymbol{q}).$$

Bringing together these expressions, we obtain the formula

$$[\overset{*}{U}{}_\nu^-(\boldsymbol{k}),\ U_\mu^+(\boldsymbol{q})] = [U_\nu^-(\boldsymbol{k}),\ \overset{*}{U}{}_\mu^+(\boldsymbol{q})] = \left(\frac{k_\nu k_\mu}{m^2} - g_{\nu\mu}\right)\delta(\boldsymbol{k}-\boldsymbol{q}), \qquad (14)$$

which is relativistically symmetric and compatible with the subsidiary conditions (3.22). Going over to the coordinate representation, we obtain

$$\begin{aligned}
&[\overset{*}{U}{}_\nu^-(x),\ U_\mu^+(y)] = [U_\nu^-(x),\ \overset{*}{U}{}_\mu^+(y)] = \left(g_{\nu\mu} + \frac{1}{m^2}\frac{\partial^2}{\partial x^\nu\,\partial x^\mu}\right) iD^-(x-y),\\
&[\overset{*}{U}_\nu(x),\ U_\mu(y)] = \left(g_{\nu\mu} + \frac{1}{m^2}\frac{\partial^2}{\partial x^\nu\,\partial x^\mu}\right) iD(x-y).
\end{aligned} \qquad (15)$$

It is not difficult to verify that the resulting commutation relations are compatible both with the field equations and with the subsidiary conditions. Thus, operating on both sides of (15) with the Klein–Gordon operator $\Box_x - m^2$, we obtain the identity $0 = 0$, since according to (3)

$$(\Box - m^2)\,D(x) = 0.$$

8.4. The electromagnetic field—difficulties of quantization. In the quantization of the electromagnetic field, it is necessary to satisfy simultaneously the requirements of positive energy density, the Lorentz subsidiary condition, and the condition of transversality. Moreover, the whole formulation should have the property of being relativistically covariant.

We encountered a similar situation in the quantization of the vector-meson field. The difference is that vector mesons can exist in three spin states, while photons, because of their transversality, only in two spin states. This in turn is because in contrast to mesons, the mass of the photons equals zero. The first circumstance leads to the fact that the components of the potential of the electromagnetic field contain, to an even greater extent than in the case of the vector field, "superfluous" variables, since we have four components and only two states in which real photons may exist.

The second difference ($m = 0$) makes it impossible to apply to the electromagnetic field the procedure used for the quantization of the vector field. Indeed, the quantization of a hypothetical vector field with vanishing mass (i.e., of a field which differs from the electromagnetic field by not having a gauge transmormation and therefore having three components) already meets with considerable difficulties. In attempting to carry out such a quantization we obtain meaningless expressions, first of all when we diagonalize the energy-momentum by means of the substitution (3.24), and second in setting up commutation relations for components of the potential A_ν in (14), (15), because of the presence of the vanishing mass in the denominators of the above expressions.

Therefore, the quantization of the electromagnetic field is carried out in the following manner. We regard the components of the vector potential as independent quantities, thereby giving up the Lorentz condition in operator form. Then, in accordance with the general prescription for quantization, starting with the structure of the energy-momentum four-vector expressed in terms of its longitudinal, transverse, and timelike components $a_\mu(\boldsymbol{k})$,

$$P^\nu = \int d\boldsymbol{k}\, k^\nu \left(-a_\mu^+(\boldsymbol{k})\, a_\mu^-(\boldsymbol{k})\right), \tag{16}$$

we arrive at the commutation relations

$$[a_\nu^-(\boldsymbol{k}),\ a_\mu^+(\boldsymbol{q})] = -g_{\nu\mu}\delta(\boldsymbol{k}-\boldsymbol{q}). \tag{17}$$

The quantization condition (17) allows us to regard the operators as creation and annihilation operators for four independent kinds of photons—two transverse kinds, and "longitudinal" and "timelike" kinds. However, the following difficulty arises in such a quantization. The component a_0 satisfies the relation

$$[a_0^-(\boldsymbol{k}),\ a_0^+(\boldsymbol{q})] = -\delta(\boldsymbol{k}-\boldsymbol{q}),$$

a comparison of which with (7.19) shows that the creation and annihilation operators for the "timelike" photons behave as if they had interchanged places, which occurs as a result of the term $a_0^+ a_0^-$ in (16) having a negative sign. Such a commutation relation is incompatible with the assumption that the metric in the Hilbert space of states is positive definite (see Assignment O12).

The difficulty outlined above may be eliminated by means of a formal approach (the Gupta–Bleuler method) which is based on the fact that this difficulty is due entirely to the "timelike" photons that do not actually exist, but the appearance of which in the intermediate steps of the argument is connected with the transition from observable quantities (the vectors $\boldsymbol{E}$ and $\boldsymbol{H}$) to the nonobservable four-potential A_ν, which has been made in order that the theory may be relativistically symmetric and covariant.
covariant.

8.5. The electromagnetic field—its quantization according to Gupta and Bleuler. We refer the reader for details to the literature (the "*Introduction*", Section 12.2; Akhiezer and Berestetskii (1969), Section 5.2) and present here only the fundamental idea of the Gupta–Bleuler method and the final formulas.

A central part is played here by the Lorentz condition. In the course of the quantization (17) we assumed that the components of the potential A_ν are independent, and consequently we may not impose the Lorentz condition on the operators. it may easily be seen that it is likewise impossible to impose the Lorentz condition on the admissible states

$$\partial A(x)\,\Phi = 0 \qquad (\partial A \equiv \partial^\nu A_\nu)$$

for such a condition contradicts, for example, the definition of the vacuum state. Indeed, by setting $\Phi = \Phi_0$ we have

$$\partial A(x)\,\Phi_0 = \partial A^+(x)\,\Phi_0 = 0.$$

Multiplying on the left by $A_\mu^-(y)$, we obtain

$$A_\mu^-(y)\,\partial^\nu A_\nu^+(x)\,\Phi_0 = \frac{\partial}{\partial x_\nu}\left(A_\mu^-(y)\,A_\nu^+(x)\right)\Phi_0 =$$

$$= \frac{\partial}{\partial x_\mu}\left[A_\mu^+(x)\,A_\nu^-(y) - ig_{\mu\nu}D_0^+(x-y)\right]\Phi_0 = -\,i\partial_\nu D_0^+(x-y)\,\Phi_0 \neq 0.$$

Here D_0^+ is the commutation function of the frequency components of the electromagnetic field, i.e., the positive-frequency part of the commutation function D_0 to be introduced below in (22).

Therefore we shall formulate the quantum Lorentz condition as a condition on admissible states:

$$\partial A^{-}(x)\,\Phi = \frac{\partial A_{\nu}^{-}(x)}{\partial x_{\nu}}\,\Phi = 0, \quad \overset{*}{\Phi}\,(\partial A^{+}) = 0. \tag{18}$$

The *weakened* condition (18) guarantees that the Lorentz condition is satisfied for expectation values

$$\langle \partial A \rangle = \overset{*}{\Phi}\,\partial A\,(x)\,\Phi = 0,$$

which is quite sufficient to establish correspondence with the classical field.

As pointed out in Section 4, the scalar function $\chi(x) = \partial A(x)$ in the diagonal gauge herein considered satisfies the free equation (4.15) even in the presence of interaction with sources. Therefore the weakened Lorentz condition (18) retains its form also in switching on interaction of the quantum electromagnetic field with the conserved current of charged particles.

Transition to the momentum representation in terms of variables in the local reference frame gives (taking into account that $k_0 = |\boldsymbol{k}|$) instead of (18) the following:

$$F^{-}(\boldsymbol{k})\,\Phi = 0, \quad \overset{*}{\Phi} F^{+}(\boldsymbol{k}) = 0, \tag{19}$$

where

$$F^{\pm}(\boldsymbol{k}) = a_{0}^{\pm}(\boldsymbol{k}) - a_{3}^{\pm}(\boldsymbol{k}).$$

From this it follows that the combination $a_3^+ a_3^- - a_0^+ a_0^-$ does not contribute to the admissible states:

$$\langle a_3^+ a_3^- - a_0^+ a_0^- \rangle = \overset{*}{\Phi}\,(a_3^+ - a_0^+)\,a_3^- \Phi = 0,$$

as a result of which the total contribution to the energy and momentum of admissible states from the longitudinal and timelike pseudophotons is equal to zero.

It is not difficult to verify that admissible states compatible with the conditions (18), (19) have the following structure:

$$\Phi = \Phi_{\text{tr}}^{0} + \sum_{n} c_n \prod_{1 \leqslant m \leqslant n} [a_3^+(\boldsymbol{k}_m) - a_0^+(\boldsymbol{k}_m)]\,\Phi_{\text{tr}}^{n}, \tag{20}$$

where Φ^0_{tr}, Φ^n_{tr} are amplitudes of real photon states which do not contain pseudophotons.

We shall now show that in the expansion over the local reference frame (4.23) only the transverse components

$$A^{tr}_\nu(\boldsymbol{k}) = e^1_\nu(\boldsymbol{k})\,a_1(\boldsymbol{k}) + e^2_\nu(\boldsymbol{k})\,a_2(\boldsymbol{k}).$$

have significance. We present the right-hand side of (4.23) in the form

$$A^-_\nu(\boldsymbol{k}) = A^{tr,-}_\nu(\boldsymbol{k}) + \left(\frac{k_\nu}{|\boldsymbol{k}|} - \delta_{\nu 0}\right) a^-_3(\boldsymbol{k}) + \delta_{\nu 0} a^-_0(\boldsymbol{k}) =$$

$$= A^{tr,-}_\nu(\boldsymbol{k}) + k_\nu \Lambda(\boldsymbol{k}) + \delta_{\nu 0} F^-(\boldsymbol{k}).$$

Here the term $k_\nu \Lambda = k_\nu a^-_3(\boldsymbol{k})/|\boldsymbol{k}|$ has the same structure as a 4-gradient and can be made to vanish by a gauge transformation, while $F^-(\boldsymbol{k})\Phi = 0$ according to (19). Therefore,

$$A^-_\nu(\boldsymbol{k})\,\Phi = A^{tr,-}_\nu(\boldsymbol{k})\,\Phi,$$

and, consequently,

$$\overset{*}{\Phi} A_\nu \Phi = \overset{*}{\Phi} A^{tr}_\nu \Phi.$$

It may be shown that the more general assertion

$$\overset{*}{\Phi} K(A)\,\Phi = \overset{*}{\Phi} K(A^{tr})\,\Phi = \overset{*}{\Phi}{}^0_{tr} K(A^{tr})\,\Phi^0_{tr}, \tag{21}$$

holds, where $K(A)$ is an operator which depends polynomially on A, and Φ^0_{tr} was introduced in (20). Thus, in accordance with (21) the expectation value of an arbitrary potential A over admissible states is equal to the expectation value of the transverse potential A^{tr} over the real physical states.

We note also that to carry out this procedure in a consistent manner requires the introduction of the indefinite metrics in that part of the Hilbert subspace which contains timelike pseudophotons. We cannot go into details here.

In conclusion we list the formulas for the quantized electromagnetic field in the Lorentz gauge: the commutator of the field functions,

$$[A_\nu(x),\ A_\mu(y)] = i g_{\nu\mu} D_0(x-y), \tag{22}$$

where

$$D_0(x) = D(x)|_{m=0} = -i(2\pi)^{-3}\int e^{ikx}\delta(k^2)\,\varepsilon(k^0)\,dk; \tag{23}$$

the field equations,

$$\Box A_\nu = 0; \tag{24}$$

the Lorentz condition on admissible states,

$$\left(\frac{\partial A^-(x)}{\partial x}\right)\Phi = 0; \tag{25}$$

and the four-momentum expectation value over admissible states,

$$\langle P_\nu\rangle = \left\langle \int d\boldsymbol{k}\, k_\nu \sum_{\sigma=1,\,2} a_\sigma^+(\boldsymbol{k})\, a_\sigma^-(\boldsymbol{k})\right\rangle. \tag{26}$$

9. SPINOR FIELDS

9.1. Quantization of the Dirac field. It may be seen from the structure of the energy-momentum 4-vector of the spinor field,

$$P^\nu = \int d\boldsymbol{k}\, k^\nu [\overset{*}{a}{}_s^+(\boldsymbol{k})\, a_s^-(\boldsymbol{k}) - \overset{*}{a}{}_s^-(\boldsymbol{k})\, a_s^+(\boldsymbol{k})] \tag{5.21}$$

that the independent amplitudes a_s $(s = 1, 2)$ may be straightforwardly quantized.

As we have mentioned earlier, the requirement that the expectation value of the energy operator P^0 should be positive leads to the demand that the spinor field must be quantized in accordance with Fermi–Dirac rules. Therefore the Fourier amplitudes of the spinor field $a_s(\boldsymbol{k})$ should be regarded as operators satisfying the Fermi–Dirac commutation relations:

$$[\overset{*}{a}{}_s^-(\boldsymbol{k}),\ a_r^+(\boldsymbol{q})]_+ = [a_s^-(\boldsymbol{k}),\ \overset{*}{a}{}_r^+(\boldsymbol{q})]_+ = \delta_{sr}\delta(\boldsymbol{k}-\boldsymbol{q}) \tag{1}$$

(all the other anticommutators are zero).

The commutation relations for the operators $\psi(x)$ will be obtained from the above with the aid of formulas for the Fourier transforms,

$$\psi^{\pm}(x) = (2\pi)^{-3/2} \int dk\, e^{\pm ikx} a_s^{\pm}(k)\, v^{s,\,\pm}(k), \tag{2}$$

$$\overline{\psi}^{\pm}(x) = (2\pi)^{-3/2} \int dk\, e^{\pm ikx} \overset{*}{a}{}_s^{\pm}(k)\, \bar{v}^{s,\,\pm}(k) \tag{3}$$

which follow from (5.9), (5.13), (5.14) and the formulas for summation over the spin index (5.20). We have

$$[\psi_\alpha^-(x),\ \overline{\psi}_\beta^+(y)]_+ = (2\pi)^{-3} \int dk\, e^{ik(y-x)} \sum_\sigma v_\alpha^{\sigma,\,-}(k)\, \bar{v}_\beta^{\sigma,\,+}(k) =$$

$$= \frac{1}{(2\pi)^3} \int dk\, e^{ik(y-x)} \frac{(\hat{k}+m)_{\alpha\beta}}{2k_0} =$$

$$= \frac{1}{(2\pi)^3} \int dk\, (\hat{k}+m)_{\alpha\beta}\, \delta(k^2-m^2)\, \theta(k^0)\, e^{ik(y-x)} =$$

$$= (i\gamma^\nu\, \partial/\partial x^\nu + m)_{\alpha\beta} \frac{1}{i} D^-(x-y).$$

Analogously it may be shown that

$$[\psi^+(x),\ \overline{\psi}^-(y)]_+ = (i\hat{\partial}_x + m)(1/i)\, D^+(x-y).$$

Introducing the notation

$$S_{\alpha\beta}^{\pm}(x) = (i\hat{\partial}+m)_{\alpha\beta}\, D^{\pm}(x), \quad S = S^+ + S^-,$$

we obtain

$$[\psi(x),\ \overline{\psi}(y)] = \frac{1}{i} S(x-y), \tag{4}$$

where

$$S(x) = (i\hat{\partial}+m)\, D(x) = \frac{i}{(2\pi)^3} \int e^{-ikx} \delta(k^2-m^2)\, \varepsilon(k^0)\, (\hat{k}+m)\, dk. \tag{5}$$

The commutation relations are compatible with the field equations, since

$$(i\hat{\partial}-m)\, S(x) = (\Box - m^2)\, D(x) = 0.$$

Writing the Lagrangian in the normal form

$$\mathscr{L} = (i/2){:}[\overline{\psi}(x)\, \gamma^\nu \partial_\nu \psi - (\partial_\nu \overline{\psi})\, \gamma^\nu \psi(x)]{:} - m{:}\overline{\psi}\psi{:}, \tag{6}$$

we obtain the energy-momentum tensor

$$T^{\mu\nu} = (i/2) : \left[\bar{\psi}\gamma^{\mu} \frac{\partial \psi}{\partial x_{\nu}} - \frac{\partial \bar{\psi}}{\partial x_{\nu}} \gamma^{\mu}\psi \right] :, \tag{7}$$

the spin density tensor

$$S^{\lambda(\mu\nu)} = \frac{1}{4} :\bar{\psi}\gamma^{\lambda}\sigma^{\mu\nu}\psi: + \frac{1}{4} :\bar{\psi}\sigma^{\mu\nu}\gamma^{\lambda}\psi: \tag{8}$$

and the current vector

$$J^{\nu}(x) = :\bar{\psi}(x)\gamma^{\nu}\psi(x):. \tag{9}$$

As a result of the transition to the momentum representation by (2) and (3), the corresponding integrals of motion, in accordance with (5.21), (5.27), and (5.28), take the following form: the energy-momentum 4-vector is

$$P^{\nu} = \int dk\, k^{\nu} [\overset{*}{a}{}^{+}_{s}(k)\, a^{-}_{s}(k) + a^{+}_{s}(k)\, \overset{*}{a}{}^{-}_{s}(k)], \tag{10}$$

the charge is

$$Q = \int dk [\overset{*}{a}{}^{+}_{s}(k)\, a^{-}_{s}(k) - a^{+}_{s}(k)\, \overset{*}{a}{}^{-}_{s}(k)], \tag{11}$$

and the spin component along the direction of motion is

$$\Delta S_3 \sim (^1/_2) [\overset{*}{a}{}^{+}_{1}(k)\, a^{-}_{1}(k) - \overset{*}{a}{}^{+}_{2} a^{-}_{2} - a^{+}_{1}\overset{*}{a}{}^{-}_{1} + a^{+}_{2}\overset{*}{a}{}^{-}_{2}]. \tag{12}$$

From (10)–(12) it follows that the operators $\overset{*}{a}{}^{+}_{s}(k)$ and $a^{-}_{s}(k)$ are respectively the creation and annihilation operators for particles of momentum k, energy $k_0 = \sqrt{k^2 + m^2}$, charge $+1$, and component of spin along the direction of k equal to $\frac{1}{2}(s = 2)$. The operators $a^{+}_{s}(k)$ and $\overset{*}{a}{}^{-}_{s}(k)$ correspond to particles which differ from those described above by the sign of the charge (-1) and by the sign of the spin (i.e., $\frac{1}{2}$ $(s = 2)$ and $-\frac{1}{2}$ $(s = 1)$).

9.2. The spinor field with zero mass. The spinor field with zero mass, which corresponds to the neutrino, is of particular interest.* Setting $m = 0$ in the Dirac equation, we obtain

$$i\gamma^{\nu}\partial_{\nu}\psi(x) = 0.$$

*In recent years wide discussion on the possibility of the neutrino having nonzero rest mass has been going on. Recent experiments at the ITEP (Moscow) yield evidence for the mass of the electron neutrino being of the order of 30 eV.

This equation decomposes into two independent two-component equations. To demonstrate this, we make use of the fact that in the absence of the mass term, the operator of the Dirac equation anticommutes with the matrix γ_5.
γ_5.

By introducing the projection operators

$$P_{\pm}=\frac{1\pm\gamma_5}{2},$$

we obtain

$$P_{\pm}\hat{\partial}\psi=P_{\pm}\gamma^{\nu}\,\partial_{\nu}\psi=\gamma^{\nu}P_{\mp}\,\partial_{\nu}\psi,$$

i.e., two separate equations for $\psi_{\pm}=P_{\pm}\psi$

$$i\hat{\partial}\psi_{+}=0,\quad i\hat{\partial}\psi_{-}=0. \tag{13}$$

In the standard representation,

$$\psi_{\pm}=\frac{1}{2}\begin{pmatrix}\psi_1\mp\psi_3\\ \psi_2\mp\psi_4\\ \mp\psi_1+\psi_3\\ \mp\psi_2+\psi_4\end{pmatrix}.$$

Thus, each of the functions ψ_+, ψ_- contains only two independent components and in the "split" form

$$\psi_{+}=\begin{pmatrix}\varphi_{(+)}\\ -\varphi_{(+)}\end{pmatrix},\quad \psi_{-}=\begin{pmatrix}\varphi_{(-)}\\ \varphi_{(-)}\end{pmatrix}$$

can be expressed in terms of the two-component spinors

$$\varphi_{(+)}=\frac{1}{2}\begin{pmatrix}\psi_1-\psi_3\\ \psi_2-\psi_4\end{pmatrix},\quad \varphi_{(-)}=\frac{1}{2}\begin{pmatrix}\psi_1+\psi_3\\ \psi_2+\psi_4\end{pmatrix}.$$

Using the "split" form (Appendix II.7) for the Dirac matrices, we obtain instead of (13) the following:

$$\left(\frac{\partial}{\partial x^0}\pm\boldsymbol{\sigma}\frac{\partial}{\partial \mathbf{x}}\right)\varphi_{(\pm)}(x)=0. \tag{14}$$

Equations of this form were first proposed by Weyl. To understand the physical meaning of the two-component functions φ_α [$\alpha = (+), (-)$], we go over to the momentum representation

$$\varphi_\alpha = \varphi_\alpha^+ + \varphi_\alpha^-, \quad \varphi_\alpha^\pm (x) = (2\pi)^{-3/2} \int d\boldsymbol{p}\, \tilde{\varphi}_\alpha^\pm (\boldsymbol{p})\, e^{\pm ipx}.$$

The *Weyl equations* assume the form

$$(p_0 \mp \boldsymbol{\sigma p})\, \tilde{\varphi}_{(\pm)} (\boldsymbol{p}) = 0. \tag{15}$$

As was established in Section 5.3 (see (5.24)), the matrix vector σ describes the spin of a fermion. From equations (15) it follows (see also Assignment O16) that the spin of the neutrino can be directed either along its momentum or in the opposite direction. Therefore each of the four-component momentum amplitudes $\omega_\pm$ contains only one spin state.

The doubled value of the spin component of a fermion along its momentum, $(\boldsymbol{\sigma p})/|\,\boldsymbol{p}\,|$, is called its *helicity*. It has been experimentally established that the spin of a neutrino is antiparallel to its momentum, i.e., the neutrino helicity is equal to -1 (which holds both for the electron and for the muon neutrino). As will be shown below, the function ψ_- describes particles with negative helicity, and ψ_+ those with positive helicity.

Therefore we define the field operator for the neutrino as follows:

$$\nu (x) = P_- \psi (x).$$

It satisfies the massless Dirac equation

$$i\hat{\partial}\nu (x) = 0$$

and the subsidiary condition

$$P_+ \nu (x) = \frac{1+\gamma_5}{2} \nu (x) = 0. \tag{16}$$

The corresponding conjugate function $\bar{\nu} = \overset{*}{\nu}\gamma^0$ obeys the conjugate equation

$$i\partial_\mu \bar{\nu} (x)\, \gamma^\mu = 0$$

and the subsidiary condition

$$\bar{\nu} (x)\, P_- = \bar{\nu} (x)\, \frac{1-\gamma_5}{2} = 0.$$

The Lagrangian of the quantized free neutrino field is written in the form

$$\mathscr{L}(x) = \frac{i}{2} :[\bar{\nu}(x)\gamma^\mu \nu_{;\,\mu}(x) - \bar{\nu}_{;\,\mu}(x)\gamma^\mu \nu(x)]:. \tag{17}$$

The expressions for the energy-momentum tensor and the spin density tensor in the configuration representation are obtained from formulas (7), (8) by substituting $\nu(x)$ for $\psi(x)$. In the momentum representation the formula for P^μ is obtained from (10) by dropping the summation over the spin index:

$$P^\mu = \int d\boldsymbol{p}\, p^\mu [\overset{*}{a}{}^+(\boldsymbol{p})\, a^-(\boldsymbol{p}) + a^+(\boldsymbol{p})\, \overset{*}{a}{}^-(\boldsymbol{p})]. \tag{18}$$

We shall consider the spin of the neutrino field separately. For this purpose we shall proceed from an expression of the form (5.25) for the spin vector density

$$(^1/_2): \{\overset{*}{a}{}^+(\boldsymbol{p})\, a^-(\boldsymbol{p})\, \overset{*}{v}{}^+(\boldsymbol{p})\, \sigma_3 v^-(\boldsymbol{p}) + \overset{*}{a}{}^-(\boldsymbol{p})\, a^+(\boldsymbol{p})\, \overset{*}{v}{}^-(\boldsymbol{p})\, \sigma_3 v^+(\boldsymbol{p})\}:, \tag{19}$$

which is valid in the reference frame where $p_1 = p_2 = 0$, i.e., when the x_3-axis is directed along the momentum. In this frame the normalized spinors satisfying the equations

$$\hat{p} v^\pm(\boldsymbol{p}) = \overset{*}{v}{}^\pm(\boldsymbol{p})\, \hat{p} = 0$$

and subsidiary conditions

$$(1+\gamma^5)\, v^\pm(\boldsymbol{p}) = \overset{*}{v}{}^\pm(\boldsymbol{p})\,(1+\gamma^5) = 0,$$

may be chosen in the form (the standard representation of the Dirac matrices is implied)

$$v^\pm(\boldsymbol{p}) = (2)^{-1/2} \begin{pmatrix} w(\boldsymbol{p}) \\ w(\boldsymbol{p}) \end{pmatrix} \equiv v(\boldsymbol{p}), \quad \overset{*}{v}(\boldsymbol{p}) = (2)^{-1/2} (\overset{*}{w}, \ \overset{*}{w}).$$

Here $w(\boldsymbol{p})$ are two-component spinors satisfying the Weyl equation (15):

$$(p_0 + \sigma_3 p_3)\, w(\boldsymbol{p}) = 0,$$

and have the form

$$w(\boldsymbol{p}) = \begin{pmatrix} 0 \\ 1 \end{pmatrix}.$$

Here we have taken into account that for a massless neutrino $p_0 = p_3$. It is also obvious that

$$\overset{*}{w}(\mathbf{p}) = (0,\ 1),$$

and therefore

$$\overset{*}{v}{}^{\pm}(\mathbf{p})\,\sigma_3 v^{\mp}(\mathbf{p}) = -1.$$

Inserting this result into (19), we find

$$\Delta S_3 \sim -\frac{1}{2}\overset{*}{a}{}^{+}(\mathbf{p})\,a^{-}(\mathbf{p}) + \frac{1}{2}a^{+}(\mathbf{p})\,\overset{*}{a}{}^{-}(\mathbf{p}). \tag{20}$$

The first term of this expression corresponds to the left-handed neutrino with helicity equal to -1, and the second to the right-handed antineutrino with helicity equal to $+1$.

The expression (20) corresponds to (12) when $a_1 = 0$, $a_2 = a$. If instead of the function $\nu = \psi_-$ we were to examine the function ψ_+, we would obtain the analog of (12) for $a_2 = 0$, which describes particles of positive helicity and antiparticles with negative helicity.

Note that in the consideration of weak-interaction processes, in which left-handed neutrinos are involved, sometimes the following representation for the Dirac matrices ($n = 1, 2, 3$) is utilized:

$$\Gamma_0 = \Gamma^0 = \begin{pmatrix} 0 & I \\ I & 0 \end{pmatrix}, \quad \Gamma^n = -\Gamma_n = \begin{pmatrix} 0 & \sigma_n \\ -\sigma_n & 0 \end{pmatrix},$$

$$\Gamma_5 = \Gamma^5 = \begin{pmatrix} -I & 0 \\ 0 & I \end{pmatrix} = -i\Gamma^0\Gamma^1\Gamma^2\Gamma^3 = i\Gamma_0\Gamma_1\Gamma_2\Gamma_3. \tag{21}$$

Here we have made use of the *split* form of writing in terms of 2×2 matrices **I**, **0** and Pauli matrices (see Appendix II).

In this representation the projection operators $P_\pm$ are diagonal, and consequently the spinor with negative helicity has the simpler form

$$\nu(x) = P_-\psi(x) = \begin{pmatrix} \psi_1 \\ \psi_2 \\ 0 \\ 0 \end{pmatrix}.$$

The prepresentation (21) may be appropriately called the *helicity* representation of the Dirac matrices. It is connected with the standard representation (DII.5), (DII.7) by the transformation

$$\Gamma = O^{-1}\gamma O, \quad O = \frac{1}{\sqrt{2}}\begin{pmatrix} I & I \\ -I & I \end{pmatrix}.$$

9.3. Charge conjugation. As is easily seen from the formulae of Section 2 for complex fields of integral spin describing charged particles, the transformation

$$\begin{aligned} \varphi_a(x) &\rightarrow \varphi'_a(x) = \overset{*}{\varphi}_a(x), \\ \overset{*}{\varphi}_a(x) &\rightarrow \overset{*}{\varphi}{}'_a(x) = \varphi_a(x) \end{aligned} \tag{22}$$

to the new functions φ', $\overset{*}{\varphi}{}'$ leaves the expressions for all the dynamical quantities (with the exception of the current and charge) unchanged and changes the sign of J_ν and Q. Therefore a transformation of the type (22) is called the transformation of charge conjugation, or just *charge conjugation*.

The charge conjugation of a spinor field is more complicated than (22) because of the matrix character (multicomponent nature) of spinor wave functions; it is a matrix transformation of the following form:

$$\psi'(x) = C\bar{\psi}(x)^{\mathrm{T}}, \quad \bar{\psi}' = \psi^{\mathrm{T}}\overset{\mathrm{T}}{C}{}^{-1} = (C^{-1}\psi)^{\mathrm{T}}, \tag{23}$$

for the consistency of which we must demand that

$$\overset{\mathrm{T}}{C}\gamma_0\overset{+}{C}\gamma_0 = 1. \tag{24}$$

The transformation inverse to (23) is correspondingly given by

$$\psi(x) = C\bar{\psi}'(x)^{\mathrm{T}}, \quad \bar{\psi} = \psi'^{\mathrm{T}}\overset{\mathrm{T}}{C}{}^{-1} = (C^{-1}\psi')^{\mathrm{T}}.$$

We thus see that charge conjugation defined by (23), in addition to the obvious property that the twice iterated transformation is the identity transformation, also has the property of "mirror" symmetry, i.e., its form coincides with that of the inverse transformation.

We shall determine the explicit form of the matrices C by requiring that the Lagrangian of the free field and, consequently, the energy-momentum four-vector should not change their form, while the current four-vector should change its sign; i.e., that the following relations should hold:

$$\mathscr{L}(\psi) = \mathscr{L}(\psi'), \quad T(\psi) = T(\psi'), \quad J_\nu(\psi) = -J_\nu(\psi').$$

For this it will be sufficient (see Assignment O17) to demand that the following two relations should hold:

$$:\bar{\psi}_1(x)\gamma_\nu\psi_2(x): = -:\bar{\psi}_2'(x)\gamma_\nu\psi_1'(x): \tag{25}$$

and

$$:\bar{\psi}(x)\psi(x): = :\bar{\psi}'(x)\psi'(x):, \tag{26}$$

where ψ_1 and ψ_2 are equal either to the function ψ itself or to its derivatives $\partial_\mu\psi$, and they transform according to (23).

Let us now examine the restrictions imposed on the matrix C by the conditions (25) and (26). Inserting (23) into the left hand side of (25), we have

$$:\bar{\psi}_1(x)\gamma_\nu\psi_2(x): = :\overset{T}{\psi_1'}(x)\overset{T}{C}{}^{-1}\gamma_\nu C\overset{T}{\bar{\psi}_2'}(x):.$$

Taking into account the fact that the quantized spinors ψ' and $\bar{\psi}'$ anti-commute, we hence obtain the relation

$$:\bar{\psi}_1\gamma_\nu\psi_2: = -:\bar{\psi}_2'\overset{T}{C}\overset{T}{\gamma}_\nu C^{-1}\psi_1':.$$

Comparing with (25) we obtain the first condition on the matrix C:

$$\overset{T}{C}\overset{T}{\gamma}_\nu C^{-1} = \gamma_\nu,$$

or, equivalently, in the transposed form

$$\overset{T}{C}{}^{-1}\gamma_\nu C = \overset{T}{\gamma}_\nu. \tag{27}$$

Similarly, substituting (23) into (26), we obtain

$$:\bar{\psi}\psi: = :\overset{T}{\psi'}\overset{T}{C}{}^{-1}C\overset{T}{\bar{\psi}'}: = -:\bar{\psi}'\overset{T}{C}C^{-1}\psi':,$$

from which we find the second condition on the matrix C: $\overset{T}{C}C^{-1} = -1$; i.e.,

$$\overset{T}{C} = -C. \tag{28}$$

Utilizing (28), we may write (27) in the form

$$C^{-1}\gamma^{\nu}C = -\overset{T}{\gamma^{\nu}}. \tag{29}$$

It is not difficult to show (see Assignment O17) that from (28), (29), and (24) it follows that the charge-conjugation matrix is unitary:

$$C\overset{+}{C} = 1.$$

The formulas (29) and (30) allow us to determine the explicit form of the matrix C in any fixed representation of the Dirac matrices. We suggest that the reader verify on his own that both in the standard representation (AII.5) and in the helicity representation (21) the matrix C equals the product of the "zeroth" and the "second" Dirac matrices, i.e.,

$$C_{\text{stand}} = \gamma^0\gamma^2 = \alpha_2, \quad C_{\text{helic}} = \Gamma^0\Gamma^2.$$

We point out that generally, in going over from one representation of the Dirac matrices γ to some other one $\tilde{\gamma}$, i.e.

$$\gamma^{\nu} = O\tilde{\gamma}^{\nu}O^{-1}$$

we shall obtain the new matrix $\tilde{C}$ for the charge conjugation from the relation

$$C = O\tilde{C}\overset{T}{O},$$

where $\overset{T}{O}$ is a matrix transposed with respect to O.

The physical meaning of the operation of charge conjugation C consists in the substitution of an antiparticle for a particle. Therefore, for example, invariance under the operation C (which takes place in strong and electromagnetic interactions) leads to the following: if a certain process Π (which represents an arbitrary combination of strong and electromagnetic interactions) is realized, then the process $C\Pi$ which differs from Π in that all particles are replaced by antiparticles may take place and is realized with the same probability as Π.

9.4. The CPT theorem. To complete the consideration of quantized free fields, we recall one of the most important assertions of local quantum field theory—the so-called *CPT* theorem that formulates the invariance property under three discrete transformations:

the charge conjugation *C*,
the space inversion *P*,
the inversion of the time axis *T*.

Besides the operation *C* considered above, we have previously encountered the operation *P* of reflection of all the three spatial coordinate axes (see Section 1 and Appendix II). The third operation—inversion (or reflection) of the time axis—has the form

$$x \to x', \quad x_0' = -x_0, \quad \boldsymbol{x}' = \boldsymbol{x}.$$

Unlike the continuous transformations entering into the Poincaré group, the operations *P* and *T* do not correspond to any real motion of a physical system. They do, however, connect various physical processes with each other, as does the operation *C*.

We shall return to this question at the end of the section, and for the time being we shall examine these transformations formally. Let us consider the simple example of a complex scalar field described by the operators

$$\varphi(x) = \varphi^+(x) + \varphi^-(x), \quad \overset{*}{\varphi}(x) = \overset{*}{\varphi}{}^+(x) + \overset{*}{\varphi}{}^-(x).$$

The operations *P* and *T* change the sign of the argument in the configuration representation. For instance,

$$PT\varphi^{\pm}(x) = PT\varphi^{\pm}(x_0, \boldsymbol{x}) = P\varphi^{\pm}(-x_0, \boldsymbol{x}) = \varphi^{\pm}(-x_0, -\boldsymbol{x}) = \varphi^{\pm}(-x),$$

while the operation *C* replaces particles by antiparticles:

$$C\varphi^{\pm}(x) = \overset{*}{\varphi}{}^{\pm}(x), \quad C\overset{*}{\varphi}{}^{\pm}(x) = \varphi^{\pm}(x).$$

Therefore

$$CPT\varphi^{\pm}(x) = \overset{*}{\varphi}{}^{\pm}(-x), \quad CPT\overset{*}{\varphi}{}^{\pm}(x) = \varphi^{\pm}(-x).$$

In the same way as to transformations belonging to the Lorentz group *L* there correspond unitary operators U_L which transform fields and state vectors (see, e.g., (6.25)), to the transformations *C, P, T* in the quantum case there also correspond the operators U_C, U_P, U_T. *Thus, for the complex scalar field,*

$$\varphi_C(x) = U_C^{-1}\varphi(x)\,U_C = \overset{*}{\varphi}(x),$$
$$\varphi_P(x) = U_P^{-1}\varphi(x)\,U_P = \varphi(x_0, -\boldsymbol{x})$$

and so on.

The CPT theorem asserts* that the local Lagrangian of a system of quantum fields constructed according to "normal rules" turns out to be invariant:

$$\mathscr{L}_{CPT}(x') = U_C^{-1}U_P^{-1}U_T^{-1}\mathscr{L}(x')\,U_T U_P U_C = \mathscr{L}(x), \quad x' = -x$$

under the simultaneous application of all three operations. The "normal rules" involve the quantization rules of fields in accordance with the Pauli theorem and the choice of coupling constants in accordance with the condition of Hermiticity of the Lagrangian.

Let us now discuss the physical consequences of the *CPT* theorem. The operation *C* replaces particles by antiparticles. The operation *P* changes the directions in the 3-dimensional space, i.e., replaces all three-vectors (momenta, components of the electric field, etc.) by their negatives. The operation *T* exchanges initial and final states and vice versa; the directions of the momenta and spins are also changed. Thus, from the *CPT* theorem follows the equality of the squares of the matrix elements of the processes Π and *CPT*Π, while the process *CPT*Π is obtained from Π by the substitution of antiparticles for all the particles, by the inversion of all the spins, and by the substitution of final states for the initial ones. In particular, from the *CPT* theorem the equality of the masses and lifetimes of particle and antiparticle follows, as well as the fact that their magnetic moments differ in sign only.

---

*The reader may find a more precise formulation in terms of modern axiomatic field theory in the book by Bogolubov, Logunov and Todorov (1969), Chapter 5.

CHAPTER III

INTERACTING FIELDS

10. THE INTERACTION OF FIELDS

10.1. Interaction between particles. One of the most important physical features of microscopic particles is their ability to interact and to transform mutually into each other. According to present-day conceptions all the variety of different kinds and forms of interaction of matter can be reduced to four types of *elementary* (or *fundamental*) *interactions* between particles:

(1) gravitation,
(2) electromagnetism,
(3) weak interactions,
(4) strong interactions.

Gravitation is a universal interaction. It is an inherent quality of all forms of matter, of all microscopic particles.

Electrically charged particles partake directly in electromagnetic interactions.

Both gravitational and electromagnetic interactions are *long-range*. Their intensity decreases weakly at large distances ($\sim r^{-1}$); as a result of this it is just these interactions that play a determining role in macrophysics.

At the same time, as pointed out in Section 1, in the microcosm the gravitational interaction is much weaker than the electromagnetic interaction. Therefore we shall not consider it.

Let us first examine *electromagnetic interactions*. It is well known that the Coulomb interaction of two charged particles which represents action at a distance (*actio in distans*) is the result of the interaction of each of the particles with the electromagnetic field. After quantization the latter is represented by its quanta—by photons—and the *elementary act* of interaction of a charged particle Q with the electromagnetic field manifests itself as an act of emission or of absorption of a γ-quantum by this particle:

$$Q \to Q+\gamma \quad \text{or} \quad Q+\gamma \to Q. \tag{1}$$

We shall agree to represent the combination of two such mutually converse processes by a double arrow or by the following picture-diagram:

$$Q \longleftrightarrow Q + \gamma, \tag{2}$$

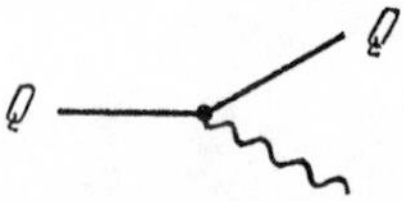

In the latter the solid lines correspond to the charged particle Q, while the wavy line corresponds to the γ-quantum. The Coulomb interaction of two particles Q_1 and Q_2 is represented by the diagram

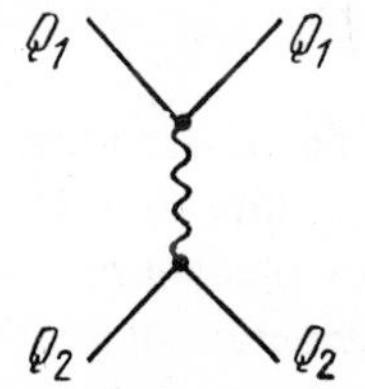

$$Q_1, Q_2 \longrightarrow Q_1, Q_2, \tag{3}$$

which describes the exchange of a γ-quantum between the particles Q_1 and Q_2. We shall adopt the convention that in such diagrams time "flows" horizontally from left to right.

The diagram describing the annihilation or creation process of a pair will then look like the following:

$$Q + \bar{Q} \longleftrightarrow \gamma, \tag{4}$$

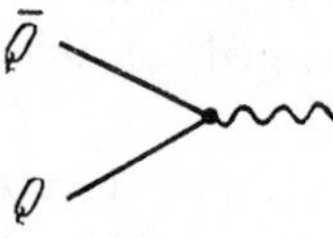

In such a "transfer" to the opposite side of (2) the particle Q must be replaced by its antiparticle $\overline{Q}$.

If we now turn the picture (3) 90° clockwise, we obtain

$$Q_2 + \bar{Q}_2 \longleftrightarrow Q_1 + \bar{Q}_1, \tag{5}$$

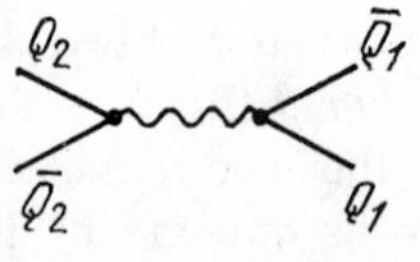

—the diagram which describes the process of transformation of the pair (Q_2, $\overline{Q}_2$) into the pair (Q_1, $\overline{Q}_1$) by means of one-photon annihilation.

As we shall see below (see Section 10.3), to the elementary electromagnetic interaction (2), (4), there corresponds a term in the Lagrangian of the form

$$\mathscr{L}(x) \sim eA_\mu(x)\, J_Q^\mu(x), \tag{6}$$

where A_μ is the potential of the electromagnetic field, and J_Q^μ is the field current of the particles Q. For example, in the electron–positron case

$$\mathscr{L}(x) = eA_\mu(x)\,\bar{\psi}(x)\,\gamma^\mu\,\psi(x). \tag{7}$$

A typical example of the *weak interaction* is the β-decay of the neutron:

$$n \to p + e^- + \bar{\nu}_e, \tag{8}$$

which we shall present in a "more symmetric" form

$$n + \nu_e \longleftrightarrow p + e^-$$

or

$$n + \bar{p} \longleftrightarrow e + \nu_e$$

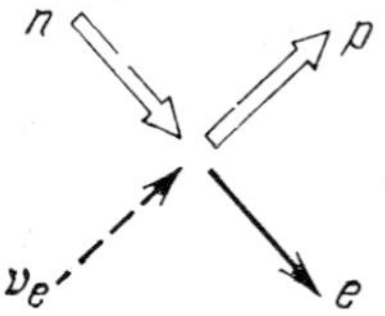

. (9)

As a rule, elementary weak-interaction processes have a four-fermion structure similar to (9). A second example is the decay of a muon ($\mu \to e^- + \nu_\mu + \bar{\nu}_e$), which corresponds to the interaction

$$\mu + \nu_e \longleftrightarrow e^- + \nu_\mu$$

or

$$\mu + \bar{\nu}_\mu \longleftrightarrow e + \nu_e$$

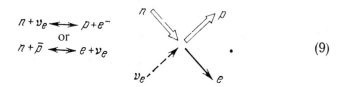

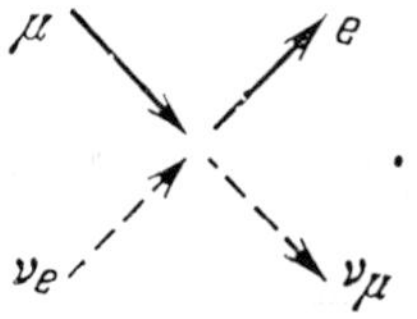

. (10)

Weak interactions describing processes of the type (9), (10) are represented in the Lagrangian by terms which have the structure (current $\times$ current):

$$\mathscr{L} \sim G J_w^\nu(x)\,\overset{+}{J}{}_\nu^w(x), \tag{11}$$

where

$$J_\nu^w(x) = \bar{p}(x)\,O_\nu n(x) + \bar{\nu}_e(x)\,O_\nu e(x) + \bar{\nu}_\mu(x)\,O_\nu \mu(x) + \dots \tag{12}$$

is the weak current, O_ν are certain matrices, G is the Fermi constant, and the operator field functions are denoted by the same symbols as the particles corresponding to them. The numerical value of the Fermi constant may be expressed in terms of the fundamental constants c, $\hbar$, and some mass squared. If we choose for this purpose the mass of the nucleon M, we have

$$\frac{G}{\hbar c} \cdot \frac{M^2c^2}{\hbar^2} \simeq 1.0 \times 10^{-5}. \tag{13}$$

The fact that in the system $\hbar = c = 1$ ($[G] = m^{-2}$) the Fermi constant has a dimension differing from zero means that the intensity of the weak interaction depends significantly on the energy. At energies of the order of magnitude of 1 GeV (i.e., of the order of magnitude of the nucleon mass) in the center-of-mass system the intensity is described by the numerical relation (13). However, at energies within the range 10^2–10^3 GeV it increases up to values of the order of magnitude of 1, and the Fermi interaction ceases to be weak.

Strong interactions embrace a large group of particles with relatively large masses ($\geq m_\pi$) which are called by the generic name *hadrons*. These interactions are short-ranged:

$$F_{\text{strong}} \sim \exp(-r/r_{\text{nucl}});$$

they manifest themselves only at small distances characterized by the range of action of nuclear forces,

$$r_{\text{nucl}} \simeq \frac{\hbar}{m_\pi c} \simeq 1.4 \text{ fm} = 1.4 \times 10^{-13} \text{ cm}, \tag{14}$$

which is close in value to the Compton wavelength of the pion.

Strong interactions are responsible for the nuclear forces which bind protons and neutrons together in atomic nuclei. They are called strong because in the microcosm they lead to significant quantitative effects. At the same time, at small distances (when $r < r_{\text{nucl}}$) they are large in absolute value and are considerably greater than electromagnetic interactions. We shall now elucidate this idea.

There exist various ways of comparing the intensities of different interactions. One of them is based on the comparison of energy effects. Thus, electromagnetic interactions may be characterized by the binding energy of the s-electron in the hydrogen atom: $E_{\text{el}} \simeq 10$ eV. Strong interactions are responsible for the binding energy of nucleons in the nucleus: $E_{\text{nucl}} \simeq 10$ MeV $= 10^6\, E_{\text{el}}$.

The second method consists in comparing numerical dimensionless quantities which describe the intensities of the various interactions. The Lagrangian of the electromagnetic interaction between the Dirac field and the electromagnetic field (7) contains the electron charge e, which is a measure of the interaction intensity. The Coulomb interaction of two charged particles is proportional to e^2 and may be characterized by the dimensionless combination

$$\alpha \equiv \frac{e^2}{4\pi\hbar c} \simeq \frac{1}{137} \tag{15}$$

—the so-called *fine-structure constant.* This same quantity (more precisely, $\alpha/\pi \simeq \frac{1}{430}$) represents the expansion parameter in electrodynamic perturbation theory. Therefore the electromagnetic interactions are considered to be of the order of magnitude of 10^{-2}–10^{-3}. The relative splittings of masses within the isotopic multiplets

$$\frac{M_n - M_p}{M_p} \simeq \frac{939.55 - 938.26}{938.26} \simeq 1.38 \times 10^{-3},$$
$$\frac{m_{\pi^\pm} - m_{\pi^0}}{m_{\pi^0}} \simeq \frac{139.57 - 134.96}{136.96} \simeq 3.4 \times 10^{-2},$$

are just of this order of magnitude, and they are regarded as electromagnetic effects.

In the framework of the traditional description of the interaction of pions and nucleons by the Yukawa mechanism

$N \longleftrightarrow N + \pi$, 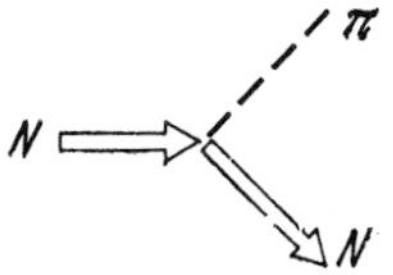

(16)

the interaction strength may be characterized, with the aid of the so-called pseudoscalar coupling

$$\mathscr{L}(x) = g\overline{\Psi}(x)\gamma^5\boldsymbol{\tau}\Psi(x)\boldsymbol{\pi}(x), \tag{17}$$

by the coupling constant g. The numerical value of this combination, analogous to the fine-structure constant, is

$$\frac{g^2}{4\pi\hbar c} \simeq 14.7, \tag{18}$$

which is not small compared to unity.

At the same time the modern description of strong interactions by the gauge quark–gluon mechanism results in a term in the Lagrangian describing the quark–gluon interaction,

$$\mathscr{L}_{\text{QCD}} = g_s \bar{q}_a(x)\gamma^\mu B_\mu^{ab}(x) q_b(x), \tag{17a}$$

involving the constant of chromodynamic interaction, numerically estimated to be

$$\alpha_s = \frac{g_s^2}{4\pi\hbar c} \sim 0.1. \tag{18a}$$

10.2. Interaction Lagrangians. In the study of a system of connected fields describing interaction processes of the corresponding particles, one starts from the total Lagrangian of the system, which is usually represented by the sum

$$\mathscr{L}_{\text{total}} = \mathscr{L}_0 + \mathscr{L}_{\text{int}}$$

of the free Lagrangian $\mathscr{L}_0$ (which is itself the sum of the Lagrangians entering into the system of free fields) and the interaction Lagrangian $\mathscr{L}_{\text{int}}$. Examples of $\mathscr{L}_{\text{int}}$ are given above by the formulas (7), (11), (17). We shall now formulate some of the general properties which the interaction Lagrangians of fields must possess.

The Lagrangian $\mathscr{L}_{\text{int}}$ must, first, represent a local Hermitian Lorentz-invariant construction consisting of the field functions of various fields and their derivatives. $\mathscr{L}_{\text{int}}$ must, besides this, reflect the symmetry properties and the discrete conservation laws characteristic of the given interaction mechanism.

Limiting ourselves to the simplest polynomial (to be more precise, multilinear) forms, we obtain, for example, that the interaction of a scalar field φ with a spinor field ψ, if parity is conserved, may be constructed to be the type scalar × scalar $(\bar{\psi}(x)\psi(x)\varphi(x))$ or of the type vector × vector $(\bar{\psi}(x)\gamma^\nu\psi(x)\varphi_{;\nu}(x))$. In a similar manner the simplest form of interaction between the pseudoscalar isovector pion field $\pi(x)$ and the nucleon field $\psi(x)$ described by (17) has the form pseudoscalar × pseudoscalar in the configuration space and at the same time the structure isovector × isovector in the isotopic space. The alternative form constructed to be of the type (pseudovector × pseudovector) × (isovector × isovector) is

$$\bar{\Psi}(x)\,\gamma^\nu\gamma^5\boldsymbol{\tau}\Psi(x)\,\boldsymbol{\pi}_{;\,\nu}(x) \tag{19}$$

and represents the so-called "pseudovector" variant of the Yukawa interaction. Here we point out the fact that isotopic symmetry represents an example of broken symmetry—it exists only in strong interactions. In a similar way there is symmetry under space inversions in strong and electromagnetic interactions. Weak interactions do not conserve space parity. The weak current is a linear combination of a vector and a pseudovector.

At the same time such laws as the conservation laws of the electric and baryon charges are obeyed in all known interactions. Together with Lorentz invariance, they are considered to be absolute.

In accordance with the conservation law of electric charge, the Lagrangian should be invariant under the phase transformation of the complex-conjugated field functions which describe charged particles:

$$\varphi \to e^{i\alpha}\varphi, \quad \overset{*}{\varphi} \to e^{-i\alpha}\overset{*}{\varphi}, \tag{20}$$

and also of all other fields which carry the given charge.

Analogous limitations are imposed by the conservation laws of the baryon charge, of strangeness, charm, etc.

Interactions of fields are usually introduced in accordance with experimentally observed processes. Such reasoning does not, of course, have a sufficiently firm theoretical basis and may lead to errors. The question remains open, in particular, of the existence of the intermediate vector boson which would make the 4-fermion Fermi interaction (11) a consequence of the trilinear interaction $J^{\mu}_{\text{weak}}(x)B_{\mu}(x)$, where B_{μ} is a vector field—i.e., in a manner similar to the transition from (3) to (2), the four-fermion vertex presented in (9), (10) is split and represented in the form

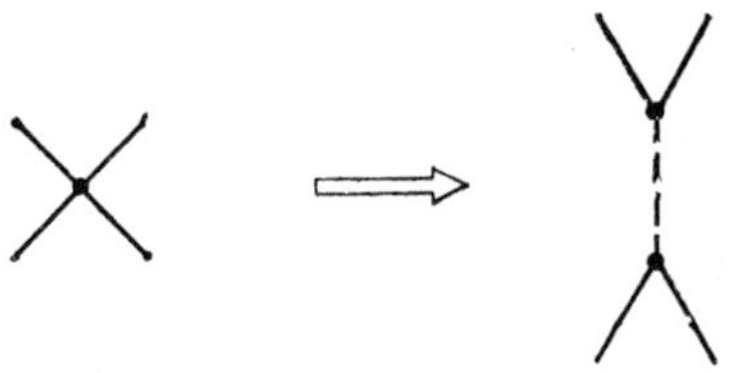

In quantum field theory an important role is played by simple models involving a small number of quantized fields connected with each other by interactions simple in form.

Below we shall often deal with the model of a scalar real field with a quartic interaction the total Lagrangian of which has the form

$$\mathcal{L}(x) = \frac{1}{2}\,\partial_{\mu}\varphi\partial^{\mu}\varphi - \frac{m^2}{2}\,\varphi^2(x) - h\varphi^4(x). \tag{21}$$

Here we shall regard the latter term containing the coupling constant h as the interaction Lagrangian and call the model (21) the φ^4 model. In the Lagrangian (21) the field φ may be considered to be either a scalar or a pseudoscalar. Therefore the φ^4 model is close to a model of pion–pion interaction decribed by the Lagrangian

$$\mathscr{L}(x)=\frac{1}{2}\partial_\mu\boldsymbol{\varphi}(x)\,\partial^\mu\boldsymbol{\varphi}(x)-\frac{m^2}{2}\boldsymbol{\varphi}(x)\,\boldsymbol{\varphi}(x)-h\,(\boldsymbol{\varphi}\boldsymbol{\varphi})^2, \tag{22}$$

where φ is the isovector introduced in (3.10). The interaction term $h(\varphi\varphi)^2$ describes several processes simultaneously:

$$\begin{gathered}\pi^\pm+\pi^\pm\to\pi^\pm+\pi^\pm\\ \pi^++\pi^-\leftrightarrow\pi^0+\pi^0\\ \pi^0+\pi^0\to\pi^0+\pi^0\end{gathered} \tag{23}$$

As it was pointed out above, we have no serious basis for regarding the interaction (23) as a fundamental interaction. In the quark–gluon model of strong interactions (see Section 33 below), for instance, pion–pion interactions (23) as well as the pion–nucleon interaction (16), (17) turn out to be derivatives of the more fundamental quark–gluon interaction.

Despite such a general unsatisfactory state of knowledge of the structure of interactions of fields, there exist profound arguments which allow one to determine the general structure of electromagnetic interactions. We shall now present these arguments.

10.3. The electromagnetic field as a gauge field. The starting point is a sort of singling out of the electromagnetic field, contrasting it with all other fields, which we shall call *fields of matter*.

The field functions of electrically charged fields of matter u, like the quantum-mechanical ψ-function, are not observable and enter into dynamical quantities in terms of bilinear forms in u and $\overset{*}{u}$. (On the contrary, the field strengths of the electromagnetic field are directly measurable.) Because of this, the field functions of matter are determined only up to a phase factor. In the preceding (Sections 2.5 and 6.5) this arbitrariness permitted us to introduce a gauge transformation of the form (20) with the phase α independent of the coordinates, and ultimately to obtain expressions for the charge operator.

Let us now demand that the theory remain invariant also when the parameter of the gauge transformation depends on x

$$u(x)\to u'(x)=e^{i\alpha(x)}u(x),\quad \overset{*}{u}(x)\to\overset{*}{u}{}'(x)=e^{-i\alpha(x)}\overset{*}{u}(x), \tag{24}$$

i.e., that the relative phase $\alpha(x)-\alpha(x')$ at two different points of the space-time x and x' be absolutely arbitrary. The *local phase transformation* (24) expresses the possibility that the arbitrariness in the choice of the phase factor of the wave function of the field of matter may be of a local nature.

It is not difficult to verify that the Lagrangians of the above-considered complex fields (see, for example, (3.16), (5.5)) are not invariant under (24), since the field gradients, besides the phase factor, get an extra term proportional to the gradient of the function $\alpha(x)$. Invariance may be restored by introducing an additional vector field A_ν which transforms together with (24) according to the law

$$A_\nu(x) \to A'_\nu(x) = A_\nu(x) + f_{;\nu}(x). \tag{25}$$

All the derivatives $\partial_\nu u$, $\partial_\nu u^*$ in the Lagrangian must then be replaced by the operators

$$\begin{aligned} D_\nu(A)\, u &\equiv (\partial_\nu - ieA_\nu(x))\, u(x), \\ \overset{*}{D}_\nu(A)\, \overset{*}{u} &\equiv (\partial_\nu + ieA_\nu(x))\, \overset{*}{u}(x), \end{aligned}$$

which depend on A. The expressions (26) are referred to as *covariant derivatives*.

If the condition $\alpha(x) = ef(x)$ is satisfied, the covariant derivatives under the simultaneous transformations (24), (25) transform in a manner similar to (24), i.e., they change their phase:

$$\begin{aligned} D_\nu(A')\, u'(x) &= e^{ief(x)} D_\nu(A)\, u(x), \\ \overset{*}{D}_\nu(A')\, \overset{*}{u}'(x) &= e^{-ief(x)} \overset{*}{D}_\nu(A)\, \overset{*}{u}(x). \end{aligned} \tag{27}$$

Besides the substitution of covariant derivatives for the ordinary ones, it is also necessary to introduce into the Lagrangian the Lagrangian $\mathscr{L}_0$ of the free field A_ν:

$$\mathscr{L}\left(u, \overset{*}{u}, \partial u, \partial \overset{*}{u}\right) \to \mathscr{L}' = \mathscr{L}\left(u, \overset{*}{u}, D(A)\, u, \overset{*}{D}(A)\, \overset{*}{u}\right) + \mathscr{L}_0(A), \tag{28}$$

which in turn should be invariant under (25). It should be chosen in the form (4.12):

$$\mathscr{L}_0(A) = -\frac{1}{4} F_{\mu\nu} F^{\mu\nu}, \tag{29}$$

with

$$F_{\mu\nu} = \partial_\mu A_\nu - \partial_\nu A_\mu \tag{4.1}$$

so the field A_ν of necessity turns out to be massless. The constant e in (26) and (27) may be identified with the electric charge, and the vector field A_ν is to be identified with the electromagnetic field. Thus, the electromagnetic field introduced through the covariant derivatives D, $\overset{*}{D}$ plays the role of a *compensating* field. fields which compensate the change of gauge of the fields of matter are known also as *gauge fields*. As the group composed of the transformations (25) is of an Abelian (commutative) nature, the electromagnetic field is an Abelian gauge field.

The transition in the free Lagrangian from the ordinary to the covariant derivatives in accordance with the rule (28), (26) leads to the appearance of an interaction between the initial field u with the electromagnetic field A.

The interaction introduced with the aid of covariant derivatives is also called the *minimal electromagnetic interaction*.

In the case of spinor fields, when $\mathscr{L}_0$ is linear in the derivatives, the minimal electromagnetic interaction is linear in the gauge field A and assumes the form (7).

For fields of integral spin the minimal interaction in the usual formulation contains quadratic terms. Thus, to the Lagrangian of the complex scalar field there corresponds

$$\mathscr{L}_{\text{inter}}(x) = ie\left[\overset{*}{u}(x)\, u_{;\nu}(x) - \overset{*}{u}_{;\nu}(x)\, u(x)\right] A^\nu(x) + e^2 \overset{*}{u}(x)\, u(x)\, A^\nu(x)\, A_\nu(x). \tag{30}$$

Thus, we have verified that the electromagnetic field provides invariance of the total Lagrangian under local phase transformations which represent generalizations of the "global" phase transformations leading to the conservation of electric charge. Such a viewpoint establishes connections between the properties of fields in space-time and the so-called intrinsic symmetries. A most important example of intrinsic symmetry which directly generalizes the charge symmetry is isotopic symmetry.

The requirement of invariance under the local isotopic transformations leads to the necessity of introducing a new compensating vector field—the Yang–Mills field.

11. NON-ABELIAN GAUGE FIELDS

11.1. The Yang–Mills field. The requirement of invariance under local phase transformations belonging to the continuous symmetry group leads to the appearance of a compensating vector field. Such a field was first introduced for the non-Abelian group of isotopic transformations by Yang and Mills in 1953.

The non-Abelian gauge fields subsequently acquired the name of Yang–Mills fields. In modern physics such fields are widely employed in models of weak and strong interactions (see Sections 32, 33). In these models use is made of groups of appropriately large size. To facilitate the discussion, we shall restrict ourselves to examining the simplest non-Abelian group—the $SU(2)$ group (the notation and terminology of the theory of continuous groups is explained in Appendix III) which we shall conventionally call "isotopic", not attributing any physical meaning to this term.

Consider a multicomponent field function φ_a ($a = 1, \ldots, A$) which transforms according to a certain irreducible representation of the gauge group G (i.e., it forms an isotopic multiplet). The infinitesimal transformation

$$\varphi_a \to \varphi_a' = \varphi_a + ig\alpha_k (T_k)_{ab} \varphi_b \qquad (k = 1, \ldots, K) \tag{1}$$

contains K independent real infinitesimal parameters α_k, from which the finite common multiplier g is factored out for convenience. The number K is called the dimensionality of the group G. In the case of (10.24) it is equal to 1. For the group $SU(2)$ under consideration $K = 3$; for the group $SU(3)$ of unitary transformations $K = 8$, etc. The Hermitian generators T_k are matrices of the representation to which the multiplet φ_a belongs. For the isodoublet ($A = 2$) they are proportional to the Pauli matrices τ_k (compare (AI.5)), for the isotriplet ($A = 3$) the matrices T may be expressed in terms of the unit antisymmetric tensor ε (see (AI.15)):

$$T_k = t_k, \quad (t_k)_{ab} = t_{kab} = -i\varepsilon_{kab} \tag{2}$$

and so on.

It may be useful to point out that the commutators of the matrices T of an arbitrary representation have the following universal form:

$$[T_k, T_l] = -(t_n)_{kl} T_n = i\varepsilon_{kln} T_n \tag{3}$$

and contain the matrix elements of the matrix t of the so-called adjoint representation as structure consants. The dimension of this representation is equal to the dimension of the group—in the case under consideration it is equal to three.

Equation (1), which involves the parameters α_k independent of the coordinates, is analogous to (2.27) and describes the so-called global gauge transformation. Passing over to the local transformation of the type (10.24), i.e. considering the parameters α to depend on x

$$\alpha_k \to \alpha_k(x),$$

by analogy with electrodynamics we replace the ordinary derivatives of φ by the covariant ones

$$\partial_\nu \varphi_a \to (D_\nu(B)\varphi)_a = \partial_\nu \varphi_a - igB_\nu^k(x)(T_k)_{ab}\varphi_b, \tag{4}$$

which contain the new compensating vector field B_ν. This field, contrary to the electromagnetic field A_ν, has group indices (the upper ones) and transforms according to a more complicated law

$$B_\nu^a \to (B_\nu^a)' = B_\nu^a + ig\alpha_b(x)\, t_{bac} B_\nu^c + \partial_\nu \alpha_a(x), \tag{5}$$

which, besides the gauge covariant $\partial_\nu \alpha$, also involves isotopic rotation. The generators of this rotation, $\varepsilon = it$, are matrices of the adjoint representation.

Note that because of the gauge arbitrariness here, as in electrodynamics (see Section 4.3), of the four components of the vector potential B_ν only two "three-dimensionally transverse" components correspond to physical degrees of freedom.

With the aid of the formulas (3) it is now possible to demonstrate that the covariant derivative $D_\nu \varphi$ transforms under gauge transformations (1), (5) like φ:

$$[D_\nu(B')\varphi']_a = [D_\nu(B)\varphi]_a + ig\alpha_k(x)(T^k)_{ab}[D_\nu(B)\varphi]_b. \tag{6}$$

The tensor

$$F_{\mu\nu}^k = \partial_\mu B_\nu^k - \partial_\nu B_\mu^k + igt_{kln} B_\mu^l B_\nu^n. \tag{7}$$

turns out to be the analog of the electromagnetic field strength tensor $F_{\mu\nu}$. For the SU(2) group with the aid of (2) we also obtain

$$\boldsymbol{F}_{\mu\nu} = \partial_\mu \boldsymbol{B}_\nu - \partial_\nu \boldsymbol{B}_\mu + g[\boldsymbol{B}_\mu \times \boldsymbol{B}_\nu]. \tag{8}$$

The tensor F transforms as the adjoint representation of the gauge group

$$F_{\mu\nu}^a \to (F_{\mu\nu}^a)' = F_{\mu\nu}^a + ig\alpha_k(x)(t_k)_{ab} F_{\mu\nu}^b(x). \tag{9}$$

The invariant Lagrangian of the free Yang-Mills field has the form

$$\mathscr{L}_{YM}(B) = -\frac{1}{4} F_{\mu\nu}^a F_{\mu\nu}^a, \tag{10}$$

analogous to the Lagrangian of the electromagnetic field. However, unlike the expression (10.29), in which each of the multipliers $F_{\mu\nu}$ is gauge invariant, in the case in question only the expression (10) as a whole is invariant.

The Lagrangian (10) is similar to the transverse Lagrangian of electrodynamics (4.12), (10.29) in that it is degenerate. Therefore it must be modified for its quantization. The equations of motion

$$\partial^{\mu} F^{a}_{\mu\nu} + g\varepsilon^{abc} B^{b}_{\mu} F^{c}_{\mu\nu} = 0 \tag{11}$$

following from it are nonlinear and contain "self-interaction".

If we were to try to quantize the Yang–Mills field by the Gupta–Bleuler method by changing the Lagrangian in accordance with the prescription (4.13),

$$\mathscr{L} \to \mathscr{L} + \frac{a}{2} (\partial B)^2$$

then, as can be shown, the longitudinal function

$$\chi^a(x) = \partial B^a(x)$$

in this case would satisfy the nonlinear equation

$$\Box \chi^a(x) + g\varepsilon^{abc} B^b_{\nu}(x) \partial^{\nu} \chi^c(x) = 0. \tag{12}$$

instead of the free d'Alembert equation.

Thus, the longitudinal degree of freedom does not "split away" from the transverse ones. Hence it follows that in the process of time evolution "physical states" for which the Lorentz condition $\langle \chi \rangle = 0$ holds (on the average) may undergo transition to states having nonzero longitudinal component. Therefore, the Gupta–Bleuler method needs to be generalized. This generalization will be described in Section 19.4.

11.2. Gauge interaction of fields. As was shown in Section 10.3, the requirement that the Lagrangian of the fields of matter be invariant under the local phase transformations (10.24) leads to the necessity of introducing a compensating electromagnetic field A_{μ}, the interaction of which with all electrically charged fields of matter is described by a single constant e. The transformation (10.24) corresponds to the Abelian U(1) group of rank one.

In more complicated cases, when some simple non-Abelian group G occurs as the gauge group, the requirement of invariance of the Lagrangian also leads to a "uniform" interaction between the fields of matter and the gauge field B_ν. Thus, a certain set is formed of matter fields (with differing spins, electric charges, etc.) connected in a universal manner with the field B_ν. The terms in the interaction that appear due to making covariant the derivatives of the matter fields in accordance with the rule (4) contain the same dimensionless coupling constant g as the nonlinear components of the Yang–Milles field (see (7), (8)).

The procedure for the construction of a gauge model of interacting fields consists of the following steps:

(1) We choose a (simple) gauge group G and fix the set of matter fields $u_{(1)}, \ldots, u_{(n)}$ which transform according to different (but definite) representations of this group and are described by the free Lagrangian

$$\mathscr{L}_0(u, \partial u) = \sum_i \mathscr{L}_0^i(u_{(i)}, \partial_\nu u_{(i)}). \tag{13}$$

(2) We introduce a compensating vector field B_ν corresponding to the adjoint representation of group G and form the overall Lagrangian (the fields of matter $u_{(1)}, \ldots, u_{(n)}$ plus the Yang–Mills field B_ν) by the recipe

$$\begin{gathered}\mathscr{L}_0(u, \partial u) \rightarrow \mathscr{L}(u, B), \\ \mathscr{L}(u, B) = \mathscr{L}_0(u, D(B)u) + \mathscr{L}_{\mathrm{YM}}(B).\end{gathered} \tag{14}$$

In the first term of the right-hand side the derivatives of each of the fields $u_{(i)}$ are made covariant according to the rule (4) with the same constant g which enters into $\mathscr{L}_{\mathrm{YM}}$. Therefore the Lagrangian $\mathscr{L}(u, B)$, describing a system of $n + 1$ fields, contains in the case of a simple group only one coupling constant. This Lagrangian is invariant under the simultaneous gauge transformation (1), (5) of its fields.

As pointed out at the end of Section 11.1, the free gauge vector field is perforce massless. This circumstance imposes significant restrictions on the possibility of using gauge models of field interaction, since in the real world the only existing massless vector field is the electromagnetic field A_ν, which corresponds to the Abelian gauge group.

It turns out, however, that there exists a nontrivial way of introducing mass into a gauge field *interacting* with a scalar field of matter. This method is based on the so-called mechanism of spontaneous symmetry breaking. We shall now proceed to examine this mechanism.

11.3. Spontaneous symmetry breaking. In various areas of physics, starting with classical mechanics, examples are known of systems which exhibit

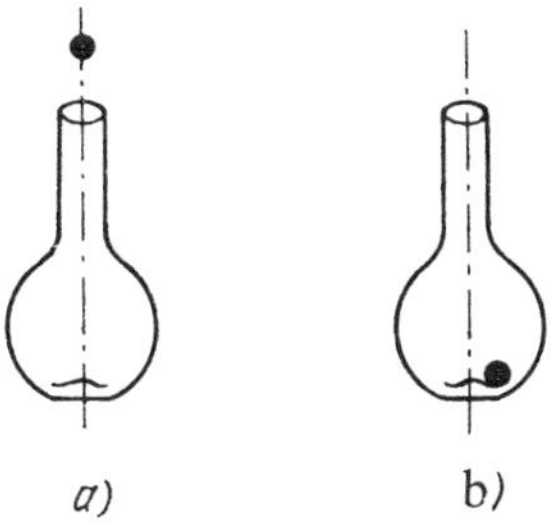

Fig. 11.1. A mechanical system demonstrating spontaneous symmetry breaking: (a) the initial state; (b) the final state.

spontaneous symmetry breaking. We mean by such a system one described by expressions (a Lagrangian, a Hamiltonian, equations of motion) possessed of some symmetry, while the real physical state of the system corresponding to a particular solution of the equations of motion does not have this symmetry. Such a situation arises when the lowest symmetric state does not have the lowest possible energy and is itself unstable. The actual cause of symmetry breaking may turn out to be an infinitesimal nonsymmetric perturbation.

A simple example is provided by a mechanical system consisting of an empty vessel with a convex bottom and a small ball inside it. We make the vessel, which has rotational symmetry, stand up in a perfectly vertical position (with respect to the Earth gravitation field) and let the ball fall into it precisely along its axis (Figure 11.1a). When the ball reaches the bottom, it will not stay on the central protuberance and will roll down to the periphery (Figure 11.1b). The final state does not have rotational symmetry. Another such example is represented by an isotropic ferromagnet which breaks the symmetry with respect to rotations in space by having a magnetization vector in a definite direction.

We shall now turn to the classical scalar field $\varphi(x)$ characterized by the Lagrangian

$$\mathscr{L} = T(\partial\varphi) - V(\varphi), \tag{15}$$

in which T is the kinetic part which depends only on the components of the gradient, while V is a potential function. For the usual free scalar field considered in Section 3, $V(\varphi)$ is a second-order parabola $m^2\varphi^2/2$, presented in Figure 11.2. The zero solution corresponds to the minimum of the potential function, and quantization of the free scalar field performed in Section 8 corresponds to the quantization of small oscillations about the position of equilibrium at $\varphi = 0$.

The Lagrangian (15) is symmetric with respect to the "reflection transformation" $\varphi \to -\varphi$, and its lower state $\varphi = 0$ is symmetric and stable.

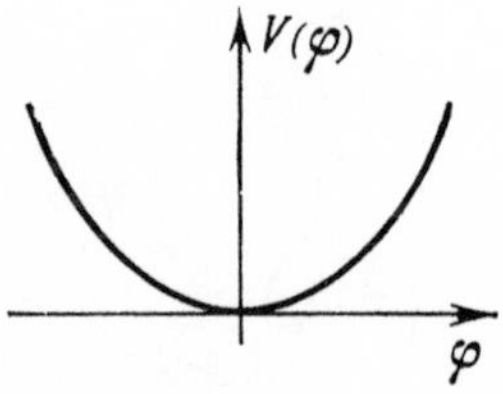

Fig. 11.2. The potential function of the free scalar field.

the introduction into $V(\varphi)$ of positive terms of the form $+\varphi^4$ (compare (10.21), (10.22)) describing self-interaction of the field does not affect these properties.

To achieve spontaneous breaking of the reflection symmetry, we assume that the dependence of V on φ has the form of an even function with two symmetric minima shown in Fig. 11.3. Such a dependence may be described by the binomial

$$V(\varphi) = -\frac{\mu^2}{2}\varphi^2 + \frac{h^2}{4}\varphi^4.$$

This expression contains a term quadratic in φ with "negative mass squared", which may suggest such hypothetical particles as tachyons—particles moving faster than light. To us, however, what is essential is that, owing to the negative sign of the first term, the point $\varphi = 0$ is not a minimum, and stable equilibrium is observed at the two points

$$\varphi = \pm\varphi_0, \quad \varphi_0 = \mu/h.$$

Both lower states have the same energy

$$V(\pm\varphi_0) = V_0 = -\mu^4/4h^2,$$

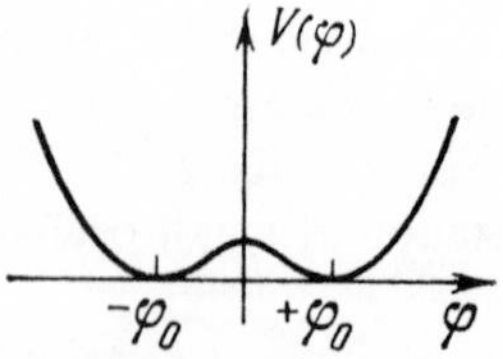

Fig. 11.3. The potential function of the scalar-field model involving spontaneous symmetry breaking.

as a result of which the corresponding quantum system has a doubly-degenerate vacuum state.

Now we must choose one of the lower states, for instance, and, through the operation of displacement by a constant, pass over to a new field function

$$\varphi(x) \to u(x) = \varphi(x) - \varphi_0,$$

which will turn out to be convenient for studying small deviations from φ_0. In this way we obtain

$$V(\varphi) = V(\varphi_0 + u) = V_0 + \mu^2 u^2 + \mu h u^3 + (h^2/4)\, u^4.$$

Neglecting the constant V_0, which only changes the energy scale, we see that the positive second term on the right-hand side corresponds to a mass of the quanta of small oscillations equal to $m = \mu/\sqrt{2}$, while the third and fourth terms giving nonlinear contributions to the equations of motion may be regarded at small h as a perturbation corresponding to the self-interaction of the field.

Note here that the field transformation of the form

$$\varphi(x) = \text{const} + u(x)$$

was introduced into quantum physics for the first time by one of the authors of this book at the time of the construction of the microscopic theory of superfluidity in 1946 (see Bogoliubov (1979), pp. 115–131), whereas the method of studying quantum systems with spontaneously broken symmetry, referred to as the method of quasiaverages, was introduced in 1960–61 (see Bogoliubov (1979), pp. 193–269).

A quite simple but physically profound generalization arises in the transition from the single-component (singlet) field to the isotopic multiplet $\varphi = (\varphi_1, \ldots, \varphi_A)$ with the potential

$$V(\varphi) = -\frac{\mu^2}{2}(\boldsymbol{\varphi\varphi}) + \frac{h^2}{4}(\boldsymbol{\varphi\varphi})^2, \quad \boldsymbol{\varphi\varphi} = \sum_a \varphi_a^2. \tag{16}$$

The discrete reflection symmetry is replaced by the continuous symmetry of isotopic rotations, similar to the symmetry of rotations in space about the vertical axis in the above example from mechanics. The minimum in the potential energy now corresponds to the relation

$$\boldsymbol{\varphi\varphi} = \varphi_0^2 = \mu^2/h^2, \tag{17}$$

defining an $(n-1)$-dimensional manifold in the n-dimensional isotopic space. It is necessary to shift the field function φ by a constant vector φ_0 satisfying the condition (17). Without loss of generality, we choose this vector in the form

$$\boldsymbol{\varphi}_0 = (\varphi_0, 0, \ldots, 0), \quad \varphi_0 = +\mu/h.$$

Setting, further,

$$\boldsymbol{\varphi}(x) = \boldsymbol{\varphi}_0 + \boldsymbol{u}(x), \tag{18}$$

we obtain for V the expression

$$V(\boldsymbol{\varphi}_0 + \boldsymbol{u}(x)) = V_0 + \mu^2 u_1^2 + \mu h u_1 (\boldsymbol{u}\boldsymbol{u}) + \frac{h^2}{4} (\boldsymbol{u}\boldsymbol{u})^2,$$

which contains no terms quadratic in $u_2, u_3, \ldots, u_n$. At the same time the structure of V with respect to the new variable u_1 is analogous to the one-component case.

Thus, as a result of spontaneous symmetry breaking, represented by the displacement (18), the shifted component u_1 "acquires" a free mass $m_1 = \mu\sqrt{2}$, while the remaining components $u_2, u_3, \ldots, u_n$ correspond to massless degrees of freedom ($m_2 = m_3 = \cdots = m_n = 0$). This latter property is invariably associated with the spontaneous breakdown of a *continuous* symmetry, because each of the massless degrees of freedom corresponds to an infinite degree of degeneracy of the vacuum state, which in our case is described by the relation (17). In the classical example the massless degree of freedom is related to the virtual motion of the ball in the final state (Figure 11.1b) within the circular hollow at the bottom of the vessel.

The assumption formulated above that there necessarily appear massless scalar mesons when continuous symmetry is spontaneously broken is known in quantum field theory as the *Goldstone theorem*, and the massless particles are called Goldstone bosons. The Goldstone theorem is the relativistic analog of the theorem on long-range action in quantum statistics (see Bogoliubov (1979), pp. 193–269, especially Chapter II).

11.4. The massive Yang–Mills field. We emphasize that spontaneous symmetry breaking signifies the breaking of the symmetry of the solution and not of the Lagrangian. The initial Lagrangian (14), (16) remains unchanged under displacement by a constant (18), which only represents the transition to new independent variables. This fact can be used for introducing a mass for the non-Abelian gauge field while not breaking the gauge symmetry of the

Lagrangian. Such a massive field cannot be free. It must interact with the scalar field which serves to implement the spontaneous symmetry breaking.

Consider a non-Abelian Yang–Mills gauge field $\boldsymbol{B}$ interacting with the scalar field $\boldsymbol{u}$ which transforms according to the triplet representation of the $SU(2)$ group. Assume that the field $\boldsymbol{u}$ obeys nonlinear equations corresponding to a Lagrangian of the type (15), (16):

$$\mathcal{L}_0(\boldsymbol{u}, \partial \boldsymbol{u}) = \frac{1}{2}\partial_\nu \boldsymbol{u}\, \partial^\nu \boldsymbol{u} + \frac{\mu^2}{2}\boldsymbol{u}\boldsymbol{u} - \frac{h^2}{4}(\boldsymbol{u}\boldsymbol{u})^2.$$

The total Lagrangian of the system of fields $\boldsymbol{u} + \boldsymbol{B}$, on the basis of (14), will be

$$\mathcal{L}(\boldsymbol{u}, \boldsymbol{B}) = \mathcal{L}_0(\boldsymbol{u}, D(\boldsymbol{B})\boldsymbol{u}) + \mathcal{L}_{YM}(\boldsymbol{B}), \tag{19}$$

where the covariant derivative $D(\boldsymbol{B})\boldsymbol{u}$ is defined by (4). With this formula we find

$$\mathcal{L}(\boldsymbol{u}, \boldsymbol{B}) = (1/2)(\partial^\nu \boldsymbol{u} + g[\boldsymbol{B}_\nu \times \boldsymbol{u}])(\partial^\nu u + g[\boldsymbol{B}^\nu \times \boldsymbol{u}]) + \\ + \frac{\mu^2}{2}\boldsymbol{u}\boldsymbol{u} - \frac{h^2}{4}(\boldsymbol{u}\boldsymbol{u})^2 + \mathcal{L}_{YM}(\boldsymbol{B}). \tag{20}$$

Consider the terms containing both fields: $\boldsymbol{u}$ and $\boldsymbol{B}$. After performing slight simplifications these terms may be represented as follows:

$$(g^2/2)\{(\boldsymbol{B}_\nu \boldsymbol{u})(\boldsymbol{B}^\nu \boldsymbol{u}) - (\boldsymbol{u}\boldsymbol{u})(\boldsymbol{B}_\nu \boldsymbol{B}^\nu\} - g([\partial_\nu \boldsymbol{u} \times \boldsymbol{u}]\boldsymbol{B}^\nu). \tag{21}$$

Let us now carry out the operation of displacement by a constant for the scalar field

$$\boldsymbol{u}(x) = \boldsymbol{u}_0 + \boldsymbol{v}(x), \tag{22}$$

where the isovector $\boldsymbol{u}_0$ is directed along the third axis:

$$u_0^a = \varphi_0 \delta_{a3}.$$

The terms quadratic in the Yang–Mills field and not containing the new scalar field obtained from (21) are:

$$-(g^2\varphi_0^2/2)(g^{\mu\nu}B_\mu^3 B_\nu^3 - \boldsymbol{B}_\nu \boldsymbol{B}^\nu) = +\frac{M^2}{2}g^{\mu\nu}(B_\mu^1 B_\nu^1 + B_\mu^2 B_\nu^2), \tag{23}$$

which are to be regarded as the mass terms of the components B^1, B^2 of the Yang-Mills field. Thus, these two components acquire the mass $M = g\varphi_0$. The component B^3 has remained massless.

The form (21) after the shift of the scalar field also involves quadratic terms which contain the vector as well as the new scalar fields:

$$-M\,(B^1_\mu\,\partial^\mu v^2 - B^2_\mu\,\partial^\mu v^1) \tag{24}$$

The presence of such "cross" quadratic terms, generally speaking, complicates the interpretation of the massless components of the scalar field v_1 and v_2 (concerning this question see also below, in Section 12.2). However, in the case under discussion there exists an important circumstance leading to a significant simplification. The cross terms may be eliminated by a gauge transformation of a special kind, namely, by fixing the gauge so that the following conditions will be satisfied:

$$v'_1(x) = v'_2(x) = 0, \quad \partial^\nu B'^3_\nu(x) = 0. \tag{25}$$

As a result, we obtain a Lagrangian depending only on the scalar field $v'_3 = v$ and the vector triplet $\boldsymbol{B}' = (B'_1, B'_2, B'_3)$:

$$\mathscr{L}(\boldsymbol{u}, \boldsymbol{B}) = \mathscr{L}(v, \boldsymbol{B}'). \tag{26}$$

The two components B'_1, B'_2 of the vector field, in conformity with (22), have equal masses:

$$M_1 = M_2 = M = g\varphi_0 = g\mu/h, \tag{27}$$

the component B'_3 remains massless, while the scalar boson v has the mass

$$m = \sqrt{2}\mu. \tag{28}$$

Thus, instead of the initial isotriplet of massless vector fields and of the triplet of scalar fields of negative squared mass, we have obtained a doublet of massive vector fields: one massless vector field and one scalar meson with mass. The massless Goldstone mesons have disappeared as a result of the gauge transformation. This process of transforming the massless gauge field into a massive one together with the simultaneous "annihilation" of the Goldstone mesons is called the *Higgs effect* (or mechanism). The surviving massive scalar meson is called the Higgs boson, and its field (in the case under discussion $v_3 = v$) is called the Higgs field.

It is not difficult to verify that the Higgs mechanism does not change the number of particles of the system as a whole. For this we recall that a massless four-vector field (for instance, the electromagnetic field) as a result of quantization describes two particles. For brevity we shall say that it has two physical degrees of freedom. A massive vector field has three degrees of freedom, as follows from Section 8.3.

Therefore the initial Lagrangian (20) has the $3 \times 2 = 6$ degrees of freedom of the three massless vector fields and the $3 \times 1 = 3$ degrees of freedom of the three scalar fields. Altogether, it has $6 + 3 = 9$ degrees of freedom. The final Lagrangian $\mathscr{L}(v_3, \boldsymbol{B})$ has the $2 \times 3 = 6$ degrees of freedom of two massive vector fields, the $1 \times 2 = 2$ degrees of freedom of the massless vector component B^3, and the single $1 \times 1 = 1$ component v_3 of the Higgs field. Altogether, it too has $6 + 2 + 1 = 9$ degrees of freedom.

We point out also that, if the above model is made somewhat more complicated, and instead of a triplet of scalar fields a quadruplet (a two-component isospinor of complex scalar fields) is dealt with, we then will obtain a system with 10 degrees of freedom. As a result of a transformation of the form (22), three Goldstone bosons are generated and all three components of the Yang-Mills field acquire mass. The three massless bosons will subsequently be absorbed by the gauge transformation. Ultimately, we obtain three massive vector fields and one Higgs boson.

Thus, the Higgs mechanism permits us to assign mass to all the components of a gauge vector field. However, an indispensable attribute of the model containing the massive Yang–Mills field is the Higgs boson. The Higgs mechanism is applied to construct unified models of the weak and electromagnetic ("electroweak") interaction (see Section 32 below).

12. QUANTUM SYSTEMS WITH INTERACTION

12.1. Formulation of the problem. In considering the interaction of quantized fields one may take as a starting point the total Lagrangian of a system of fields

$$\mathscr{L} = \mathscr{L}(u_i, \partial u_j) = F(u_i, \nabla u_j, \dot{u}_k).$$

With the aid of Noether's theorem we obtain from the formulas of Section 2 the energy-momentum tensor and the total energy (Hamiltonian)

$$H = H(u, \nabla u, \dot{u}). \qquad (1)$$

As the solutions of the equations of motion are (contrary to the preceding) as a rule not known to us, the use of four-dimensional Fourier transforms for the field functions turns out to be ineffectual. Instead we may write

$$u(x) = u^{+} + u^{-}, \quad u^{\pm}(x) = (2\pi)^{-3/2} \int e^{\mp ikx} a^{\pm}(\boldsymbol{k}, t)\, d\boldsymbol{k}. \tag{2}$$

It should be noted that, in constrast to the free-field case, the decomposition into positive- and negative-frequency parts may here turn out to be relativistically noninvariant, and that the quantities $a^{\pm}$ do not necessarily have the meaning of creation and annihilation operators of real particles.

Nevertheless, we shall assume below that the amplitudes a^{+} may be represented by a linear combination of positive time components

$$a^{+}(\boldsymbol{k}, t) = \sum_{\alpha} e^{i\omega_{\alpha}(\boldsymbol{k})t} a_{\alpha}^{+}(\boldsymbol{k}) \tag{3}$$

and are, therefore, creation operators, while the amplitudes a^{-} are annihilation operators. The Hamiltonian H may now be expressed in terms of the amplitudes $a^{\pm}$ and their time derivatives. As H does not depend on the time, we may set $t = 0$ and pass over to the amplitudes

$$a^{\pm}(\boldsymbol{k}) = a^{\pm}(\boldsymbol{k}, 0), \quad \dot{a}^{\pm}(\boldsymbol{k}) = \dot{a}^{\pm}(\boldsymbol{k}, 0). \tag{4}$$

To establish the commutation relations, we may now make use of the canonical quantization and obtain the quantum equations of motion in the Heisenberg representation:

$$i \frac{\partial u(x, t)}{\partial t} = [u, H], \tag{5}$$

as well as the simultaneous commutation relations of the type

$$\left[u(\boldsymbol{x}, 0), \left. \frac{\partial \mathscr{L}}{\partial \dot{u}(\boldsymbol{y}, t)} \right|_{t=0} \right] = i\delta(\boldsymbol{x} - \boldsymbol{y}). \tag{6}$$

Inserting (2) and (1) into (5), we arrive at quite complicated nonlinear operator relations for the amplitudes $a^{\pm}(\boldsymbol{k}, t)$. The solution of these relations is equivalent to the solution of the equations of motion.

On the other hand, using (6) and passing to the amplitudes (4), we obtain commutation relations of the following form:

$$[a^{-}(\boldsymbol{k}), \dot{a}^{+}(\boldsymbol{q})] = \delta(\boldsymbol{k} - \boldsymbol{q}). \tag{7}$$

Taken as a whole, the Schrödinger equation, the expressions for the Hamiltonian H (and for other dynamic quantities) written in terms of the time-independent operators (4), and the commutation relations (7), represent the formulation of the problem in the Schrödinger representation.

The transition from the equal-time commutation relations (7) to nonsimultaneous ones, i.e., the constructive transition from the Schrödinger to the Heisenberg representation, requires knowledge of the time dependence of the operators $a^{\pm}$ (i.e., of the explicit form of (3)), which in turn is equivalent to solving the equations of motion. Thus, although the canonical commutation relations do formally solve the problem of quantization of a system of interacting fields, they actually turn out not to have a deep physical meaning, since they absolutely do not reflect dynamics.

12.2. An illustration. This point may be illustrated by the following solvable model:

$$\mathcal{L} = \mathcal{L}_0(\varphi_1, m_1) + \mathcal{L}_0(\varphi_2, m_2) + g\varphi_1\varphi_2, \tag{8}$$

where

$$\mathcal{L}_0(\varphi, m) = (1/2)[(\partial\varphi)^2 - m^2\varphi^2].$$

The Lagrangian (8) may be diagonalized by the substitution

$$\varphi_1 = \cos\alpha \cdot u_1 + \sin\alpha \cdot u_2, \quad \varphi_2 = \cos\alpha \cdot u_2 - \sin\alpha \cdot u_1,$$

where

$$\alpha = \operatorname{arctg}\left(\frac{\sqrt{(m_1^2 - m_2^2)^2 + 4g^2} - (m_1^2 - m_2^2)}{2g}\right).$$

Here it is implied that $\Delta m^2 = m_1^2 - m_2^2 > 0$, and the choice of sign of the square root corresponds to $u_1 \to \varphi_1$ and $u_2 \to \varphi_2$ when $g \to 0$.

Performing the corresponding transformation, we have

$$\mathcal{L} = \mathcal{L}_0(u_1, \mu_1) + \mathcal{L}_0(u_2, \mu_2),$$

where the new masses may be represented in the form

$$\mu_1^2 = \frac{m_1^2 + m_2^2}{2} + \frac{\Delta\mu^2}{2}, \quad \mu_2^2 = \frac{m_1^2 + m_2^2}{2} - \frac{\Delta\mu^2}{2},$$

$$\Delta\mu^2 = \frac{(\Delta m^2)^2 + 4g^2}{g} \cdot \frac{\operatorname{tg}\alpha}{1 + \operatorname{tg}^2\alpha} = [(\Delta m^2)^2 + 4g^2]^{1/2} > 0.$$

Performing quantization of the new free fields

$$[u_k(x),\ u_j(y)] = -i\delta_{kj} D_k(x-y)$$

and returning to the initial ones, we obtain

$$\begin{aligned} i[\varphi_1(x),\ \varphi_1(y)] &= \cos^2\alpha \cdot D_1(x-y) + \sin^2\alpha \cdot D_2(x-y), \\ i[\varphi_2(x),\ \varphi_2(y)] &= \cos^2\alpha \cdot D_2(x-y) + \sin^2\alpha \cdot D_1(x-y), \\ i[\varphi_1(x),\ \varphi_2(y)] &= \sin\alpha\cos\alpha \cdot \{D_2(x-y) - D_1(x-y)\} \end{aligned} \tag{9}$$

—nonsimultaneous commutation relations which completely reflect the dynamics of the problem. Differentiating them with respect to x^0 and then assuming $x^0 = y^0 = t$, we may obtain the canonical commutation relations

$$\begin{aligned} &[\varphi_l(t,\ \boldsymbol{x}),\ \dot\varphi_k(t,\ \boldsymbol{y})] = i\delta_{kl}\delta(\boldsymbol{x}-\boldsymbol{y}), \\ &[\varphi_l(t,\ \boldsymbol{x}),\ \varphi_k(t,\ \boldsymbol{y})] = [\dot\varphi_l(t,\ \boldsymbol{x}),\ \dot\varphi_k(t,\ \boldsymbol{y})] = 0, \end{aligned} \tag{10}$$

which contain no dynamical information concerning the interaction of the fields φ_1, φ_2.

Obviously, diagonalization (similar to that just performed) of the Lagrangian at the same time leads to the explicit solution of the equations of motion. However, within the framework of realistic models such a solution, as a rule, turns out to be impossible.

Instead, we use the perturbative method, which consists in expanding the solution in a series in powers of a small parameter represented by the coupling constant. Thus the initial zeroth approximation is reduced to a system of free equations, the solutions of which are known and are represented by superpositions of plane waves. Such an approach is prevalant in the present-day quantum field theory. It is formulated as a covariant perturbation theory and is diagramatically represented by the well-known Feynman diagrams. Their exposition is presented in the following chapters.

For methodological purposes it seems expedient to consider the problem of interaction between quantum systems having an infinite number of degrees of freedom in the nonrelativistic approximation starting with the Schrödinger equation in the occupation-number represention. Such a consideration will permit us to introduce in a sufficiently simple manner such concepts as renormalization of mass and of wave functions and connects them with the idea of the "dressing" of particles due to quantum interaction effects. It will also permit us to understand the physical nature of ultraviolet divergences present in the local quantum field theory.

12.3. The Hamiltonian approach. Consider a system of N identical particles (bosons or fermions) connected by forces of interaction between pairs. In the second-quantization representation such a system is described by the Hamiltonian

$$H=\int \omega(\boldsymbol{q}) \overset{+}{a}(\boldsymbol{q})\, a(\boldsymbol{q})\, d\boldsymbol{q}+$$
$$+\frac{1}{2}\int \overset{+}{a}(\boldsymbol{q}_1) \overset{+}{a}(\boldsymbol{q}_2) V(\boldsymbol{q}_1, \boldsymbol{q}_2; \boldsymbol{p}_1, \boldsymbol{p}_2)\, a(\boldsymbol{p}_1)\, a(\boldsymbol{p}_2)\, d\boldsymbol{q}_1\, d\boldsymbol{q}_2\, d\boldsymbol{p}_1\, d\boldsymbol{p}_2. \quad (11)$$

Here a, $\overset{+}{a}$ are Bose or Fermi operators with the given commutation relations

$$[a(\boldsymbol{k}), \overset{+}{a}(\boldsymbol{q})]_{\pm}=\delta(\boldsymbol{k}-\boldsymbol{q}), \; [a, a]=[\overset{+}{a}, \overset{+}{a}]=0. \quad (12)$$

What we have obtained here is, of course, the quantum interaction problem formulated in the Schrödinger representation.

The projection of the operator H on the single-particle state $\Phi_1(\boldsymbol{k}) = \overset{+}{a}(\boldsymbol{k})\Phi_0$ will be

$$H\Phi_1(\boldsymbol{k})=\omega(\boldsymbol{k})\,\Phi_1(\boldsymbol{k}),$$

and consequently $\omega(\boldsymbol{k})$ represents the energy of a particle with momentum $\boldsymbol{k}$. Then projecting H onto the N-particle state, we see that V represents the energy of the interaction between a pair of particles with corresponding momenta (see, for example, Blokhintsev (1976), Section 118; Lipkin (1973), Chapter 5).

Formulas of the type (11), (12) are "good" in that they describe a system of N particles when the number N is arbitrary although fixed, and they do not depend explicitly on this number. The above property permits us to make an important generalization and abandon the condition $N=$ const. For this purpose it is sufficient to change the function $H(a, \overset{+}{a})$ so that it ceases to commute with the operator of the total number of particles

$$\check{N}=\int \overset{+}{a}(\boldsymbol{q})\, a(\boldsymbol{q})\, d\boldsymbol{q}$$

or, in the case the system consists of sets differing from each other but identical within each set of particles, with the operators of the numbers of particles of the corresponding types N_i.

As an example, we shall present the so-called Frölich Hamiltonian

$$H_{\mathrm{Fr}}=\sum_{s}\int E(\boldsymbol{k})\, \overset{+}{a}_s(\boldsymbol{k})\, a_s(\boldsymbol{k})\, d\boldsymbol{k}+\int \omega(\boldsymbol{q})\, b(\boldsymbol{q})\, b(\boldsymbol{q})\, d\boldsymbol{q}+H_1, \quad (13)$$

where

$$H_1 = \int dk\, dk'\, dq V_{\mathrm{Fr}}(\boldsymbol{k}, \boldsymbol{k}', \boldsymbol{q}) \overset{+}{a}_s(\boldsymbol{k})\, a_s(\boldsymbol{k}')[\overset{+}{b}(\boldsymbol{q}) + b(-\boldsymbol{q})],$$

$$V_{\mathrm{Fr}}(\boldsymbol{k}, \boldsymbol{k}', \boldsymbol{q}) = \delta(\boldsymbol{k}+\boldsymbol{q}-\boldsymbol{k}') \frac{g}{4\pi} \sqrt{\frac{\omega(\boldsymbol{q})}{\pi}}.$$

Here $\overset{+}{a}$, a are Fermi electron operators, $s = \frac{1}{2}$ is the spin index over which summation is implied, $\overset{+}{b}$, b are phonon Bose operators, E and ω are the electron and phonon energies, and g is a coupling constant in the normalization adopted in superconductivity theory.

This Hamiltonian corresponds to a system of electrons in a metal which interacts with the oscillations of the crystalline ion lattice. The term H_1 describes the excitation process of vibrations of the crystalline lattice by electrons, i.e., the creation of a quantum of these oscillations (the phonon), as well as the inverse process, i.e., the absorption of a phonon by an electron. Accordingly, the Hamiltonian H_{Fr} commutes with the operator of the total number of electrons and does not commute with the operator of the total number of phonons.

In this *Hamiltonian formulation*, as we call it, the problem is formulated as follows: the eigenvectors of the states of the Hamiltonian H are to be found, as well as its eigenvalues corresponding to the above eigenvectors.

If, together with the Hamiltonian H, there are given operators of other dynamic quantities $A(a, \overset{+}{a}, \ldots)$, then knowledge of the eigenvectors will allow us to obtain the corresponding expectation values. An essential quantity of that type is the momentum of a system. Note in this connection that the arguments $\boldsymbol{k}, \boldsymbol{q}, \boldsymbol{p}, \ldots$ of the operators a and $\overset{+}{a}$ introduced in this section do not yet have physical meaning and have been identified with the momenta of particles in a purely formal way. They do, however, acquire the meaning of momenta if one defines the momentum operator of the system. Thus, for the Frölich model

$$P_n = \int p_n \overset{+}{a}_s(\boldsymbol{p})\, a_s(\boldsymbol{p})\, d\boldsymbol{p} + \int k_n \overset{+}{b}(\boldsymbol{k})\, b(\boldsymbol{k})\, d\boldsymbol{k} \qquad (n = 1, 2, 3). \tag{14}$$

Now it is not difficult to check that

$$P_n \overset{+}{a}(\boldsymbol{p})\, \Phi_0 = p_n \overset{+}{a}(\boldsymbol{p})\, \Phi_0, \quad P_n \overset{+}{b}(\boldsymbol{k})\, \Phi_0 = k_n \overset{+}{b}(\boldsymbol{k})\, \Phi_0$$

and, generally,

$$P_n\Big[\prod_i \overset{+}{a}(p^i)\prod_j \overset{+}{b}(k^j)\,\Phi_0\Big]=\Big[\sum_i p_n^i+\sum_j k_n^j\Big]\Big[\prod_i \overset{+}{a}(p^i)\prod_j \overset{+}{b}(k^j)\,\Phi_0\Big],$$

if

$$P_n\Phi_0=0.$$

In appropriate cases one may in a similar manner introduce the operators for electric charge, spin, etc.

12.4. Diagonalization of model Hamiltonians. The process of solving the problem in the Hamiltonian formulation is reduced to the diagonalization of the Hamiltonian by means of a unitary operator transformation. A simple, exactly solvable example is given by the Hamiltonian of the so-called "static source" problem

$$H=\int\omega(\boldsymbol{k})\,\overset{+}{b}(\boldsymbol{k})\,b(\boldsymbol{k})\,d\boldsymbol{k}+\int[g(\boldsymbol{k})\,\overset{+}{b}(\boldsymbol{k})+\overset{*}{g}(\boldsymbol{k})\,b(\boldsymbol{k})]\,\omega(\boldsymbol{k})\,d\boldsymbol{k}. \quad (15)$$

Here b, $\overset{+}{b}$ are Bose operators, and $g(\boldsymbol{k})$, $\overset{+}{g}(k)$ are c-number functions which characterize the source of the field of the b-particles and are Fourier transforms of the corresponding space distributions.

The problem consists in determining the eigenfunctions and eigenvalues of the Schrödinger equation

$$H\Phi(\boldsymbol{q})=E(\boldsymbol{q})\,\Phi(\boldsymbol{q}).$$

Applying formally the operation of "displacement" of variables

$$b(\boldsymbol{k})\to B(\boldsymbol{k})=b(\boldsymbol{k})+g(\boldsymbol{k}),\quad \overset{+}{b}(\boldsymbol{k})\to\overset{+}{B}(\boldsymbol{k})=\overset{+}{b}(\boldsymbol{k})+\overset{*}{g}(k)$$

reduces the Hamiltonian (15) to the diagonal form

$$H=\int\omega(\boldsymbol{k})\,\overset{+}{B}(\boldsymbol{k})\,B(\boldsymbol{k})\,d\boldsymbol{k}-\int\omega(\boldsymbol{k})\,|g(\boldsymbol{k})|^2\,d\boldsymbol{k}.$$

Therefore the problem is reduced to seeking a unitary operator implementing the transformation $b\to B$, i.e.,

$$U^{-1}(g)\,b(\boldsymbol{k})\,U(g)=B(\boldsymbol{k}),\quad U^{-1}(g)\,\overset{+}{b}(\boldsymbol{k})\,U(g)=\overset{+}{B}(\boldsymbol{k}).$$

With the help of standard technique (see Appendix IV) we obtain

$$U(g) = \exp \int d\boldsymbol{k} [g(\boldsymbol{k}) \overset{+}{b}(\boldsymbol{k}) - \overset{*}{g}(\boldsymbol{k}) b(\boldsymbol{k})]. \tag{16}$$

To find the eigenfunctions of the Hamiltonian H we note that in terms of the operators B, $\overset{+}{B}$ they take on the form

$$\Psi_N = \overset{+}{B}(\boldsymbol{k}_1) \dots \overset{+}{B}(\boldsymbol{k}_N) \Psi_0, \tag{17}$$

ψ_0 being the vacuum of the B-field:

$$B(\boldsymbol{k}) \Psi_0 = 0. \tag{18}$$

The eigenvalues corresponding to (17) will be

$$H\Psi_N = E_N \Psi_N, \quad E_N = \omega(\boldsymbol{k}_1) + \dots + \omega(\boldsymbol{k}_N).$$

In order to express the state ψ_N in terms of eigenvectors of the field b, we insert the formulas (16) into equations (17) and (18). Multiplying further on the left by U, we obtain

$$\Phi_N = U(g) \Psi_N = \overset{+}{b}(\boldsymbol{k}_1) \dots \overset{+}{b}(\boldsymbol{k}_N) \Phi_0,$$

where the state

$$\Phi_0 = U(g) \Psi_0 \tag{19}$$

in accordance with (18), satisfies the condition $b\Phi_0 = 0$ and is, obviously, the vacuum of the b-field.

We shall now consider in greater detail the structure of the new vacuum vector $\Psi_0 = U^{-1}(g)\Phi_0$. For this purpose we shall reduce the operator U^{-1} to its normal form (see (AIV.10))

$$U^{-1}(g) = Z^{1/2} \exp\left(-\int d\boldsymbol{k}\, g(\boldsymbol{k}) \overset{+}{b}(\boldsymbol{k})\right) \exp\left(\int d\boldsymbol{q}\, \overset{*}{g}(\boldsymbol{q}) b(\boldsymbol{q})\right),$$

where

$$Z = \exp\left(-\int |g(\boldsymbol{k})|^2 d\boldsymbol{k}\right).$$

Thus, the new vacuum

$$\Psi_0 = Z^{1/2} \exp\left(-\int dk\, g(k)\, \overset{+}{b}(k)\right)\Phi_0 =$$
$$= Z^{1/2}\left\{\sum_n \frac{(-1)^n}{n!}\int dk_1 \ldots dk_n g(k_1)\ldots g(k_n)\,\overset{+}{b}(k_1)\ldots \overset{+}{b}(k_n)\right\}\Phi_0 \qquad (20)$$

is a superposition of an infinite number of eigenstates of the Hamiltonian $H_0 = \int \omega(k)\overset{+}{b}(k)b(k)\, dk$.

We further present without computations (see also Assignment N13) the solution of the problem involving the bilinear interaction of two Bose fields:

$$H = \sum_{r=1,2}\int \omega(k)\,\overset{+}{b}_r(k)\, b_r(k)\, dk +$$
$$+\int g(k)\,\omega(k)\,[\overset{+}{b}_1(k)\, b_2(k) + \overset{+}{b}_2(k)\, b_1(k)]\, dk. \qquad (21)$$

Diagonalization is achieved by means of the transformation

$$B_1(k) = U^{-1}(g)\, b_1(k)\, U(g) = \frac{1}{\sqrt{2}}[b_1(k) - b_2(k)],$$
$$B_2(k) = U^{-1}(g)\, b_2(k)\, U(g) = \frac{1}{\sqrt{2}}[b_1(k) + b_2(k)],$$

where

$$U(g) = \exp\left\{\frac{1}{\sqrt{2}}\int dk[\overset{+}{b}_2(k)\, b_1(k) - \overset{+}{b}_1(k)\, b_2(k)]\right\}. \qquad (22)$$

As a result, we obtain the diagonal expression for the Hamiltonian

$$H = \sum_r \int \Omega_r(k, g)\,\overset{+}{B}_r(k)\, B_r(k)\, dk, \qquad (23)$$

where

$$\Omega_{1,2}(k, g) = \omega(k)\,(1 \mp g(k)).$$

The eigenvalues have the form

$$H\Psi(n, m) = E(n, m)\,\Psi(n, m),$$

$$E(n, m) = \sum_{1 \leqslant i \leqslant n} \Omega_1(k_i) + \sum_{1 \leqslant j \leqslant m} \Omega_2(k_j),$$
$$\Psi(n, m) = B_1^+(\boldsymbol{k}_1) \dots B_1^+(\boldsymbol{k}_n) B_2^+(\boldsymbol{q}_1) \dots B^+(\boldsymbol{q}_m) \Psi_0,$$

while the vacuum of the fields B coincides with the vacuum of the fields b:

$$b_\nu \Psi_0 = B_\nu \Psi_0 = 0.$$

Finally, we consider the model Hamiltonian which appears in the superfluidity problem:

$$H = \int d\boldsymbol{k} \{\omega(\boldsymbol{k}) \overset{+}{b}(\boldsymbol{k}) b(\boldsymbol{k}) + g(\boldsymbol{k}) [b(\boldsymbol{k}) b(-\boldsymbol{k}) + \overset{+}{b}(\boldsymbol{k}) \overset{+}{b}(-\boldsymbol{k})]\}. \tag{24}$$

This Hamiltonian is diagonalized by the unitary transformation

$$B(\boldsymbol{k}) = \operatorname{ch}\alpha(\boldsymbol{k})\, b(\boldsymbol{k}) + \operatorname{sh}\alpha(\boldsymbol{k})\, \overset{+}{b}(-\boldsymbol{k}) = U^{-1}(\alpha)\, b(\boldsymbol{k})\, U(\alpha),$$
$$\overset{+}{B}(\boldsymbol{k}) = \operatorname{ch}\alpha(\boldsymbol{k})\, \overset{+}{b}(\boldsymbol{k}) + \operatorname{sh}\alpha(\boldsymbol{k})\, b(-\boldsymbol{k}) = U^{-1}(\alpha)\, \overset{+}{b}(\boldsymbol{k})\, U(\alpha),$$

where (compare (AIV.5.6))

$$\ln U(\alpha) = \int d\boldsymbol{k} \alpha(\boldsymbol{k}) [\overset{+}{b}(\boldsymbol{k}) \overset{+}{b}(-\boldsymbol{k}) - b(\boldsymbol{k}) b(-\boldsymbol{k})], \tag{25}$$

and $\alpha(\boldsymbol{k})$ is found from the relation

$$2g(\boldsymbol{k}) \operatorname{th}\alpha(\boldsymbol{k}) = \omega(\boldsymbol{k}) - \sqrt{\omega^2(\boldsymbol{k}) - g^2(\boldsymbol{k})}. \tag{26}$$

The ground state of the Hamiltonian (24) is determined by the expression

$$\Psi_0(\alpha) = U^{-1}(\alpha) \Phi_0,$$

where Φ_0 is the vacuum of the field b. Reducing with the aid of (AIV.12) the operator $U^{-1}(\alpha)$ to its normal form, we obtain

$$\Psi_0(\alpha) = Z^{1/2} \exp\left\{-\int \operatorname{th}\alpha(\boldsymbol{k})\, \overset{+}{b}(\boldsymbol{k})\, \overset{+}{b}(-\boldsymbol{k})\, d\boldsymbol{k}\right\} \Phi_0, \tag{27}$$

where

$$Z = \exp\left(2 \int d\boldsymbol{k} \ln \operatorname{ch}\alpha(\boldsymbol{k})\right).$$

Equation (27) contains the coherent superposition of correlated pairs of quanta of the initial field b with opposite momenta, i.e. of pairs whose total momentum equals zero.

12.5. Interaction effects. The solvable models considered in the preceding section are important mostly as examples of technique. it is true that the model (21), which is analogous to the field model (8), is related to the problem of mixing of K^0-mesons (see, for example, Lipkin (1977), Chapter 7; Feynman (1978), Section 10), while the Hamiltonian model with linear interaction (15), as was pointed out, is related to the problem of fermion–boson interaction of the Yukawa type, and its solution will be made use of in Section 13 in the analysis of the one-nucleon sector of the static nucleon model. However, for us the real value of these models here is that the solutions obtained above enable us to understand the main features of the physical structure of interaction effects.

The most important of them involves the transition to new coordinates, to a new system of excitation quanta which are described by the transformed operators B, $\overset{+}{B}$. In the more realistic models exact diagonalization turns out to be impossible; however, as a rule, the key to an approximate solution lies in the proper choice of that part of the interaction which can be diagonalized, which is the most important part from a physical point of view. The remaining part of the interaction is taken into account by perturbation theory.

As we have seen, as a result of interaction there appear new states which are (see (20)) infinite superpositions of the states of the initial Hamiltonian H_0. Thus, from the set of vectors ψ_N of the free Hamiltonian H_0 we pass over to the set of vectors Φ_N of the total Hamiltonian H.

Sometimes it turns out to be possible to establish the one-to-one correspondence $\psi_N \leftrightarrow \Phi_N$, and then there is sense in saying that interaction "dresses" the initial "bare" state and transforms it into a "dressed" or "physical" state Φ_N. Such a clothing process changes the energy of the quantum (in the relativistic case, the mass), and may also lead to a change of the norm and (as we shall see below) to a change of the coupling constant, i.e., to a screening effect.

We wish to recall here that the effect consisting in the change of the mass of a particle subjected to interaction is known, for instance, from the quantum-mechanical problem of the motion of an electron in a crystal (see Blokhintsev (1976), Section 55; Davydov (1973), Section 128), while the screening of a charge introduced into a medium takes place within the framework of classical electrodynamics.

Such effects as the changes of mass, coupling constants, and norms of states are called in quantum theory *renormalization effects.* A characteristic feature of renormalizations in quantum field theory is connected with their singular nature, which is due to the interaction being local. The model examined below of the heavy nucleon is not local, and when discussing renormalization effects we shall consider the transition to the local limit.

13. THE HEAVY-NUCLEON MODEL

13.1. Formulation of the model. Consider the problem of an interaction of the Yukawa type between spinless light bosons (mesons) and heavy fermions (nucleons). We shall consider the fermions to be so heavy that their energy does not depend on their momentum (and is equal to their mass), and we shall also neglect the possibility of reorientation of their spin (this is justified at sufficiently small momenta).

Introducing Bose operators $b(\boldsymbol{k})$, $\overset{+}{b}(\boldsymbol{k})$ for the mesons and Fermi operators $a(\boldsymbol{p})$, $\overset{+}{a}(\boldsymbol{p})$ for the nucleons, we write the system Hamiltonian in the form

$$H = H_0 + H_1, \tag{1}$$

where

$$H_0 = M \int d\boldsymbol{p}\, \overset{+}{a}(\boldsymbol{p})\, a(\boldsymbol{p}) + \int d\boldsymbol{k}\, \omega(k)\, \overset{+}{b}(\boldsymbol{k})\, b(\boldsymbol{k}), \tag{2}$$

$$H_1 = \int d\boldsymbol{p} \int d\boldsymbol{k}\, g(k)\, \omega(k)\, \overset{+}{a}(\boldsymbol{p}+\boldsymbol{k})\, a(\boldsymbol{p})\, [b(\boldsymbol{k}) + \overset{+}{b}(-\boldsymbol{k})]. \tag{3}$$

Here it is assumed that the initial Lagrangian of the meson–nucleon interaction is nonlocal:

$$\mathcal{L}_{\text{inter}}(x) = -g\overset{*}{\psi}{}^{+}(x)\, \psi^{-}(x) \int d\boldsymbol{y}\, F(\boldsymbol{x}-\boldsymbol{y})\, \varphi(y) \qquad (y_0 = x_0),$$

and the form factor F depends only on the modulus of the radius vector, while its Fourier transform is connected with the function introduced in (3) by the following relation:

$$\frac{g}{(2\pi)^3} \int e^{-i\boldsymbol{k}\boldsymbol{x}} F(\boldsymbol{x})\, d\boldsymbol{x} = gf(k) = (2\pi)^{3/2} \sqrt{2\omega^3(k)}\, g(k), \quad k = |\boldsymbol{k}|. \tag{4}$$

To the local limit $F(\boldsymbol{x}) = \delta^3(\boldsymbol{x})$ there correspond the limiting expressions

$$g(k) = \frac{g}{(2\pi)^{3/2}\sqrt{2\omega^3(k)}}, \qquad f(k) = 1. \tag{5}$$

The Hamiltonian (1) is written down in the Schrödinger representation. The operators a, $\overset{+}{a}$ satisfy the Fermi commutation relations, and b, $\overset{+}{b}$ satisfy the Bose commutation relations. The term H_1 describes the creation and annihilation of mesons on nucleons. The number of nucleons remains unchanged. In contrast to the usual expression of the type $\overset{*}{\psi}\psi\varphi$ (see Section 10.2), H_1 does not contain terms of the form $\overset{+}{a}\,\overset{+}{a}\,b$ and $aa\overset{+}{b}$ responsible for the creation and annihilation of nucleon pairs. Therefore nonlocality is not limited to the "smearing" function $f \neq 1$ and is of a more profound character. (This circumstance is considered in greater detail below in Section 15.3.) If, however, the nucleons are considered to be very heavy, i.e., $M \gg \omega(k)$ for all physically significant momenta, then real processes of pair creation and annihilation cannot take place, and we shall agree to neglect virtual effects of this form, not going into details of the legitimacy of such an approximation. It is in this sense that we shall consider the formula (5) to be the definition of the local limit.

We point out also that in accordance with the preceding, the Hamiltonian (1) commutes with the operator of the total number of nucleons N_n as well as with the operator of the total momentum

$$P_n = \int p_n\, d\boldsymbol{p}\, \overset{+}{a}(\boldsymbol{p}) a(\boldsymbol{p}) + \int k_n\, d\boldsymbol{k}\, \overset{+}{b}(\boldsymbol{k})\, b(\boldsymbol{k}) \qquad (n = 1,\ 2,\ 3) \tag{6}$$

and does not commute with the operator of the total number of mesons.

13.2. The solution in the one-nucleon sector. The static-nucleon model allows an exact solution in the so-called one-nucleon sector of the Hilbert space, consisting of the vectors ψ satisfying the relation

$$N_n \Psi = \Psi.$$

Such states with a fixed total momentum equal to $\boldsymbol{p}$ may be written in the form

$$\Psi_{\boldsymbol{p}} = \int d\boldsymbol{q}\, \overset{+}{a}(\boldsymbol{p} - \boldsymbol{q})\, \Phi(\boldsymbol{q}), \tag{7}$$

where $\Phi(\boldsymbol{q})$ is the vector of a state containing no nucleons (and containing an arbitrary number of mesons) which has the following structure:

$$\Phi(\boldsymbol{q}) = F_b(\boldsymbol{q})\,\Phi_0. \tag{8}$$

Here Φ_0 is the amplitude of the vacuum, and F_b is some functional of the creation operators $\overset{+}{b}$ of mesons of fixed total momentum $\boldsymbol{q}$.

In order to obtain the equations for the meson amplitude Φ, we compute the projection of the commutator $[a(0), H]$ onto the state Ψ_p. We have

$$[a(0),\ H_0]\,\Psi_p = \left\{M + \int d\boldsymbol{k}\,\omega(k)\,\overset{+}{b}(\boldsymbol{k})\,b(\boldsymbol{k})\right\}\Phi(\boldsymbol{p}),$$
$$[a(0),\ H_1]\,\Psi_p = \int d\boldsymbol{k}\,\tilde{B}(\boldsymbol{k})\,\Phi(\boldsymbol{p}+\boldsymbol{k}),$$

where the following notation is adopted:

$$\tilde{B}(\boldsymbol{k}) = g(k)\,\omega(k)\left[b(\boldsymbol{k}) + \overset{+}{b}(-\boldsymbol{k})\right].$$

Thus, in the one-nucleon sector the problem has been reduced to finding the solution of the equation

$$(M + W)\,\Phi(\boldsymbol{p}) + \int d\boldsymbol{k}\,\tilde{B}(\boldsymbol{k})\,\Phi(\boldsymbol{p}+\boldsymbol{k}) = E\Phi(\boldsymbol{p}), \tag{9}$$

where

$$W = \int d\boldsymbol{k}\,\omega(k)\,\overset{+}{b}(\boldsymbol{k})\,b(\boldsymbol{k})$$

is the energy operator of the mesons.

To solve this equation we make use of formula (AIV.15). Assuming in it that

$$n(\boldsymbol{k}) = \overset{+}{b}(\boldsymbol{k})\,b(\boldsymbol{k}),\quad \lambda_{\pm}(\boldsymbol{k}) = \pm g(\boldsymbol{k})\,e^{\mp i\boldsymbol{kx}},\quad \lambda^2 = -\,g^2(\boldsymbol{k}),$$

we obtain

$$[n(\boldsymbol{k}),\ e^{A(\boldsymbol{x})}] = -\left\{g(\boldsymbol{k})\left[b(\boldsymbol{k}) + \overset{+}{b}(-\boldsymbol{k})\right]e^{i\boldsymbol{kx}} + g^2(\boldsymbol{k})\right\}e^{A(\boldsymbol{x})}, \tag{10}$$

where

$$A(\boldsymbol{x}) = \int d\boldsymbol{q}\,e^{i\boldsymbol{qx}}g(q)\left[b(\boldsymbol{q}) - \overset{+}{b}(-\boldsymbol{q})\right]$$

is an anti-Hermitian operator.

Let us check that the solution of equation (9) may be written in the form

$$\Phi^{(0)}(\boldsymbol{p}) = T\left(\boldsymbol{p};\ b,\ \overset{+}{b}\right)\Phi_0, \tag{11}$$

where Φ_0 is the amplitude of the meson vacuum, and the operator T has the following structure:

$$T\left(\boldsymbol{p};\ b,\ \overset{+}{b}\right) = c\int d\boldsymbol{x}\, e^{i\boldsymbol{p}\boldsymbol{x}}\, e^{A(\boldsymbol{x})}. \tag{12}$$

Commuting T with the energy operator of the mesons W, with the aid of (10) we find

$$[T(\boldsymbol{p}),\ W] = \int d\boldsymbol{k}\,\tilde{B}(\boldsymbol{k})\,T(\boldsymbol{p}+\boldsymbol{k}) + \Delta M\cdot T(\boldsymbol{p}), \tag{13}$$

where

$$\Delta M = \int \omega(k)\,g^2(k)\,d\boldsymbol{k} = \frac{g^2}{(2\pi)^3}\int d\boldsymbol{k}\,\frac{f^2(k)}{2\omega^2(k)}. \tag{14}$$

Taking into account that $W\Phi_0 = 0$ from (13), we obtain

$$W\Phi^{(0)}(\boldsymbol{p}) + \int d\boldsymbol{k}\,\tilde{B}(\boldsymbol{k})\,\Phi^{(0)}(\boldsymbol{p}+\boldsymbol{k}) + \Delta M\Phi^{(0)}(\boldsymbol{p}) = 0.$$

Comparing this relation with (9), we see that the formulas (11), (12) do actually give the solution of equation (9) provided that

$$E = E^{(0)} = M - \Delta M.$$

Thus, (14) represents a correction to the nucleon mass due to interaction with the meson field.

The set of formulas (7), (11), (12) gives us the explicit expression for the "dressed" one-nucleon state $\Psi_p^{(0)}$. To make it completely concrete, one must define the constant c in equation (12). This is conveniently done starting with the normalization condition

$$\overset{*}{\Phi}{}^{(0)}(\boldsymbol{p})\,\Phi^{(0)}(\boldsymbol{p}') = \delta(\boldsymbol{p}-\boldsymbol{p}').$$

Substituting into this condition the representations (7) and (11), we find

$$\int \langle 0 | \overset{*}{T}(\boldsymbol{p}+\boldsymbol{q})\, T(\boldsymbol{q}) | 0 \rangle \, d\boldsymbol{q} = \delta(\boldsymbol{p}).$$

Using (12) and taking into account the anti-Hermiticity of the operator A, we obtain the following as a result of integrating over $\boldsymbol{q}$, $\boldsymbol{x}$, and $\boldsymbol{y}$:

$$|c|^2 \int d\boldsymbol{x} \int d\boldsymbol{y} \int d\boldsymbol{q}\, e^{i\boldsymbol{q}\boldsymbol{y} - i(\boldsymbol{q}+\boldsymbol{p})\boldsymbol{x}} \langle 0 | e^{A(\boldsymbol{y}) - A(\boldsymbol{x})} | 0 \rangle =$$
$$= (2\pi)^3 |c|^2 \int d\boldsymbol{x}\, e^{-i\boldsymbol{p}\boldsymbol{x}} = (2\pi)^6 |c|^2 \delta(\boldsymbol{p}).$$

Therefore we may assume $c = (2\pi)^{-3}$.

We furthermore reduce the operator T to its normal form in order to write down the right-hand side of (11) in the form of (8). Using for this purpose the Hausdorf formula (see AIV.10), we write

$$e^{A(\boldsymbol{x})} = e^{A_+(\boldsymbol{x})} e^{A_-(\boldsymbol{x})} e^{-L/2},$$

where

$$A_\pm(\boldsymbol{x}) = \mp \int d\boldsymbol{k}\, g(\boldsymbol{k})\, b^\pm(\boldsymbol{k})\, e^{\mp i\boldsymbol{k}\boldsymbol{x}}$$

and

$$L = [A_+(\boldsymbol{x}),\ A_-(\boldsymbol{x})] = \frac{g^2}{(2\pi)^3} \int \frac{d\boldsymbol{k}}{2\omega^3(k)} f^2(k).$$

Collecting the results, we represent the "dressed" one-nucleon state in the form (7):

$$\Psi_p^{(0)} = \frac{Z^{1/2}}{(2\pi)^3} \int d\boldsymbol{x}\, e^{A(\boldsymbol{x})} \int d\boldsymbol{q}\, e^{i\boldsymbol{q}\boldsymbol{x}} \overset{+}{a}(\boldsymbol{p}-\boldsymbol{q})\, \Phi_0, \tag{15}$$

where

$$Z = e^{-L}, \tag{16}$$

and the amplitude $\Psi_p^{(0)}$ satisfies the equation

$$H\Psi_p^{(0)} = (M - \Delta M)\, \Psi_p^{(0)} \tag{17}$$

and the normalization condition

$$\overset{*}{\Psi}{}_p^{(0)} \Psi_p^{(0)} = 1. \tag{18}$$

13.3. Properties of the single-nucleon solution. The formula (15) represents the state of a physical nucleon, $\Psi_p^{(0)}$, in terms of an infinite superposition of states containing a single bare nucleon and an arbitrary number of mesons. Expanding in a series the operator exponential occurring on the right-hand side of this formula, we obtain after integrating over $\boldsymbol{x}$ the following:

$$\Psi_p^{(0)} = Z^{1/2} \Big\{ \overset{+}{a}(\boldsymbol{p}) - \int d\boldsymbol{q}\, g(\boldsymbol{q}) \overset{+}{b}(\boldsymbol{q}) \overset{+}{a}(\boldsymbol{p}-\boldsymbol{q}) +$$

$$+ \frac{(-1)^2}{2!} \int d\boldsymbol{q}_1 \int d\boldsymbol{q}_2 g(\boldsymbol{q}_1) g(\boldsymbol{q}_2) \overset{+}{b}(\boldsymbol{q}_1) \overset{+}{b}(\boldsymbol{q}_2) \overset{+}{a}(\boldsymbol{p}-\boldsymbol{q}_1-\boldsymbol{q}_2) + \ldots \Big\} \Phi_0. \tag{19}$$

In this case the probability amplitude that the physical nucleon with momentum $\boldsymbol{p}$ is in the state of a bare nucleon is equal to $Z^{1/2}$, the probability amplitude for it to be "made up" of a bare nucleon with momentum $\boldsymbol{p} - \boldsymbol{q}$ and a meson with momentum $\boldsymbol{q}$ is equal to $Z^{1/2} g(\boldsymbol{q})$, and so on.

It is important to point out that the amplitudes of the higher-order terms are factored out with respect to the momenta of the mesons. Infinite superpositions of boson states of this type are referred to as *coherent states.* Such states Φ_{coh} may be defined by the operator relation

$$b(0)\Phi_{\text{coh}} = \beta \Phi_{\text{coh}}, \tag{20}$$

where β is a c-number. In other words, coherent states are eigenstates of the annihilation operator. This property holds both in the momentum and in the configuration representation. From a physical point of view the factorizability of the correlation field functions b following from (20) is important. Exactly this property allows one to connect Φ_{coh} with the phenomenon of optical coherence (see, for instance, Klauder and Sudarshan (1970), Chapter 7).

It is not difficult to show that the amplitude $\Psi_p^{(0)}$ describes the ground state in the single-nucleon region. Note for this that the action of the annihilation operator on this amplitude gives

$$b(\boldsymbol{k}) \Psi_p^{(0)} = -g(\boldsymbol{k}) \Psi_{p-k}^{(0)}, \tag{21}$$

which at $\boldsymbol{k} = 0$ transforms into (20). Hence it may be seen that the amplitude $\Psi_p^{(0)}$ satisfies the condition

$$B(0)\,\Psi_p^{(0)} \equiv (b(0)+g(0))\,\Psi_p^{(0)}=0$$

(here we have made use of the notation from Section 12.4). Thus, the annihilation operator of the "dressed" quantum of the field B actually does annihilate the amplitude $\Psi_p^{(0)}$. The upper index (0) now acquires a clear meaning.

It is interesting to construct Bose excitations of the state $\Psi_p^{(0)}$. The simplest of these states corresponds to the system meson+nucleon. We look for such a state $\Psi_p^{(1)}(\boldsymbol{k})$ with total momentum $\boldsymbol{p}+\boldsymbol{k}$ in the form

$$\overset{+}{b}(\boldsymbol{k})\,\Psi_p^{(0)}+c(\boldsymbol{k})\,\Psi_{p+k}^{(0)}.$$

It is not difficult to verify (see Assignment N17) that this expression, when $c(\boldsymbol{k})=g(\boldsymbol{k})$, is an eigenvector of the Hamiltonian, i.e.,

$$H\Psi_p^{(1)}(\boldsymbol{k})=(M-\Delta M+\omega(\boldsymbol{k}))\,\Psi_p^{(1)}(\boldsymbol{k}), \tag{22}$$

where

$$\Psi_p^{(1)}(\boldsymbol{k})=\overset{+}{b}(\boldsymbol{k})\Psi_p^{(0)}+g(\boldsymbol{k})\,\Psi_{p+k}^{(0)}. \tag{23}$$

It is interesting to point out that the role of the creation operator of the physical meson is here played by the expression

$$\check{B}^{+}(\boldsymbol{k})=\overset{+}{b}(\boldsymbol{k})+g(\boldsymbol{k})\,\check{I}_{k}, \tag{24}$$

where $\check{I}_k$ is an operator that increases the momentum by $\boldsymbol{k}$, i.e., that is defined by the relation

$$\check{I}_k\Psi_p=\Psi_{p+k}.$$

It can also be seen that (20) may be rewritten in the form

$$\check{B}(\boldsymbol{k})\,\Psi_p^{(0)}=0, \tag{25}$$

where

$$\check{B}(\boldsymbol{k})=b(\boldsymbol{k})+g(\boldsymbol{k})\,\check{I}_{-k}. \tag{26}$$

Note also that from the fact established above that (23), describing the physical state meson + nucleon, is an eigenvector of the Hamiltonian, it directly follows that in the model under consideration there exists no meson–nucleon scattering.

13.4. Transition to the local limit. Let us consider the properties of the solution in the transition to the local limit (5). This limit turns out to be singular. To study the structure of the singularities arising, we introduce a subsidiary parameter characterizing the extent to which the form factor $f(k)$ is "smeared out". For this purpose we define for form factor in the following way:

$$f(k) = e^{-k^2/2\Lambda^2}; \qquad k = |\boldsymbol{k}|. \tag{27}$$

The parameter Λ plays the part of the cutoff momentum in the momentum representation. In the configuration representation the length $r_\Lambda = \hbar/\Lambda$ characterizes the region of smearing out of the nucleon. To pass over to the local limit, one must let Λ tend to ∞. For further discussion we shall need the following two integrals:

$$\Delta M = \frac{g^2}{(2\pi)^3} \int d\boldsymbol{k} \frac{f^2(k)}{2\omega^2(k)} = \frac{g^2}{4\pi^2} \int_0^\infty \frac{k^2 dk}{k^2 + \mu^2} e^{-k^2/\Lambda^2}$$

and

$$L = \frac{g^2}{(2\pi)^3} \int d\boldsymbol{k} \frac{f^2(k)}{2\omega^3(k)} = \frac{g^2}{4\pi^2} \int_0^\infty \frac{k^2 dk}{(k^2 + \mu^2)^{3/2}} e^{-k^2/\Lambda^2},$$

in which we use the relativistic expression for the energy of mesons of mass μ. In the limit $\Lambda \gg \mu$ we obtain

$$\Delta M \simeq \frac{g^2}{4\pi^2} \Lambda, \quad L \simeq \frac{g^2}{8\pi^2} \ln \frac{\Lambda^2}{\mu^2}. \tag{28}$$

Thus, in the local limit both integrals diverge. The integral ΔM, which according to (17) gives the field contribution to the nucleon mass, diverges linearly in Λ, i.e., is inversely proportional to the "dimension" of the nucleon r_Λ. The second integral according to (15) enters into the norm of the "dressed" state. On the basis of (19) we conclude that the probability for the "dressed" nucleon to be in the bare state in the local limit is equal to

$$w_0 = Z = e^{-L} \simeq \left(\frac{\Lambda}{\mu}\right)^{-g^2/4\pi^2},$$

i.e., it tends to zero exponentially. The probability of existence of the state containing a "bare" nucleon plus one meson,

$$w_1 = Z \int d\boldsymbol{q}\, g^2(\boldsymbol{q}) = ZL = \frac{g^2}{8\pi^2}\left(\frac{\mu}{\Lambda}\right)^{g^2/4\pi^2} \ln\frac{\Lambda^2}{\mu},$$

also tends to zero exponentially, and so on.

Thus, in the local limit the connections between the "bare" and "dressed" masses as well as between the eigenfunctions of the free and total Hamiltonians become singular.

CHAPTER IV

THE SCATTERING MATRIX

14. THE SCATTERING MATRIX

14.1. Perturbation theory. In studying interaction problems in quantum mechanics one starts with the Schrödinger equation

$$i\frac{\partial\psi(t)}{\partial t} = H\psi(t),$$

in which the total system Hamiltonian H is represented as a sum of the free-motion Hamiltonian H_0 and the interaction Hamiltonian H_1:

$$H = H_0 + H_1. \tag{1}$$

Since, as a rule, it is impossible to obtain the exact solution of the Schrödinger equation, one usually uses perturbation theory. In this case in the initial approximation the interaction term is neglected, and the exactly solvable problem is obtained:

$$i\frac{\partial\psi_0(t)}{\partial t} = H_0\psi_0(t). \tag{2}$$

In the theory of quantized fields, as well as in nonrelativistic quantum mechanics, exact solutions may be obtained only in the case of certain quite simple models. Therefore, here one also usually resorts to perturbation theory based on the initial approximation of noninteracting particles. In this case it becomes necessary to start by considering corresponding idealized free fields and to regard the interaction as a certain additional factor that causes little change in the properties of the dynamical system, a factor that one may "switch on" and "switch off". At first sight such a formulation of the problem does not seem objectionable. Indeed, elementary particles interact intensely only if they come sufficiently close together (in collision processes). Therefore it seems natural that at large distances the interaction between fields is insignificant and to a certain approximation one may neglect them and consider real particles as free ones. However (as is well known, for example, from classical electrodynamics) even free particles interact with the fields they themselves create. In the quantum case this leads to the perturbation of the lowest (vacuum) state—compare, for example, the problem of the static source in Section 12.4. Sometimes it is said that particles interact with the vacuum (of free fields) as with a kind of physical medium within which these particles move. Despite the possible weakness of the interaction, owing to the small dimensions of the particles the interaction effects turn out to be large (in the limit of pointlike particles, infinite).

Therefore using the conception of "bare" (i.e., not interacting even with the vacuum) particles turns out to be unsatisfactory, and it thus seems highly desirable to deal with real interacting particles without introducing the artificial conception of fictitious free fields. Such a program has been to a certain extent realized in the so-called axiomatic construction of the theory of quantized fields.

However, putting it into effect constructively encounters serious difficulties of a mathematical nature. The point is that at present the only sufficiently developed method of investigating interacting quantum fields is the perturbation-theory method. This method has as a starting point the approximation of noninteracting fields described by linear equations. All nonlinear contributions (cubic, trilinear, etc. in the Lagrangian), including the terms describing the interaction between fields and nonlinear effects due to self-interaction (see Sections 10.2 and 19.4), are considered here as small perturbations. The interaction constants (in other words, the coupling constants) occurring in these terms are considered to be small parameters, in successive powers of which perturbation theory is constructed. Here it must be emphasized that use of perturbation theory assumes tactitly the possibility of using expansions in power series in the coupling constant (or constants). Such an assumption actually involves two hypotheses:

(a) The existence of a weak-coupling limit for physically consistent solutions. We shall call solutions possessing such a limit adiabatic solutions.

(b) The analyticity (or sufficiently weak nonanalyticity) in the coupling constant of the adiabatic solution in the neighborhood of zero.

The use of these hypotheses is historically based on quantum electrodynamics, i.e., on the model of quantum field theory which describes the interaction between electrically charged particles and the electromagnetic field. In the framework of this model the coupling constant is the electric charge e, while the parameter of the perturbation expansion is proportional to the fine-structure constant, which in the system of units adopted herein takes on the form $\alpha = e^2/4\pi$.

The actual expansion parameter is of the order of magnitude of 10^{-3} ($\sim\alpha/\pi$—see, e.g., (30.5)). Therefore, the higher-order terms of perturbation theory in quantum electrodynamics numerically represent very small corrections to the main expressions obtained from the first nonvanishing terms of the perturbation expansion.

Such expressions, which had been calculated already in the late 1920s, turned out to be very close to the experimental values, and consequentially the possibility of using the perturbation theory expansion in quantum electrodynamics was automatically confirmed. Thus, hypotheses (a) in quantum electrodynamics always seemed to be quite natural. Owing to this fact, it was taken to be valid also for other types of interaction. For the same reasons, the validity of hypothesis (b) in quantum electrodynamics was not called in question either.

From a modern point of view the plausibility of hypotheses a) and b) for strong interactions is disputable. These hypotheses contradict certain solvable quantum-field models which have been discovered recently. Nevertheless our program will continue in constructing the apparatus of the quantum field perturbation theory designed to calculate matrix elements and transition probabilities. As we shall see later, this apparatus functions excellently in quantum electrodynamics (see Section 30 below) as well as in the realm of weak interactions (Section 32) and, at least partially, in the realm of strong interactions (Section 33).

14.2. The interaction representation. In the case when no interaction is present, the set of quantum fields is described by the analog of equation (2) for the state amplitude Ψ:

$$i\frac{\partial\Psi(t)}{\partial t}=H_0\Psi(t). \tag{3}$$

By carrying out the integration formally we obtain

$$\Psi(t)=e^{-iH_0t}\Phi. \tag{4}$$

The constant amplitude Φ was introduced in Section 6. In the case where interaction is present, (4) will no longer satisfy the equation of motion

$$i\frac{\partial\Psi(t)}{\partial t}=(H_0+H_1)\Psi(t). \tag{5}$$

However, we may generalize it by assuming Φ to depend on time:

$$\Psi(t)=e^{-iH_0t}\Phi(t). \tag{6}$$

On substituting (6) into (5) we obtain

$$i\frac{\partial\Phi(t)}{\partial t}=e^{iH_0t}H_1e^{-iH_0t}\Phi(t). \tag{7}$$

Let us discuss the meaning of the equation obtained above. The density $\mathcal{H}(\boldsymbol{x})$ of the interaction Hamiltonian

$$H_1=\int\mathcal{H}(\boldsymbol{x})\,d\boldsymbol{x},$$

(for economy of space we omit the lower index 1 under the integral sign) occurring on the right-hand sides of (5) and (7) is usually a polynomial function of the field operators in the Schrödinger representation:

$$\mathcal{H}(\boldsymbol{x})=\prod_\alpha u_\alpha(\boldsymbol{x})=\prod_\alpha u_\alpha(x)\,|_{x^0=0}.$$

But, in accordance with (6.20),

$$e^{iH_0t}\mathcal{H}(\boldsymbol{x})\,e^{-iH_0t}=\prod_\alpha e^{iH_0t}u_\alpha(0,\ \boldsymbol{x})\,e^{-iH_0t}=\prod_\alpha u_\alpha(x),$$

where $u_\alpha(x)$ are field operators in the Heisenberg representation with respect to the Hamiltonian of free motion. Therefore, the application of the operator $e^{iH_0t}\cdots e^{-H_0t}$ corresponds to replacing the operator functions of free fields in the Schrödinger representation $u(\boldsymbol{x})=u(0,\boldsymbol{x})$ by free-field operator functions in the Heisenberg representation $u(t,\ \boldsymbol{x})=u(x)$.

For the new state amplitude $\Phi(t)$ we now have, in accordance with (7), the following:

$$i\frac{\partial\Phi(t)}{\partial t}=H(t)\,\Phi(t), \tag{8}$$

where the space density $\mathcal{H}(x)$ of the operator

$$H(t)=\int\mathcal{H}(x)\,d\boldsymbol{x} \tag{9}$$

depends on the free fields in the Heisenberg representation. Such a representation of the Schrödinger equation is called the *interaction representation* (or sometimes the Dirac picture).

On expressing in this representation the expectation value of the dynamic operator B,

$$\bar{B}_t = \overset{*}{\Psi}(t)\, B\Psi(t) = \overset{*}{\Phi}(t)\, e^{iH_0t} Be^{-iH_0t}\Phi(t),$$

we see that it may be written as the expectation value over the amplitudes $\Phi(t)$ of the interaction representation of the expression

$$e^{iH_0t}Be^{-iH_0t} = B_{int}(t), \tag{10}$$

which it is natural to regard as the interaction representation for the dynamical variable B.

The transformation for the interaction Hamiltonian which was performed earlier may be written in the form

$$e^{iH_0t}H_1e^{-iH_0t} = H_{int}(t) = H(t). \tag{11}$$

Thus, in the interaction representation the operators for the dynamic quantities must be regarded as functions of field operators $u(x)$ in the Heisenberg representation for the *free* fields.

14.3. The scattering matrix. We note that with the aid of (8) we may already introduce into our discussion a very important quantity characterizing the system, the so-called *scattering matrix* or S-matrix. Let us study a process at the beginning and at the end of which we have only particles that are widely separated from each other and may be considered to be free. To calculate the probability amplitude for the scattering and for the mutual transformations of particles occuring in this process, we shall examine the situation in which the interaction $H_1(t)$ is adiabatically switched on in the infinitely remote past and is adiabatically switched off in the infinitely remote future. By denoting the amplitude of the initial state by $\Phi(-\infty)$ and the amplitude of the final state by $\Phi(\infty)$, we may establish a connection between them by means of the relation

$$\Phi(\infty) = S\Phi(-\infty), \tag{12}$$

in which the operator S is called the scattering operator. Evidently, the S operator may be defined by its matrix elements (transition amplitudes)

$$S_{\alpha\beta} = \langle \beta \mid S \mid \alpha \rangle,$$

where $\mid \alpha \rangle$ is a certain initial state which is characterized by the total set of quantum numbers $\alpha = \{\alpha_1, \alpha_2, \ldots,\}$, which represent both discrete and continuous sets of values, and $\mid \beta \rangle$ is the corresponding final state. The square of the modulus $\mid S_{\beta\alpha} \mid^2$ determines the probability of transition from $\mid \alpha \rangle$ to $\mid \beta \rangle$.

Thus, the operator S may conventionally be represented in the form of a matrix of infinite dimension composed of the elements $S_{\beta\alpha}$. It is usually called the *scattering matrix*.

To obtain the explicit expression for the operator S, we return to the Schrödinger equation in the interaction representation and proceed to construct its solution as an expansion in powers of the interaction H_1 starting with the initial approximation, which we choose to be the function $\Phi(t_0)$ of the initial state at a certain "initial" moment of time $t = t_0$. We obtain

$$\Phi(t) = S(t, t_0)\, \Phi(t_0), \tag{13}$$

where

$$S(t, t_0) = 1 - i\int_{t_0}^{t} H(t')\, dt' + (-i)^2 \int_{t_0}^{t} H(t')\, dt' \int_{t_0}^{t'} H(t'')\, dt'' + \ldots$$
$$\ldots + (-i)^n \int_{t_0}^{t} H(t_1)\, dt_1 \int_{t_0}^{t_1} H(t_2)\, dt_2 \ldots \int_{t_0}^{t_{n-1}} H(t_n)\, dt_n + \ldots \tag{14}$$

The terms of the expansion obtained allow an extremely interesting rearrangement. We shall demonstrate it taking for example the third term of the right-hand side. On changing the notation of the integration variables by $t' \leftrightarrow t''$ as well as reordering the integrations, we obtain (Figure 14.1)

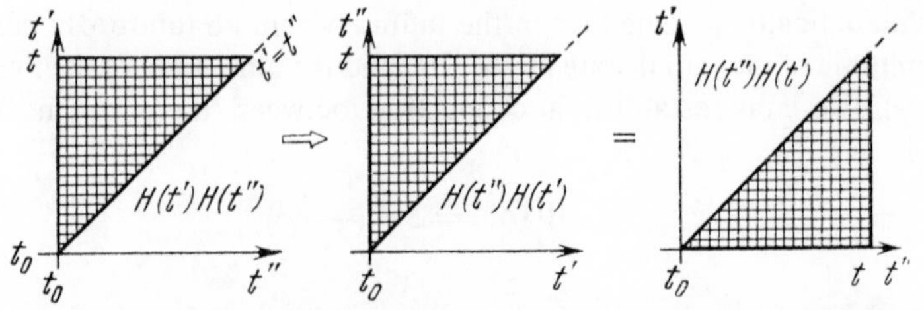

Figure 14.1. Transformation of the region of integration when the order of integration is changed in the second order of perturbation theory for the scattering matrix (see (14)).

$$\int_{t_0}^{t} H(t')\,dt' \int_{t_0}^{t'} H(t'')\,dt'' = \int_{t_0}^{t} dt' \int_{t'}^{t} dt'' H(t'')\,H(t').$$

Then representing the initial expression as half the sum of the initial and transformed ones, we find

$$\int_{t_0}^{t} H(t')\,dt' \int_{t_0}^{t'} H(t'')\,dt'' = \frac{1}{2}\int_{t_0}^{t} dt' \left\{\int_{t_0}^{t'} dt'' H(t')\,H(t'') + \int_{t'}^{t} dt'' H(t'')\,H(t')\right\}.$$

The expression in the braces may be written down in a more compact form

$$\{\ldots\} = \int_{t_0}^{t} dt'' T\,\{H(t')\,H(t'')\}$$

with the aid of the new symbol

$$T\,\{H(t')\,H(t'')\} \equiv \begin{cases} H(t')\,H(t'') & t' > t'', \\ H(t'')\,H(t') & t'' > t' \end{cases} \tag{15}$$

—the symbol denoting the *time-ordered product*. The common n-th term on the right-hand side of (14) may be reduced to the form

$$\frac{(-i)^n}{n!}\int_{t_0}^{t} dt_1 \ldots \int_{t_0}^{t} dt_n T\,\{H(t_1)\,H(t_2)\ldots H(t_n)\},$$

where the symbol $T\{H(t_1)\ldots H(t_n)\}$ for any arrangement of the time arguments is equal to the product of the Hamiltonians taken in an order corresponding to nonincreasing time arguments from left to right:

$$T\,\{H(t_1)\ldots H(t_n)\} = H(t_\alpha)\,H(t_\beta)\ldots H(t_\sigma)\,H(t_\rho), \qquad t_\alpha \geqslant t_\beta \geqslant \ldots \geqslant t_\sigma \geqslant t_\rho. \tag{16}$$

Representing each of the terms of the right-hand side of (14) with the aid of the T-product and factorizing out the symbol T from under the integration signs, we obtain

$$S(t_1, t_0) = T\left\{1 + \frac{1}{i}\int_{t_0}^{t_1} H(t)\,dt + \ldots \frac{1}{i^n n!}\left[\int_{t_0}^{t_1} H(t)\,dt\right]^n + \ldots\right\} =$$

$$= T\left\{\exp\left[\frac{1}{i}\int_{t_0}^{t_1} H(t)\,dt\right]\right\} = T\left\{\exp\left[\frac{1}{i}\int_{t_0}^{t_1} dt \int d\mathbf{x}\, \mathscr{H}(t, \mathbf{x})\right]\right\} \quad (17)$$

—the so-called *time-ordered exponential.*

14.4. Time-ordered products. The expression (17) contains time-ordered products of $\mathscr{H}(t, \mathbf{x}) = \mathscr{H}(x)$. Such products of operator expressions depending on points in four-dimensional space-time represent direct generalizations of the definitions (16). We, however, shall proceed differently and use the T-product of the local operator field functions.

We define the time-ordered product of n field operators $u_1(x_1)$, $u_2(x_2), \ldots, u_n(x_n)$ as the product of these operators taken in an order corresponding to the time arguments increasing from right to left, with due regard being paid to the overall sign, which may change if some of the operators are quantized in accordance with Fermi–Dirac rules. Here it is necessary to make an important comment. The time sequence of two points x_i, x_j is not a relativistically invariant concept in the case when they have a spacelike relation to one another ($x_i \sim x_j$). Therefore the new definition will be Lorentz-invariant only for operators which (anti)commute outside of the light cone:

$$u_i(x_i)\, u_j(x_j) = \eta u_j(x_j)\, u_i(x_i), \qquad x_i \sim x_j, \quad \eta = \pm 1 \quad (18)$$

We shall call such operators local field operators. It is clear, in particular, that the frequency parts of the operators $u^{\pm}$ are not local. We have by definition for the local field operators $u_1, \ldots, u_n$

$$T(u_1(x_1)\, u_2(x_2) \ldots u_n(x_n)) = \eta u_{i_1}(x_{i_1}) \ldots u_{i_n}(x_{i_n}),$$
$$x_{i_1} \gtrsim x_{i_2} \gtrsim \ldots \gtrsim x_{i_n}. \quad (19)$$

Here the symbol $x \gtrsim y$ signifies that the point x lies in the upper light cone of the point y or has a spacelike relation to it, while $\eta = (-1)^p$, where p is the parity of the permutation of the Fermi operators under the transition from the order $(1, 2, \ldots, n)$ to the order $(i_1, i_2, \ldots, i_n)$.

Now it is possible to introduce the time-ordered product of the local operator expressions $A_1(x_1), \ldots, A_n(x_n)$:

$$T\,(A_1(x_1)\ \ldots\ A_n(x_n)) = \eta A_{i_1}(x_{i_2})\ \ldots\ A_{i_n}(x_{i_n}),$$
$$x_{i_1} \gtrsim x_{i_2} \gtrsim \ldots \gtrsim x_{i_n}. \tag{20}$$

We define the local operator expression $A_u(x)$ with respect to the field $u(x)$ as an operator expression depending on the field operator function $u(x)$ and its derivatives taken as a whole ($u = u^+ + u^-$) at the point x and (anti)commuting with the field operator when the arguments are separated by a spacelike interval:

$$\{A_u(x),\ u(y)\}_{\pm} = 0, \qquad x \sim y.$$

Such operators are usually reduced to normal products of the operator functions $u(x)$. They possess the property of mutual local (anti)commutativity,

$$A_i(x)\,A_j(y) = \eta A_j(y)\,A_i(x), \qquad x \sim y, \tag{21}$$

as a result of which the definition (20) is self-consistent.

Here, indeed, it is necessary to make an essential qualification. The problem is that (anti)commutators of the type (21) usually have singularities at $x = y$. Therefore the time-ordered products (19), (20) turn out to be indefinite when two or more arguments ($x_1, \ldots, x_n$) coincide. This indefiniteness may be dealt with in various ways; see Section 16 below.

Henceforth we shall call operators $A(x)$ containing an even number of Fermi operators *local Bose operators* or, simply, local operators, while operator expressions in which an odd number of Fermi operators occurs will be called *local Fermi operators.*

The Lagrangian $\mathscr{L}(x)$, the energy-momentum tensor, and the current four-vector are examples of local operator expressions.

14.5. The time-ordered exponential. Returning to the scattering matrix (12), we see that in order to obtain S from the operator $S(t_1, t_0)$, t_1 must be made to tend towards $+\infty$, and t_0 towards $-\infty$. Taking into account that within the framework of models in which the interaction Lagrangian contains no derivatives of functions, the Hamiltonian differs from the Lagrangian function only by its sign, i.e.,

$$H(t) = -L(t) = -\int \mathscr{L}(x)\,d\boldsymbol{x},$$

we obtain

$$S = S(\infty, -\infty) = T \exp\left(i \int dx \mathscr{L}(x)\right) = Te^{i\mathscr{A}}. \tag{22}$$

Here $\mathscr{A}$ is the part of the action of the system which contains the interaction, i.e., the action of the interaction terms.

Here the following warning seems appropriate. In considering expressions similar to the one occurring on the right-hand side of (22), a temptation may arise to pull out from inside the T-product a factor of the type $\mathscr{A}$ which does not depend explicitly on the time. It is necessary to bear in mind that, in accordance with the definitions introduced (compare with (17)), such a factor must always be represented as an integral of a local-operator expression, and the symbol T must be placed under the integral sign. For example, the second term of the expansion (22) will be

$$T(\mathscr{A}^2) \equiv \int dx \int dy T(\mathscr{L}(x), \mathscr{L}(y)). \tag{23}$$

We may point out, also, that although we have obtained the T-exponential of the Lagrangian (22) with the aid of the Schrödinger equation and the T-exponential of the Hamiltonian (17), the expression (22) turns out to be valid also when the interaction Lagrangian contains derivatives of the field functions. In this case it turns out that the S-matrix may be defined in the form (22) by proceeding from a series of general requirements, such as causality, unitarity, relativistic invariance, and the correspondence principle (see Section 15 below). At the same time it turns out to be difficult to introduce the Hamiltonian operator H (for details see the *Introduction*, Chapter VII). Therefore we shall henceforth make use of (22) without corresponding reservations.

15. GENERAL PROPERTIES OF THE S-MATRIX

15.1. The scattering matrix as a functional. As was pointed to in Section 14, one may, by proceeding from the original idea of Heisenberg, introduce the S-matrix (corresponding to an infinitely large time interval from $t_1 = \infty$ to $t_0 = -\infty$) directly without reference to the Schrödinger equation, the Hamiltonian, and the operator $S(t_1, t_0)$ for finite t_1 and t_0. Instead of these properties and construction, in order to make the form of the S-matrix concrete, the following explicitly formulated physical conditions are used:

(a) causality,
(b) unitarity,
(c) relativistic covariance,
(d) the correspondence principle;

these conditions are accepted as initial axioms. Such a way of constructing the S-matrix, dating from the work of Stueckelberg and of Bogoliubov at the beginning of the 1950s, is conventionally called *axiomatic* (see the *Introduction*, Sections 20, 52, as well as the book of Bogoliubov, Logunov, and Todorov (1969)). In Section 16 we shall discuss the content and the formulation of the axioms, and present an expression for the S-matrix similar to (14.22) along with the corresponding comments.

It will then turn out to be necessary for us to define the S-matrix as a functional of certain subsidiary classical functions. The point is that the usual scattering matrix described by (14.22) and corresponding to the interaction of quantum wave fields throughout the whole space-time does not depend on any ordinary functions, and its matrix elements therefore do not involve any functional dependences. The introduction into the interaction Lagrangian of an unquantized function transforms the S-matrix and its matrix elements into functionals. In calculating the functional derivatives of these functionals one may introduce into consideration the space-time properties of various quantities involved in the theory (for instance, the causality condition). The "classical fields" will then play the part of subsidiary variables in the intermediate computations and are usually eliminated from the final expressions by means of some corresponding limiting process.

Such a subsidiary classical field may, for example, be introduced in the form of a fixed external field $u_{\text{ext}}(x)$ or a fixed external current of some particles, $J_{\text{ext}}(x)$. Thus, in studying the behavior of charged particles in an external electromagnetic field one introduces into the interaction Lagrangian the external unquantized electromagnetic potential

$$\mathscr{L} \to \mathscr{L}(A_{\text{ext}}) = e :\bar{\psi}(x)\gamma^{\nu}\psi(x)\left(A_{\nu}(x) + A_{\nu}^{\text{ext}}(x)\right):. \qquad (1)$$

The scattering matrix then turns out to be a functional

$$S \to S(A_{\text{ext}}).$$

A physically less clear, but technically more convenient method consists in introducing a classical functional argument into the interaction Langrangian in the form of a "function of the interaction range" $g(x)$. This function allows us to formalize the operations of "switching on" and "switching off" the interaction. To describe this operation mathematically, we introduce the function $g(x)$ which assumes values in the range $[0, 1]$ and which represents the intensity of switching on the interaction. In the regions where $g(x) = 0$ the interaction is absent, it is completely switched on only where $g = 1$, and where $0 < g < 1$ it is switched on only partially. Substituting the product $\mathscr{L}(x)g(x)$ for the actual interaction Lagrangian $\mathscr{L}$, we arrive at the interaction "switched on with an intensity $g(x)$".

Now let g differ from zero only within a certain finite space-time region. In this case in the sufficiently remote past and future the fields are free, and therefore the initial and final states of the dynamic system may be characterized by the ordinary constant state amplitudes introduced in Chapter II. These two quantities $\Phi(-\infty)$ and $\Phi(\infty)$ are related to some operator S which transforms $\Phi(-\infty)$ into $\Phi(\infty)$ and depends on the behavior of the function g. By fixing the initial state amplitude $\Phi(-\infty) = \Phi$, we may regard the final amplitude as a functional of g

$$\Phi(\infty) = \Phi(g) = S(g)\,\Phi. \tag{2}$$

According to this definition it is natural to interpret $S(g)$ as the scattering matrix for the case when the interaction is switched on with intensity g. The real case of interaction switched on completely throughout the whole space-time should in this scheme be considered by means of the limiting process in which the region where $g = 1$ expands and in the limit embraces the whole space-time. The usual scattering matrix S may then be defined in the form

$$S = S(1). \tag{3}$$

We shall often deal with the concept of the functional derivative, which is a natural extension of the concept of a partial derivative. As is well known, the partial derivative of a function of n variables $z_1, \ldots, z_n$ can be defined as the coefficient of dz_i in the sum

$$df = \sum_i R_i\, dz_i, \tag{*}$$

representing the differential of this function.

Now consider a certain functional $I(u)$ the variation of which is $\delta I(u)$, and defined as the main part of the increment $I(u + \delta u) - I(u)$; this variation may be represented by an integral of the form

$$\delta I(u) = \int_G R(x, u)\, \delta u\, dx.$$

Here R is a functional of u, depending on the position of the point x in the region G in the same manner as, in the sum (*), R_i was a function of $z_1, \ldots, z_n$ depending on the index i. By analogy with the definition presented above of the partial derivative, we then introduce the concept of the functional derivative of the functional $I(u)$ with respect to u at the point x, and define it by the relation

$$\frac{\delta I\,(u)}{\delta u\,(x)} = R\,(x,\; u).$$

In the same way it is possible to introduce also higher-order functional derivatives, it being easy to see that functional derivatives possess the same essential properties as ordinary ones do.

15.2. Relativistic covariance and unitarity. We shall now formulate a number of essential physical conditions which must be satisfied by the S-matrix. An important physical requirement, as usual, is the condition of *relativistic covariance.* For its explicit formulation consider the transformation P from the inhomogeneous Lorentz group, i.e., from the Poincaré group:

$$x \rightarrow x' = Px. \tag{4}$$

In the absence of interaction, the transformation law of the state amplitude corresponding to transformation (4), according to (6.22), had the form

$$\Phi' = U_P \Phi. \tag{5}$$

On the other hand, now, when $\Phi = \Phi(g)$, it is necessary also to take into account that the function $g(x)$ itself, which may be considered as a certain "classical scalar field", undergoes the transformation (4); since the interaction region described by the function g remains unchanged under that transformation, the transition to new coordinates in the argument of g gives

$$g\,(x) \rightarrow Pg\,(x) = g\,(P^{-1}x). \tag{6}$$

Therefore the transformation law of the amplitude $\Phi(g)$ has the form

$$\Phi\,(g) \rightarrow \Phi'\,(Pg) = U_P \Phi\,(g). \tag{7}$$

For reasons of relativistic covariance it is necessary to demand that the transformation law (2) of the initial amplitude into the final one shall not depend on the reference frame, i.e., that

$$\Phi'\,(g) = S\,(g)\,\Phi'.$$

Inserting here (5) and (7) with a displaced argument ($Pg \rightarrow g$), we find, allowing for (2),

$$U_P S\,(P^{-1}g)\,\Phi = S\,(g)\,U_P \Phi.$$

Owing to the arbitrariness of the initial state amplitude Φ, this expression may be written in the operator form

$$U_P S(P^{-1}g) = S(g)\, U_P$$

or, displacing the arguments by P, multiplying on the right by U_P^{-1}, and taking into account the unitarity of U_P,

$$S(Pg) = U_P S(g)\, U_P^{-1} = U_P S(g) \overset{+}{U}_P. \tag{8}$$

It is this formula that expresses the condition of covariance of the operator $S(g)$.

Let us now formulate another essential requirement, the requirement that the norms of the wave functions be conserved. Applied to the case under consideration, it implies that we should demand that

$$\overset{*}{\Phi}(g)\, \Phi(g) = \overset{*}{\Phi}\Phi,$$

whence, under the condition of reversibility of the operator, it follows that

$$\overset{+}{S}(g)\, S(g) = S(g)\, \overset{+}{S}(g) = 1, \tag{9}$$

i.e., that the operator $S(g)$ must be unitary.

In accordance with what was said in Section 14 the relation obtained may also be written in the form

$$\sum_{\gamma} S_{\beta\gamma}(g)\, \overset{+}{S}_{\gamma\alpha}(g) = I_{\beta\alpha}, \tag{9'}$$

where the symbol $\sum_{\gamma}$ stands for summation over the discrete and integration over the continuous quantum numbers characterizing the state $|\,\gamma\rangle$, while the symbol $I_{\beta\alpha}$ represents the product of the corresponding Kronecker symbols and Dirac δ-functions.

15.3. The condition of causality. We must also guarantee that the condition of causality is satisfied, i.e., require that the change in the interaction law in any space-time region can influence the evolution of the system only at subsequent times. To formulate the condition of causality explicitly, we shall consider the case when the space-time region G in which the function g differs from zero may be divided into two separate subregions G_1 and G_2 such that all points of one of them (G_1) lie in the past with respect to a certain time

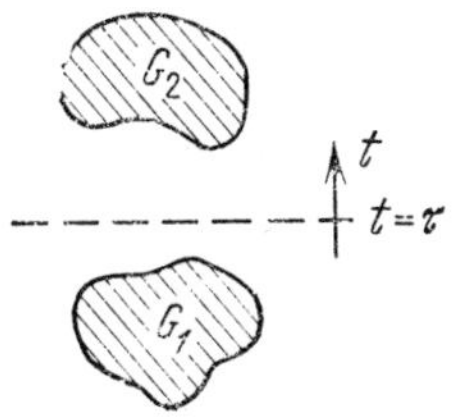

Fig. 15.1. Illustration of the mutual positions of interaction regions for the formula (13).

instant $t = \tau$, while all points of the other (G_2) lie in the future with respect to $t = \tau$(Figure 15.1). The function g may in this case be reresented as a sum of two functions

$$g(x) = g_1(x) + g_2(x), \tag{10}$$

one of which, g_1, differs from zero only in G_1, while the second, g_2, differs from zero only in G_2.

At time τ one may define a state characterized by the amplitude Φ_τ, which, by considerations of causality, should not depend on the interaction in the region G_2 and may therefore be written in the form

$$\Phi_\tau = S(g_1)\,\Phi, \tag{11}$$

where $S(g_1)$ is the scattering matrix for the case when the interaction is switched on with an intensity g_1. The final state $\Phi(g)$ can now be obtained from Φ_τ with the aid of the operator $S(g_2)$ which describes the interaction in the region G_2:

$$\Phi(g) = S(g_2)\,\Phi_\tau. \tag{12}$$

Comparing (10)–(12) with (2), we find that

$$S(g_1 + g_2) = S(g_2)\,S(g_1) \quad \text{for} \quad G_2 > G_1. \tag{13}$$

(The inequality $G_2 > G_1$ denotes that all points of the region G_2 lie at time later than all points of the region G_1.) The relation (13) represents the formulation of the principle of causality for the case $G_2 > G_1$.

Note, furthermore, that if $G_1 \sim G_2$, i.e., all points of the region G_1 have a spacelike relation to all points of the region G_2 (Figure 15.2), then the time order of these regions may be changed by means of the corresponding Lorentz transformation. Therefore

$$S(g_1)\,S(g_2) = S(g_2)\,S(g_1) \quad \text{for} \quad G_1 \sim G_2. \tag{14}$$

The formulas (13), (14) represent formulations of the condition of causality written down for the operator $S(g)$ as a whole. For theoretical applications the differential formulation of the condition of causality turns out to be more effective.

15.4. The differential condition of causality. To obtain the differential condition of causality, consider two cases which differ from each other by the form of the interaction in region G_2 and which are described by the same function in G_1 (Figure 15.3), i.e.,

$$g'(x) = g_2'(x) + g_1(x), \quad g''(x) = g_2''(x) + g_1(x).$$

In constructing the expression $S(g'')\overset{+}{S}(g')$ we see that it does not depend on the behavior of the functions g'' and g' in the region G_1. Indeed, since according to (13)

$$\overset{+}{S}(g_1 + g_2) = \overset{+}{S}(g_1)\,\overset{+}{S}(g_2) \quad \text{for} \quad G_2 > G_1,$$

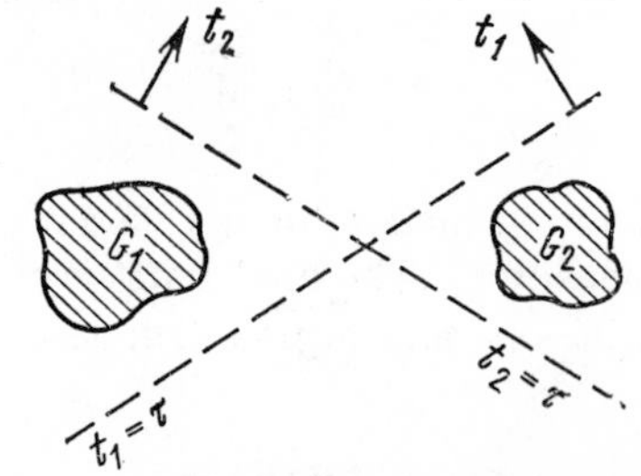

Fig. 15.2. Relative position of interaction regions corresponding to (14).

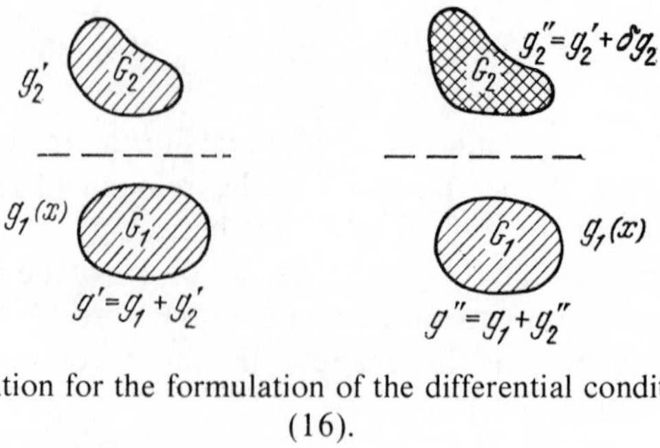

Fig. 15.3. Illustration for the formulation of the differential condition of causality (16).

and taking into account the unitarity properties of the matrix $S(g_1)$, we obtain

$$S(g'')\overset{+}{S}(g') = S(g_2'')\, S(g_1)\, \overset{+}{S}(g_1)\, \overset{+}{S}(g_2') = S(g_2'')\, \overset{+}{S}(g_2').$$

Thus the product $S(g'')\overset{+}{S}(g')$ does not in fact depend on the behaviour of the function g in the region G_1. This is due to the fact that the dependence on the state of the system prior to the time t contained in $S(g'')$ is eliminated by the corresponding part of the operator $\overset{+}{S}(g)$. Therefore, in the more general case, we shall also adopt the following formulation of the condition of causality:

> If there are two functions $g''(x)$ and $g'(x)$ which coincide with each other for x^0 smaller than a certain t, then the product $\overset{+}{S}(g'')S(g')$ must not depend on the simultaneous variation of g' and g'' by the same value in the region $x^0 < t$.

Let us now formulate the condition of causality in differential form. if we set $g'(y) = g(y)$ and $g''(y) = g(y) + \delta g(y)$, where $\delta g(y)$ is an infinitesimal variation of the function g which differs from zero only for $y^0 > t$, then the matrix $S(g'')$ can be represented in the form

$$S(g'') = S(g) + \delta S(g),$$

where

$$\delta S(g) = \int\limits_{y^0 > t} \frac{\delta S}{\delta g(y)}\, \delta g(y)\, dy.$$

It may now be seen that the product

$$S(g'')\, \overset{+}{S}(g') = S(g)\, \overset{+}{S}(g) + \delta S(g)\, \overset{+}{S}(g) = 1 + \delta S(g)\, \overset{+}{S}(g)$$

does not depend on the behavior of the function g for $x^0 < t < y^0$.

By going over to the variational derivative, we may therefore formulate the condition of causality as the condition that the expression

$$H(y;\, g) = i\frac{\delta S(g)}{\delta g(y)}\, \overset{+}{S}(g) \tag{15}$$

is independent of the behavior of the function $g(x)$ at the point x for $x < y$. By considerations of covariance, it also follows from this that the operator (15) cannot depend on the behaviour of the function $g(x)$ at $x \sim y$ either.

The condition of causality may evidently be written in the form

$$\frac{\delta}{\delta g(x)}\left(\frac{\delta S(g)}{\delta g(y)}\overset{+}{S}(g)\right)=0 \quad \text{for} \quad x \lesssim y. \tag{16}$$

This relation represents the *formulation of the principle of causality in differential form.*

We thus have conditions of relativistic covariance, unitarity, and causality which together provide a sufficient basis for the construction of the S-matrix.

16. THE AXIOMATIC S-MATRIX

16.1. Expansion in powers of the interaction. We now proceed with constructing the operator $S(g)$ satisfying the conditions of covariance (15.8), unitarity (15.9), and causality (15.16). For the time being we shall completely disregard the arguments presented in Section 14 and the formula (14.22) obtained therein on the basis of the Schrödinger equation.

We shall seek $S(g)$ in the form of a formal functional expansion in powers of g:

$$S(g)=1+\sum_{n\geqslant 1}\frac{1}{n!}\int S_n(x_1, \ldots, x_n)\, g(x_1)\ldots g(x_n)\, dx_1\ldots dx_n, \tag{1}$$

in which $S_n(x_1, \ldots, x_n)$ are operator expressions which depend on the complete field functions $u(x_1), \ldots, u(x_n)$ and their partial derivatives as a whole and not separately on their frequency parts $u^{\pm}$, while the Fermi field operators occur in S_n only in even combinations.

We shall call such operator expressions multilocal. For two multilocal operators $A(x_1, \ldots, x_n)$, $B(y_1, \ldots, y_m)$, the following property holds:

$$[A(x_1, \ldots, x_n),\ B(y_1, \ldots, y_m)]=0,$$

when all the x_i are spacelike with respect to all the y_j.

In the case under consideration the condition

$$[S_n(x_1, \ldots, x_n),\ S_m(y_1, \ldots, y_m)] = 0 \quad (x_i \sim y_j) \tag{2}$$

has an important physical meaning. If the two functions g_1 and g_2 are localized in space-time regions such that any point of one region is spacelike with respect to all the points of the other region, then $S(g_1)$ commutes with $S(g_2)$. This essentially expresses the fact (which is also a manifestation of the principle of causality) that a signal cannot propagate with a speed greater than that of light.

Returning to (1), we see that without loss of generality the S_n may be considered to be symmetric functions of their arguments $x_1, \ldots, x_n$, since the weight functions $g(x_1), \ldots, g(x_n)$ enter into the expressions in a symmetric manner.

16.2. Conditions for S_n. To determine the concrete form of the functions S_n, it is necessary to substitute the expansion (1) into the conditions (15.8), (15.9), (15.16). The condition of covariance leads to the linear condition

$$U_P S_n(x_1, \ldots, x_n) \overset{+}{U}_P = S_n(Px_1, \ldots, Px_n). \tag{3}$$

while the condition of unitarity provides the nonlinear relations

$$\begin{aligned} S_n(x_1, \ldots, x_n) + \overset{+}{S}_n(x_1, \ldots, x_n) + \\ + \sum_k P(x_1, \ldots, x_k \mid x_{k+1}, \ldots, x_n) S_k(x_1, \ldots, x_k) \times \\ \times \overset{+}{S}_{n-k}(x_{k+1}, \ldots, x_n) = 0. \end{aligned} \tag{4}$$

Here $\overset{+}{S}_n$ is the Hermitian conjugate of the operator S_n, and the symbol $P(x_1, \ldots, x_k | x_{k+1}, \ldots, x_n)$ denotes the sum over all the $n!/k!(n-k)!$ ways of dividing the set of points $x_1, \ldots, x_n$ each of which contributes equally since the functions S_k are symmetric in their arguments. For example,

$$\begin{aligned} P(x_1, x_2 \mid x_3) S_2(x_1, x_2) S_1(x_3) = S_2(x_1, x_2) S_1(x_3) + \\ + S_2(x_1, x_3) S_1(x_2) + S_2(x_2, x_3) S_1(x_1). \end{aligned}$$

The result obtained on applying the condition of causality takes on the form

$$H_n(y; x_1, \ldots, x_n) = 0, \tag{5}$$

if for at least one $x_j (j = 1, \ldots, n)$ we have $y \gtrsim x_j$. Here H_n are coefficients of the expansion of the operator (15.15):

$$H(y; g) = \sum_{n \geqslant 0} \frac{1}{n!} \int H_n(y; x_1, \ldots, x_n) g(x_1) \ldots g(x_n) \, dx_1 \ldots dx_n, \tag{6}$$

related to S_n by means of the nonlinear relations

$$H_n(y;\ x_1,\ \ldots,\ x_n) = i \sum_{0\leqslant k\leqslant n} P(x_1,\ \ldots,\ x_k\,|\,x_{k+1},\ \ldots,\ x_n)\times$$
$$\times S_{k+1}(y;\ x_1,\ \ldots,\ x_k)\overset{+}{S}_{n-k}(x_{k+1},\ \ldots,\ x_n). \tag{7}$$

The conditions (2) and (3) determine the general properties of local commutativity and invariance of the operator functions S_n, while the nonlinear conditions (4) and (5) may be regarded as recurrence relations allowing one to obtain each of the functions S_n in sequence in terms of the preceding S_k $(1 \leq k \leq n - 1)$.

Let $S_1(x)$ be the initial quantity. From the condition of unitarity (4) we obtain

$$S_1(x) = i\mathscr{L}(x), \tag{8}$$

where $\mathscr{L}$ is a Hermitian operator. The formula (3) for $n = 1$ gives the condition of covariance for $\mathscr{L}$, and the formula (2) for $n = m = 1$ leads to the *condition of local commutativity* for $\mathscr{L}$:

$$[\mathscr{L}(x),\ \mathscr{L}(y)] = 0 \qquad (x \sim y). \tag{9}$$

The condition of local commutativity represents a particular case of the condition (2). It imposes significant restrictions on the structure of the operator $\mathscr{L}(x)$. In the axiomatic construction this operator plays the role of the interaction Lagrangian. Owing to the condition (9), $\mathscr{L}(x)$ should depend on an even number of Fermi fields, and each of these fields must enter into $\mathscr{L}(x)$ as a whole and not only in its separate frequency components. In Section 14.3 we referred to such operator constructions dependent on the fields at point x as local. Thus the condition of local commutativity leads to the requirement that the interaction Lagrangian be local.

16.3. Determination of the explicit form of S_2 and S_3. To determine S_2, we write the condition of causality (5) for $n = 1$:

$$S_2(x,\ y) + S_1(x)\overset{+}{S}_1(y) = S_2(x,\ y) + \mathscr{L}(x)\mathscr{L}(y) = 0.$$

Thus,

$$S_2(x,\ y) = -\mathscr{L}(x)\mathscr{L}(y) \quad \text{for} \quad x \gtrsim y.$$

By virtue of the symmetry of S_2 we have also

$$S_2(x, y) = -\mathcal{L}(y)\mathcal{L}(x) \quad \text{for} \quad y \gtrsim x.$$

The regions within which the two preceding formulas are defined overlap for $x \sim y$. The requirement that they be compatible leads us to the condition of local commutativity (9). At the same time these relations define S_2 everywhere except at the point $x = y$. We have

$$S_2(x, y) = i^2 T(\mathcal{L}(x)\mathcal{L}(y)).$$

As is seen, the behavior of S_2 for coinciding arguments is not determined by the recurrence relations. Therefore it is possible to write

$$S_2(x, y) = i^2 T(\mathcal{L}(x)\mathcal{L}(y)) + i\Lambda_2(x, y). \tag{10}$$

The operator Λ_2 introduced here satisfies

$$\Lambda_2(x, y) = 0 \quad \text{for} \quad x \neq y \tag{11}$$

and it is therefore called *quasilocal.* In the more general case a quasilocal operator depending on n arguments is defined by the condition that

$$\Lambda_n(x_1, \ldots x_n) = 0 \quad \text{everywhere, except at the point} \quad x_1 = x_2 = \ldots = x_n. \tag{12}$$

Inserting (10) into the condition of unitarity (15.9), we find that the operator $i\Lambda_2$ must be anti-Hermitian, while Λ_2 is correspondingly Hermitian:

$$\overset{+}{\Lambda}_2(x, y) = \Lambda_2(x, y).$$

Thus, the axioms used so far allow us to define the operator S_2 in terms of the first-order operator S_1 to the accuracy of the quasilocal anti-Hermitian operator $i\Lambda_2$.

It may be shown (see the *Introduction*, Section 21) that such a situation is typical also for the higher-order S_n $(n \geq 3)$. Thus, for example,

$$\begin{aligned} S_3(x, y, z) = i^3 T(\mathcal{L}(x)\mathcal{L}(y)\mathcal{L}(z)) + \\ + i^2 T(\mathcal{L}(x)\Lambda_2(y, z)) + i^2 T(\mathcal{L}(y)\Lambda_2(x, z)) + \\ + i^2 T(\mathcal{L}(z)\Lambda_2(x, y)) + i\Lambda_3(x, y, z), \end{aligned} \tag{13}$$

where Λ_3 is a Hermitian quasilocal operator.

16.4. The general form of $S(g)$. In the general case S_n is expressed in terms of a sum, the components of which are of three different types. There is, first, the T-product of $\mathscr{L}(x_1), \ldots, \mathscr{L}(x_n)$, then the sum of the various T-products of $\mathscr{L}(x_j)$ and the quasilocal operators $\Lambda_1, \ldots, \Lambda_{n-1}$, and finally the "new" quasilocal operator Λ_n:

$$S_n(x_1, \ldots, x_n) = i^n T(\mathscr{L}(x_1)\mathscr{L}(x_2)\ldots\mathscr{L}(x_n)) + \\ + \prod_T \{\mathscr{L}, \Lambda_1, \Lambda_2, \ldots, \Lambda_{n-1}\} + i\Lambda_n(x_1, \ldots, x_n), \tag{14}$$

Comparing (8), (10), (13), (14) with the expansion of the T-exponential obtained earlier in Section 14, we see that the expression

$$S_n(x_1, \ldots, x_n) = i^n T(\mathscr{L}(x_1)\ldots\mathscr{L}(x_n)),$$

which corresponds to

$$S(g) = T\left[\exp\left(i\int \mathscr{L}(x) g(x)\, dx\right)\right], \tag{15}$$

(and, in the limit $g \to 1$, also to (14.22) in which the operator $\mathscr{L}$ is the interaction Lagrangian), is admissible in the sense that it satisfies all the requirements imposed on S_n. However, this expression *is not the most general one.* To obtain expressions for $S_1, S_2, \ldots, S_n$, one must specify in addition to the local operator $\mathscr{L}(x)$ a sequence of quasilocal operators $\Lambda_2(x_1, x_2), \ldots, \Lambda_n(x_1, x_2, \ldots, x_n)$.

We have thus arrived at results which at first glance seem somewhat strange. To completely determine the matrix $S(g)$, it turns out to be insufficient to specify the interaction Lagrangian, but it is also necessary to specify in addition an infinite sequence of quasilocal operators Λ_n. It may, however, be shown that the expression for $S(g)$ for $\Lambda_2, \ldots, \Lambda_n$ different from zero can be reduced to the form

$$T\left(\exp\left(i\int \mathscr{L}(x;\ g)\, dx\right)\right), \tag{16}$$

in which the "Lagrangian" $\mathscr{L}(x; g)$ is defined by the relation

$$\mathscr{L}(x;\ g) = \mathscr{L}(x) g(x) + \\ + \sum_{\nu \geqslant 1} \frac{1}{(\nu+1)!} \int \Lambda_{\nu+1}(x, x_1, \ldots, x_\nu) g(x) g(x_1)\ldots g(x_\nu)\, dx_1 \ldots dx_\nu. \tag{17}$$

In view of the localized nature of the functions Λ_ν, all integrations disappear here, and $\mathscr{L}(x; g)$, in fact, depends on the field functions $u(x)$ at the point x, and is therefore a local operator. In addition to the field operators u the expression $\mathscr{L}(x; g)$ also depends on the functions $g(x)$, which may be regarded as a "classical" field. Consequently, (16) satisfies all the conditions imposed on $S(g)$, and may be regarded as the expression for the scattering matrix. By expanding (16) in a power series in g, we obtain the expressions (14) for $S_n(x_1, \ldots, x_n)$ which satisfy all the necessary conditions.

Passing over to the physical limit $g = 1$, we obtain

$$S = S(1) = T\left(\exp\left(i \int \mathscr{L}(x;\ 1)\, dx\right)\right), \tag{18}$$

where

$$\mathscr{L}(x,\ 1) = \mathscr{L}(x) + \sum_{\nu} \frac{1}{(\nu+1)!} \int \Lambda_{\nu+1}(x,\ x_1,\ \ldots,\ x_\nu)\, dx_1 \ldots dx_\nu. \tag{19}$$

Therefore, the actual scattering matrix $S(1)$ is completely characterized by the interaction Lagrangian of the system $\mathscr{L}(x;\ 1)$, which in perturbation theory is sometimes represented as a series.

17. WICK'S THEOREMS

17.1. Reduction to the normal form. Before proceeding to evaluate the matrix elements of the S-matrix it is necessary to become familiar with a number of properties of an algebraic nature characteristic of operator expressions constructed of quantized free fields. In view of the evident fact that for subsequent calculation of matrix elements it is convenient to represent operator expressions in their normal form, the important question arises as to the method of reduction to normal form.

To start with, we shall consider the technique for reducing to the normal form ordinary products (the first Wick's theorem) and then chronological products (the second Wick's theorem) occurring in the scattering matrix.

Clearly, for reducing to the normal form operator expressions polynomially dependent on the field operators, it suffices to know how to reduce products of the type $A_1(x_1), \ldots, A_n(x_n)$ in which $A(x)$ are "linear operators", i.e., linear combinations of the corresponding $u^+(x)$, $u^-(x)$. In each particular case the direct transformation of such a product does not present any real difficulties. One simply consecutively shifts u^+ to the left and u^- to the right, and applies the commutation relations at each shift. Nevertheless, in view of the great number of terms appearing, it becomes expedient even for small n to have a set of prescriptions so the reduction to the normal form becomes as automatic as possible.

Such a set of prescriptions follows from an important theorem established by Wick. We shall now present the formulation of this theorem.

17.2. Wick's first theorem. Assume we have the product of two linear operators $A(x)B(y)$. In this case, in order to reduce the product to the normal form, it is obviously sufficient to carry out not more than one displacement of the operators $u^-(x)$, $u^+(y)$, and therefore as a result we shall obtain terms with the "right order" in the sequence of positive- and negative-frequency parts of wave functions, and a term in the commutation functions which no longer contains any operator expressions.

Thus, the product under discussion may differ from the normal product $:A(x)B(y):$ only by a *c*-number (compare with the computation in Section 8.2) which we shall call *a pairing* and which we shall denote by bracketing below the line:

$$A(x)\,B(y) = :A(x)\,B(y): + \underbrace{A(x)\,B(y)}. \tag{1}$$

As for the vacuum state the expectation value of the normal product is always equal to zero, we may define the pairing as the vacuum expectation value of the ordinary product:

$$\underbrace{A(x)\,B(y)} = \langle A(x)\,B(y)\rangle_0. \tag{2}$$

As an example, let us consider the real scalar field. Starting with the commutation relations for the frequency parts,

$$\varphi^-(x)\,\varphi^+(y) - \varphi^+(y)\,\varphi^-(x) = -iD^-(x-y),$$

we obtain

$$\varphi(x)\,\varphi(y) = :\varphi(x)\,\varphi(y): - iD^-(x-y). \tag{3}$$

Thus,

$$\begin{gathered}\underbrace{\varphi(x)\,\varphi(y)} = \langle\varphi(x)\,\varphi(y)\rangle_0 = \underbrace{\varphi^-(x)\,\varphi^+(y)} = -iD^-(x-y),\\ \underbrace{\varphi^+(x)\,\varphi^-(y)} = 0.\end{gathered} \tag{4}$$

Analogously, for the electromagnetic field we shall obtain

$$\underbrace{A_\nu(x)\,A_\mu(y)} = \langle A_\nu(x)\,A_\mu(y)\rangle_0 = ig_{\nu\mu}D_0^-(x-y). \tag{5}$$

Let us also consider a fermion field, for which

$$\psi_{\alpha}^{-}(x)\,\bar{\psi}_{\beta}^{+}(y) = -\,\bar{\psi}_{\beta}^{+}(y)\,\psi_{\alpha}^{-}(x) - iS_{\alpha\beta}^{-}(x-y), \tag{6}$$

and therefore

$$\underline{\psi_{\alpha}(x)\,\bar{\psi}_{\beta}(y)} = -\,iS_{\alpha\beta}^{-}(x-y), \tag{7}$$

as well as

$$\underline{\bar{\psi}_{\alpha}(x)\,\psi_{\beta}(y)} = -\,iS_{\beta\alpha}^{+}(y-x) \tag{8}$$

and

$$\underline{\psi(x)\,\psi(y)} = \underline{\bar{\psi}(x)\,\bar{\psi}(y)} = 0. \tag{9}$$

To formulate Wick's theorem, we shall need also to define the concept of *a normal product with pairing*. We shall define it as follows:

$$\begin{aligned} &:A_1(x_1)\,A_2(x_2)\,\ldots\,\underline{A_j(x_j)\,\ldots\,A_i(x_i)}\,\ldots\,A_n(x_n): \equiv \\ &\equiv \eta\,\underline{A_j(x_j)\,A_i(x_i)}\!:\,A_1(x_1)\,\ldots\,A_{j-1}A_{j+1}\,\ldots\,A_{i-1}A_{i+1}\,\ldots\,A_n:, \end{aligned} \tag{10}$$

where $\eta = (-1)^p$ and p is the parity of the Fermi permutations in the transition from the order $1, \ldots, j-1, j, j+1, \ldots, i-1, i, i+1, \ldots, n$ to the order $j, i, 1, \ldots, j-1, j+1, \ldots, i+1, \ldots, n$.

In a completely analogous manner we define a normal product with an arbitrary number of pairings. Namely, we shall consider the expression

$$:A_1(x_1)\,A_2(x_2)\,A_3(x_3)\,\ldots\,A_k(x_k)\,\ldots\,A_{n-1}(x_{n-1})\,A_n(x_n):$$

(with A_1 paired with A_k and A_2 paired with A_{n-1}) to be equal to the product of all the pairings with the normal product of the remaining unpaired operators and with the number $\eta = (-1)^p$, i.e., to the expression

$$\eta\,\underline{A_1A_k}\,\underline{A_2A_{n-1}}\;:A_3\,\ldots\,A_{k-1}A_{k+1}\,A_{n-2}\,\ldots\,A_{n-2}A_n:,$$

where p is the parity of the permutation to which the Fermi operators are subjected in the process of bringing the pairings outside the normal product.

Thus, for example,

$$:\overline{\psi}_{\alpha}(x)\,\overline{\psi}_{\beta}(y)\,\psi_{\gamma}(z)\,\psi_{\delta}(t): = -\underbrace{\overline{\psi}_{\alpha}(x)\,\psi_{\gamma}(z)}\,\underbrace{\overline{\psi}_{\beta}(y)\,\psi_{\delta}(t)} =$$

$$= S^{+}_{\gamma\alpha}(z-x)\,S^{+}_{\delta\beta}(t-y).$$

From the above definition it follows immediately that the normal product with pairing is linear with respect to its factors, and that when they are transposed within such a product, this product is multiplied by $\eta = (-1)^p$, where p is the parity of the permutations of the Fermi operators.

We can now give a simple formulation of Wick's theorem. According to this theorem

The ordinary product of linear operators is equal to the sum of all the corresponding normal product with all possible pairings including the normal product without pairing:

$$A_1 \ldots A_n = :A_1 \ldots A_n: + \sum_{i \neq j} :A_1 \ldots \underbrace{A_i \ldots A_j} \ldots A_n: +$$

$$+ \sum_{i,j,k,l} :A_1 \ldots A_i \ldots A_j \ldots A_k \ldots A_l \ldots A_n: + \ldots \tag{11}$$

It may be easily seen that this theorem is also applicable to the case when some of the factors themselves appear as normal products:

$$A_1 \ldots A_\nu \; :A_{\nu+1} \ldots A_\mu: \ldots A_k \ldots :A_\rho \ldots A_n:.$$

In this case Wick's theorem is formulated in exactly the same way as for the "pure" product $A_1 \ldots A_n$, with the one obvious difference that we now need not take into account pairings between those factors that belong to the same normal product. For example, the pairings between $A_\nu, \ldots, A_\mu$ and the pairings between $A_\rho, \ldots, A_n$ should not be taken into account, and so on.

17.3. Time-ordered pairing. In the course of evaluating matrix elements of the scattering matrix we shall have to deal with chronological products of Lagrangians. For this we shall have to express the T-products of local operator expressions through normal products of the corresponding operators of free fields.

The prescription for the reduction of time-ordered products is given by Wick's theorem for T-products, which is the analog of Wick's theorem for ordinary products. Before beginning the proof of this theorem we shall introduce the important concept of the time-ordered pairing of operators.

To do this, we examine (14.19) in the case of two field operators:

$$T(u_1(x)\,u_2(y)) = \begin{cases} u_1(x)\,u_2(y), & x^0 > y^0, \\ \eta u_2(y)\,u_1(x), & y^0 > x^0. \end{cases}$$

In accordance with the definition of ordinary pairing (1), the right-hand side can be transformed to the form

$$T(u_1(x)\,u_2(y)) = \begin{cases} :u_1(x)\,u_2(y): + \underline{u_1(x)\,u_2(y)}, & x^0 > y^0, \\ :u_1(x)\,u_2(y): + \eta\underline{u_2(y)\,u_1(x)}, & y^0 > x^0. \end{cases} \tag{12}$$

From this it may be seen that in any arbitrary case $T(u_1(x)u_2(y))$ differs from $:u_1(x)u_2(y):$ by a c-number, which we shall call the *time-ordered pairing* of the operators u_1 and u_2, i.e., by definition

$$T(u_1(x)\,u_2(y)) = :u_1(x)\,u_2(y): + \overline{u_1(x)\,u_2}(y) \tag{13}$$

and

$$\overline{u_1(x)\,u_2}(y) = \begin{cases} \underline{u_1(x)\,u_2}(y), & x^0 > y^0, \\ \eta\underline{u_2(y)\,u_1}(x), & y^0 > x^0. \end{cases} \tag{14}$$

First of all, let us note an important property of time-ordered pairing. It is permissible to interchange the order of factors within a time-ordered pairing just as it within a normal product:

$$\overline{u_1(x)\,u_2}(y) = \eta\overline{u_2(y)\,u_1}(x),$$

which directly follows from (14).

We shall now define time-ordered pairings for the operators of the principal wave fields. To do this, we note that by evaluating the vacuum expectation value of (13) and taking into account the fundamental property of the normal product and the normalization of the amplitude of the vacuum state, we obtain

$$\langle T\,(u_1(x)\,u_2(y))\rangle_0 = \overline{u_1(x)\,u_2}(y). \tag{15}$$

Thus, the time-ordered pairing of two operators is equal to the vacuum expectation value of the T-product of these operators. We have in the following particular cases: for the scalar field,

$$\begin{aligned} i\overline{\varphi(x)\,\varphi}(y) &= i\,\langle T\varphi(x)\,\varphi(y)\rangle_0 = \\ &= \theta\,(x^0-y^0)\,D^-(x-y) - \theta\,(y^0-x^0)\,D^+(x-y); \end{aligned} \tag{16}$$

for the electromagnetic field,

$$\begin{aligned} i\overline{A_\nu(x)\,A_\mu}(y) &= \langle TA_\nu(x)\,A_\mu(y)\rangle_0 = \\ &= -g_{\nu\mu}\theta\,(x^0-y^0)\,D_0^-(x-y) + g_{\nu\mu}\theta\,(y^0-x^0)\,D_0^+(x-y); \end{aligned} \tag{17}$$

for the spinor field,

$$\begin{aligned} i\overline{\psi_\alpha(x)\,\bar\psi_\beta}(y) &= i\,\langle T\psi_\alpha(x)\,\bar\psi_\beta(y)\rangle_0 = \\ &= \theta\,(x^0-y^0)\,S^-_{\alpha\beta}(x-y) - \theta\,(y^0-x^0)\,S^+_{\alpha\beta}(x-y); \end{aligned} \tag{18}$$

and so on (for the definition of the function θ see (AV.3)).

As will be established below in Section 18.2, the right-hand sides of these relations are solutions of the corresponding inhomogeneous equations, i.e., they represent Green's functions.

Note now that these right-hand sides of the relations of the type (14) were defined only for $x^0 > y^0$ and $x^0 < y^0$. From the above-established property of covariance of T-products it now follows that they are defined everywhere except for $x = y$. The rules of integration of these expressions in the infinitesimal neighborhood of the point $x = y$ may be fixed arbitrarily. Thus, for example, to the right-hand side of each of them an arbitrary coefficient function of the quasilocal operator $P(\partial_x)\delta(x-y)$ may be added, where $P(\partial_x)$ is some polynomial in $\partial_x = \partial/\partial_x$. This necessity for an additional definition of pairing within the infinitesimal neighborhood of the point $x = y$ results from an ambiguity contained in the T-product. Actually, the T-products are given by the formal definition (14.20) and (12) only when the values of their arguments do not coincide.

17.4. Wick's second theorem. This theorem for time-ordered products states that

The T-product of n linear operators is equal to the sum of their normal products with all possible time-ordered pairings (including the term with no pairings).

The proof, in fact, reduces to the proof of Wick's theorem for ordinary products. Indeed, according to the definition (14.20) a T-product is always equal to a certain ordinary product:

$$T(A_1(x_1) \dots A_n(x_n)) = \eta A_{j_1}(x_{j_1}) \dots A_{j_n}(x_{j_n}).$$

By applying the first Wick's theorem to this ordinary product, we see that it is equal to the sum of the normal products of the operators $A_{j_1}, \dots, A_{j_n}$ with all possible ordinary pairings. But since the order in the sequence $A_{j_1}, \dots, A_{j_n}$ is already time-ordered, the ordinary pairings will coincide with the time-ordered ones, i.e., the given T-product is equal to the sum of the normal products of the operators $A_{j_1}, \dots A_{j_n}$ with all possible time-ordered pairings, multiplied by η.

As has been noted before, linear operators within a time-ordered pairing, as well as within a normal product, may be commuted (taking into account possible changes of sign). As a result, we may reestablish the normal order of the factors 1, 2, ... within normal products with all possible time-ordered pairings, at the same time leaving out the factor η. This completes the proof of the theorem.

Let us now consider the T-product of several normal products of linear field operators:

$$T(:A_1(x) A_2(x) \dots A_n(x): \dots :D_1(z) \dots D_m(z):). \tag{19}$$

T-products of just this form will be needed for the evaluation of T-products of local operators, since, by definition, the local operator $\mathscr{A}(x)$ is expressed as a linear combination of terms of the type $:A_1(x)A_2(x) \dots A_n(x):$. The fomulation of the second Wick's theorem for T-products of the type (19) has the sole peculiarity that mutual time-ordered pairings of operators occurring within the same normal products should not be taken into account.

17.5. Wick's third theorem. For applications the following assertion turns out to be useful:

The vacuum expectation value of the time-ordered product of $n + 1$ linear operators A, $B_1, \ldots, B_n$ is equal to the sum of n vacuum expectation values of these same time-ordered products with all possible pairings of one of these operators (for instance, A) with all the remaining ones, i.e.,

$$\langle TAB_1 \ldots B_n \rangle_0 = \sum_i \langle T\overline{AB_1 \ldots B_i} \ldots B_n \rangle_0. \tag{20}$$

This assertion may be called *Wick's theorem for vacuum expectation values* or Wick's third theorem.

We draw attention to the fact that the right-hand side of (20), unlike Wick's first two theoroms, does not contain expressions with more than one pairing.

Nevertheless, the validity of (20) follows directly from the second Wick's thoerem. As the vacuum expectation value of the normal product of an arbitrary nonzero number of unpaired operators vanishes, the left-hand part of (20) equals the sum of all possible variants of the "total mutual pairings" within the product of operators

$$AB_1 \ldots B_n, \tag{21}$$

i.e., of pairings, when all the operators are mutually paired with each other. In a precisely analogous way any of the terms of the sum in the right-hand part of (20)—for example, the first one—may be represented in the form

$$\overline{AB_1} \langle TB_2B_3 \ldots B_n \rangle_0$$

and is equal to the product of the pairing AB_1 with the sum of all possible total pairings of the operators $B_2, \ldots, B_n$.

Performing summation over i in the right-hand part of (20), we obtain the sum of all possible total pairings of operators (21). The third of Wick's theorems has thus been proved.

CHAPTER V

THE FEYNMAN RULES AND DIAGRAMS

18. GREEN'S FUNCTION FOR FREE FIELDS

At the end of the preceding chapter it was established that the operator terms of the S-matrix are reduced by Wick's second theorem to normal products with time-ordered pairings, which represent, as will now shown, Green's functions for free fields.

18.1. Green's function for a scalar field. First we shall show that the time-ordered pairing of operators of a scalar field may be expressed in terms of the Green's function of a specific form of the equation for the scalar field.

We define the Green's function for the scalar field G as the solution of the inhomogeneous Klein–Gordon equation

$$(\Box_x - m^2)\, G(x) = -\,\delta(x). \tag{1}$$

Here and henceforth for the sake of definiteness we shall adopt the convention of choosing the sign of the δ-function on the right-hand side of the equations for Green's functions in agreement with the sign of the mass term in the left-hand side.

By using the Fourier transform we obtain for G the following formal expression:

$$G(x) = \frac{1}{(2\pi)^4} \int \frac{e^{-ikx}}{m^2 - k^2}\, dk. \tag{2}$$

This expression is not actually defined until we have specified the way in which we go around the poles at $k^2 = m^2$. This indefiniteness is related to the fact that the complete (exact) solution of equation (1) may be written as a sum of the particular solution of the inhomogeneous equation and of solutions D^+ and D^- of the homogeneous equation with arbitrary coefficients. These coefficients may be uniquely determined by specifying the rules by which we go around the two poles at $k^2 = m^2$ or by imposing boundary conditions on G.

Let us show this for the retarded Green's function satisfying the boundary condition

$$D^{\text{ret}}(x) = 0 \quad \text{for} \quad x^0 < 0. \tag{3}$$

To represent it in a form similar to (2), we note that if this function is multiplied by $\exp(-\varepsilon x^0)$, where $\varepsilon > 0$, then in accordance with (3) it will not acquire any additional singularities:

$$D^{\text{ret}}(x)\, e^{-\varepsilon x^0} = G_\varepsilon(x),$$

as a result of which it may be represented in the form of the limit

$$D^{\text{ret}}(x) = \lim_{\varepsilon \to 0} G_\varepsilon(x). \tag{4}$$

In accordance with its definition, the function G_ε satisfies the equation

$$\left\{\Delta - \left(\frac{\partial}{\partial t} + \varepsilon\right)^2 - m^2\right\} G_\varepsilon(x) = -\delta(x)$$

and therefore in the momentum representation in the limit $\varepsilon \to 0$ it has the form

$$\frac{1}{m^2 - (k^0 + i\varepsilon)^2 + \boldsymbol{k}^2} \to \frac{1}{m^2 - k^2 - 2i\varepsilon k^0}.$$

Thus, in accordance with (4), the retarded Green's function may be represented in the form

$$D^{\text{ret}}(x) = \frac{1}{(2\pi)^4} \int \frac{e^{-ikx}}{m^2 - k^2 - 2i\varepsilon k^0}\, dk. \tag{5}$$

It may easily be verified that (5) actually satisfies the condition (3). For this it is sufficient to carry out explicit integration over the variable k^0 with the help of the theory of residues. The infinitesimal additional term in the denominator indicates that both poles in the complex k^0 plane must be situated above the path of integration (Figure 18.1). Therefore for $x^0 < 0$, when the

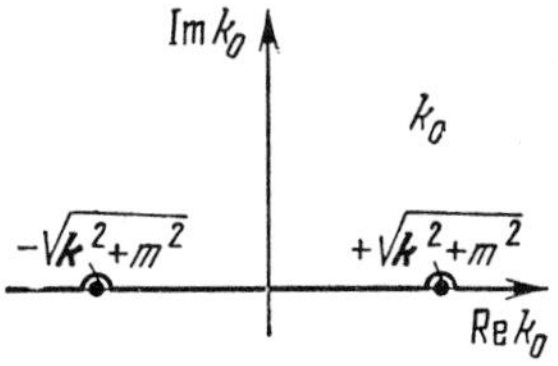

Fig. 18.1. The path of integration in the complex k^0-plane for the function D^{ret}.

contour of integration may be closed in the upper half plane with a semicircle of large radius, there are no poles inside it and we obtain zero. for $x^0 > 0$ the path of integration is closed in the lower half plane. In this case, by calculating the residues we find

$$D^{\text{ret}}(x) = \frac{1}{(2\pi)^3 i} \int \frac{e^{ik_0x_0} - e^{-ik_0x_0}}{2k_0} \bigg|_{k_0=\sqrt{\mathbf{k}^2+m^2}} e^{-i\mathbf{k}\mathbf{x}}\, d\mathbf{k} = D(x).$$

Therefore,

$$D^{\text{ret}}(x) = \theta(x^0)\, D(x). \tag{6}$$

In the same way it may be shown that the advanced Green's function, defined by the condition

$$D^{\text{adv}}(x) = 0 \quad \text{for} \quad x^0 > 0,$$

and satisfying equation (1), has the form

$$D^{\text{adv}}(x) = \frac{1}{(2\pi)^4} \int \frac{e^{-ikx}\, dk}{m^2 - k^2 + 2i\varepsilon k^0} = -\theta(-x^0)\, D(x). \tag{7}$$

18.2. The causal Green's function. The causal Green's function D^c plays a particularly important role in quantum field theory. It corresponds to the right-hand side of (17.16), and is proportional to $D^-(x-y)$ for $x^0 > y^0$, while for $x^0 < y^0$ it is proportional to $D^+(x-y)$.

It is not difficult to verify that this right-hand side may be represented in the form of the difference

$$\theta(x^0)\, D^-(x) - \theta(-x^0)\, D^+(x) = D^{\text{ret}}(x) - D^+(x)$$

of the solutions D^{ret} and D^+ of the inhomogeneous and homogeneous equations respectively, and therefore satisfies the same equation (1) as does D^{ret}. We therefore assume

$$D^c(x) = \theta(x^0)\, D^-(x) - \theta(-x^0)\, D^+(x). \tag{8}$$

To obtain an expression for the causal Green's function D^c in the momentum representation, we note that the difference $D^{\text{ret}} - D^+$ may be written in this representation as follows:

$$\frac{1}{m^2 - k^2 - 2i\varepsilon k^0} + 2\pi i\theta(-k^0)\,\delta(k^2 - m^2) =$$
$$= \frac{\mathscr{P}}{m^2 - k^2} + i\pi\delta(k^2 - m^2) = \frac{1}{m^2 - k^2 - i\varepsilon},$$

which gives

$$D^c(x) = \frac{1}{(2\pi)^4} \int \frac{e^{-ikx}}{m^2 - k^2 - i\varepsilon}\, dk \tag{9}$$

(here $\mathscr{P}$ denotes the principal value; see (AV.6)–(AV.8)). Since the commutation functions for the electromagnetic, the spinor, and the vector fields may be obtained from the Pauli–Jordan function with the aid of differential operations, the causal functions of these fields may be expressed in terms of $D^c(x)$ by the same relations

$$\begin{gathered} D_0^c(x) = D^c(x)\big|_{m=0}, \quad S_{\alpha\beta}^c(x) = (i\hat{\partial} + m)_{\alpha\beta}\, D^c(x), \\ D_{\mu\nu}^c(x) = \left(g_{\mu\nu} + \frac{1}{m^2}\partial_\mu \partial_\nu\right) D^c(x), \end{gathered} \tag{10}$$

where, for instance,

$$S_{\alpha\beta}^c(x) = \theta(x^0)\, S^-(x) - \theta(-x^0)\, S^+(x) =$$
$$= \frac{1}{(2\pi)^4} \int \frac{(m + \hat{p})_{\alpha\beta}}{m^2 - p^2 - i\varepsilon}\, e^{-ipx}\, dp. \tag{11}$$

We may now return to completing the definition of time-ordered pairings at the point $x = y$. By convention we shall agree to consider that if, when $x \neq y$, the time-ordered pairing coincides with a certain causal Green's function, then it also coincides with this function within an infinitesimal neighborhood of the point $x = y$. In other words: for the scalar field,

$$\overline{\varphi(x)\,\varphi}(y) = \langle T\varphi(x)\,\varphi(y)\rangle_0 = \frac{1}{i} D^c(x-y); \tag{12}$$

for the electromagnetic field,

$$\overline{A_\nu(x)\,A}_\mu(y) = \langle TA_\nu(x)\,A_\mu(y)\rangle_0 =$$
$$= ig_{\nu\mu} D_0^c(x-y) = \frac{g_{\nu\mu}}{(2\pi)^4 i} \int \frac{e^{ik(x-y)}}{k^2+i\varepsilon} dk; \tag{13}$$

for the vector field,

$$\overline{U_\nu(x)\,U}_\mu(y) = \langle TU_\nu(x)\,U_\mu(y)\rangle_0 = iD^c_{\nu\mu}(x-y) =$$
$$= \frac{i}{(2\pi)^4} \int \frac{g_{\nu\mu} - k_\nu k_\mu/m^2}{m^2-k^2-i\varepsilon} e^{ik(x-y)}\,dk; \tag{14}$$

and for the spinor field

$$\overline{\psi_\alpha(x)\,\bar{\psi}}_\beta(y) = \langle T\psi_\alpha(x)\,\bar{\psi}_\beta(y)\rangle_0 = \frac{1}{i} S^c_{\alpha\beta}(x-y) =$$
$$= \frac{1}{(2\pi)^4 i} \int \frac{(m+\hat{p})_{\alpha\beta}}{m^2-p^2-i\varepsilon} e^{-ip(x-y)}\,dp. \tag{15}$$

18.3. Singularities on the light cone. The Green's functions obtained above, as well as the commutation functions of the corresponding fields, may be expressed explicitly in terms of cylindrical functions and of singular functions (distributions) of the type of $\delta(x^2)$, $\theta(x^0)$, etc. We omit the corresponding computations and present the expressions for the functions of the scalar field.

The Pauli–Jordan commutation function is

$$D(x) = \frac{\varepsilon(x^0)\,\delta(\lambda)}{2\pi} - \frac{m}{4\pi\sqrt{\lambda}}\,\theta(\lambda)\,\varepsilon(x^0)\,J_1(m\sqrt{\lambda}). \tag{16}$$

Here $\lambda = x^2 = x_0^2 - \boldsymbol{x}^2$, while ε and θ are known step functions. From (16) a most important property of the commutation function follows directly—it vanishes outside of the light cone:

$$D(x)=0, \quad x^2<0. \tag{17}$$

The frequency components of the function D have the form

$$D^{\pm}(x)=\frac{\varepsilon(x^0)\,\delta(\lambda)}{4\pi}-\frac{m\theta(\lambda)}{8\pi\sqrt{\lambda}}\left[\varepsilon(x^0)\,J_1(m\sqrt{\lambda})\pm\right.$$
$$\left.\pm iN_1(m\sqrt{\lambda})\right]\mp\frac{\theta(-\lambda)\,mi}{4\pi^2\sqrt{-\lambda}}K_1(m\sqrt{-\lambda}). \tag{18}$$

Here J_1, N_1, and K_1 are respectively the Bessel, the Neumann, and the Hankel function of imaginary argument. In the neighborhood of zero they may be represented as the following expansions

$$J_1(z)=\frac{z}{2}-\frac{z^3}{16}+O(z^5),$$
$$\frac{\pi}{2}N_1(z)=-\frac{1}{z}+\frac{z}{2}\left(\ln\frac{z}{2}+1\right)+O(z^3),$$
$$K_1(z)=\frac{1}{z}+\frac{z}{2}\left(\ln\frac{z}{2}+1\right)+O(z^3).$$

The causal Green's function is

$$D^c(x)=\frac{mi}{4\pi^2}\,\frac{K_1(m\sqrt{-\lambda+i\delta})}{\sqrt{-\lambda+i\delta}}, \tag{19}$$

where

$$\sqrt{-\lambda+i\delta}=i\sqrt{\lambda} \quad \text{for} \quad \lambda>0.$$

The retarded Green's function is

$$D^{\text{ret}}(x)=\frac{\theta(x^0)\,\delta(\lambda)}{2\pi}-\frac{m\theta(x^0)\,\theta(\lambda)}{4\pi\sqrt{\lambda}}J_1(m\sqrt{\lambda}). \tag{20}$$

From the above formulas it follows that in the x-representation all of the singularities of the functions D, $D^{\pm}$, D^c, and D^{ret} are located only in the vicinity of the light cone. For example:

$$D(x) \simeq \frac{\varepsilon(x^0)\,\delta(\lambda)}{2\pi} - \frac{m^2\,\varepsilon(x^0)\,\theta(\lambda)}{8\pi} + O(\lambda), \tag{21}$$

$$D^c(x) \simeq \frac{\delta(\lambda)}{4\pi} + \frac{1}{4\pi^2 i\lambda} - \frac{m^2}{16\pi}\,\theta(\lambda) + \frac{im^2}{8\pi^2}\ln\frac{m\,|\lambda|^{1/2}}{2} + O(|\lambda|^{1/2}\ln|\lambda|). \tag{22}$$

These relations have four types of singularities: a pole $1/\lambda$, $\ln|\lambda|$, a δ-function $\delta(\lambda)$, and a discontinuity $\theta(\lambda)$. The factors $\varepsilon(x^0)$ and $\theta(x^0)$ in the individual terms do not introduce any additional singularities outside the light cone, as they occur together with $\delta(\lambda)$ or $\theta(\lambda)$. They manifest themselves only at the origin of the coordinate system.

For later use, we note that the functions for the scalar and vector fields have the following structure:

$$D_{\text{scal, vect}}(x) = m^2 F\left(m^2\lambda, \frac{x^0}{|x^0|}\right), \tag{23}$$

while the functions for the spinor field have the structure

$$S_{\text{spin}}(x) = m^3 \tilde{F}\left(m^2\lambda, \frac{x^0}{|x^0|}\right). \tag{24}$$

Owing to this, for example, singularities of the type of $\delta(\lambda)$ and λ^{-1} in (23) appear together with coefficients that do not depend on the mass m, while the singularities $\ln|\lambda|$ and $\theta(\lambda)$ appear with coefficients containing m^2. Thus, the commutation and causal functions of quantum fields are singular functions containing strong singularities on the light cone.

From a mathematical point of view such functions belong to the class of improper functions, or *distributions*. In contrast to ordinary functions, distributions are defined not by establishing the relation between the values of the function and those of its argument, but by defining rules for integrating their products with sufficiently regular functions. For example, the δ-function is characterized by the rule according to which its products with continuous functions are integrated, while the derivatives of the δ-functions are characterized respectively by the rules of integrating their products with differentiable functions, etc.

In other words, a singular function is defined by fixing the corresponding linear functional in an appropriate "linear space" of sufficiently regular functions. We cannot here go into any detailed discussion of the properties of distributions (for this purpose see the *Introduction*, Chapter III, as well as the book of Vladimirov (1976)) and so we draw attention only to the problem of multiplying distributions by one another. The point is that a distribution is defined by establishing the rules for integrating it only together with sufficiently regular functions, while from such rules the prescription for integrating the product of several distributions does not follow directly.

19. FEYNMAN DIAGRAMS

19.1. The coefficient functions. We now proceed to discuss the mathematical nature of the coefficient functions of the scattering matrix, i.e., the functions which appear as factors of the normal products of operators. These functions enter, in an obvious manner, into the matrix elements of the S-matrix to be considered in the next section. They may be obtained with the aid of Wick's second theorem and may be written in the form

$$K(x_1, \ldots, x_n) = \prod_{r,\, s} D^c_{\alpha\beta}(x_r - x_s), \tag{1}$$

where x_r and x_s take on values from the set $x_1, x_2, \ldots, x_n$. (We have in mind the coefficient function for $S_n(x_1, \ldots x_n)$.) As was established in Section 18, the factors on the right-hand side of (1) are distributions.

The problem of multiplying singular causal Green's functions represents the main difficulty of quantum field theory. This difficulty, which is called the problem of ultraviolet divergences, had retarded the development of the theory of particles for two decades when it was overcome at the end of the forties with the creation of the theory of renormalization. We shall come back to this question in the following chapter; for the present we shall bear in mind that expressions like (1) are, as a rule, defined purely formally.

We shall now turn to the construction of a convenient set of prescriptions for obtaining the individual terms in the sum contained in the operator expressions $S_n(x_1, \ldots, x_n)$. Such terms are represented as the product of a coefficient function of the form (1) with the normal product of the remaining unpaired operators:

$$K(x_1, \ldots, x_n) : \ldots u_k(x_k) \ldots u_j(x_j) \ldots : . \tag{2}$$

The construction of a particular function K corresponding to a given order of pairing of the operators may be performed with the help of the Feynman diagrams and rules.

19.2. The graphical representation of S_n. In accordance with Wick's second theorem, the operator functions $S_n(x_1, \ldots, x_n)$ may be represented by a sum of terms, each of which is a product of a certain number of pairings of the operator field functions with the normal product of the remaining unpaired operators of free fields. Each such term may be put into correspondence with a graphical picture made up according to very simple rules. This representation, which is called a *Feynman diagram,* contains, to begin with, n points $x_1, x_2, \ldots, x_n$ which are the vertices of the diagram describing the arguments of the operator $S_n(x_1, \ldots x_n)$. To each of the pairings there corresponds a line connecting the related vertices:

$$\overline{u_i(x_i)\, u_j}(x_j) \sim$$ 

.

To each unpaired operator there corresponds a line connecting the vertex under consideration with the edge of the diagram (a line with a free end):

$$:\ldots u_k(x_k) \ldots : \sim$$ 

.

The numerical and matrix factors occurring in the Lagrangian are made to correspond to the appropriate vertex. Clearly, as a result of this, each vertex, from a topological point of view, will be characterized by a complete set of lines leaving it (or entering it), each of which corresponds to one of the operators occurring in the given interaction Lagrangian. Thus, for the interaction Lagrangian in spinor electrodynamics we have

$$e : \bar{\psi}(x)\gamma^\nu \psi(x) A_\nu(x) : \sim$$ 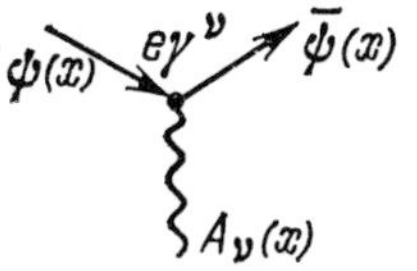

. (3)

To distinguish between various fields, we make wavy lines correspond to operators of the electromagnetic field, and solid lines correspond to operators of the Dirac field. Then, a line entering point x corresponds to the spinor $\psi(x)$, while a line leaving point x corresponds to the adjoint spinor $\bar{\psi}(x)$. Thus spinor lines (in contrast to electromagnetic ones) turn out to be directed. The graph (3) may be said to represent a visual description of the act of emission (or absorption) of a photon by an electron (or positron).

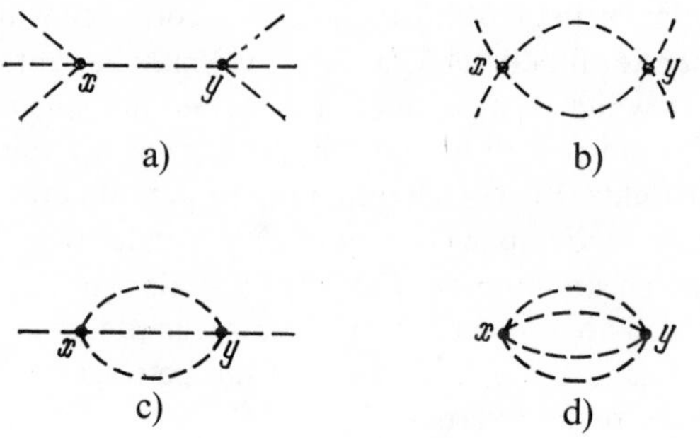

Fig. 19.1. Diagrams of second order in the ϕ^4 model.

A simpler example is given by the model of a scalar field with the fourfold interaction $h\varphi^4$. To such an interaction Lagrangian there corresponds a vertex with four identical lines:

$$h:\varphi^4(x): \quad \sim$$  (4)

As an illustration, we shall consider the term S_2 of this model:

$$S_2(x,\ y) = i^2h^2T\{:\varphi^4(x)::\varphi^4(y):\} = i^2h^2:\varphi^4(x)\varphi^4(y):+$$
$$+16i^2h^2\overline{\varphi(x)\varphi(y)}:\varphi^3(x)\varphi^3(y):+72i^2h^2[\overline{\varphi(x)\varphi(y)}]^2:\varphi^2(x)\varphi^2(y):+$$
$$+96i^2h^2[\overline{\varphi(x)\varphi(y)}]^3:\varphi(x)\varphi(y):+24i^2h^2[\overline{\varphi(x)\varphi(y)}]^4.$$

The corresponding diagrams are presented in Figure 19.1. In the more general case of $n > 2$ we obtain a set of diagrams with n vertices, involving various combinations of internal and external lines. The external lines of diagrams have a simple meaning. Figure 19.1b contains four free (external) lines which correspond precisely to the four unpaired operators occurring under the sign of the normal product. The corresponding terms of the S-matrix contribute to the matrix element of the transitions $1 \leftrightarrow 3$ and $2 \leftrightarrow 2$. Therefore external lines may be considered to be visual discriptions of the motion of the initial and final particles.

The internal lines correspond to virtual particles.

19.3. Spinor electrodynamics. More complicated cases describe the interaction of various fields having matrix structure. We shall now turn to spinor electrodynamics; see (3). In this case the operator expression of S_n will have the following form:

$$K_{\ldots\nu\ldots}(x_1, \ldots, x_n) : \ldots \bar{\psi}(x_k) \ldots A^{\nu}(x_l) \ldots \psi(x_m) \ldots : . \tag{5}$$

As is seen from this, the S-matrix transforms the state $|\,\alpha\rangle$ with a given set of quanta of the electromagnetic and spinor fields into states with other sets of quanta. Turning to (8.21), we see that when the initial state $|\,\alpha\rangle$ is a physical state (i.e., the time and longitudinal quanta either are absent in it or enter into it in certain combinations corresponding to (8.20)), then the final states $|\,\beta\rangle$ will also be physical ones.

Hence, in particular, it follows that relations like the condition of unitarity (15.9′) may be written in the form

$$\sum_{\tilde{\gamma}} S_{\beta\tilde{\gamma}} \overset{+}{S}_{\tilde{\gamma}\alpha} = I_{\beta\alpha},$$

where the summation on the left-hand side includes the states containing only physical components of the electromagnetic field. We recall that the reason for this is related to the fact that, as was emphasized in Section 4.3, the longitudinal degree of freedom $\chi(x)$ is connected neither with the transverse fields A^{tr} nor with any other ones (in the case under consideration the field ψ, $\bar{\psi}$).

The coefficient function $K_{\ldots\nu\ldots}(\ldots x \ldots)$ in (5) represents the product of time-ordered pairings of two types, which we shall by convention denote by different graphical symbols:

$\overline{A_\nu(x) A_\mu(y)}$ $\sim$ 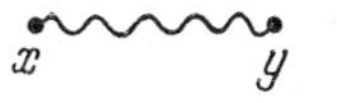 , (6)

$\overline{\psi(x)\bar{\psi}(y)}$ $\sim$  . (7)

The arrow in the representation of the second pairing serves to distinguish between the expressions $\psi(x)\bar{\psi}(y)$ and $\psi(y)\bar{\psi}(x)$.

For unpaired operators occurring under the sign of the normal product we assume

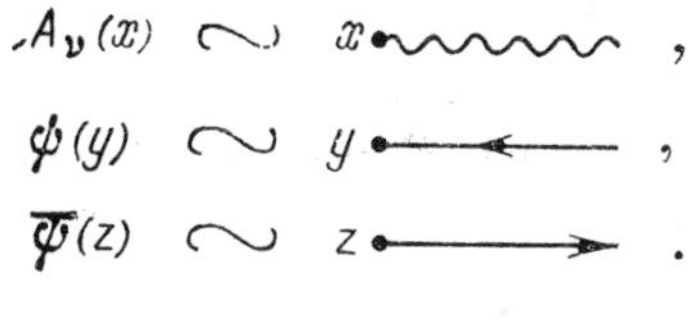

(8)

Thus, at each vertex one photon line and two spinor lines (one of them "incoming" and the other one "outgoing") will meet. Here it is necessary to recall the Dirac matrix which occurs in each of the Lagrangians $\mathscr{L}(x)$. By convention we shall make this matrix as well as the coupling constant e correspond to the vertex in a way similar to the one presented in (3):

$$e\gamma^{\nu} \sim$$ 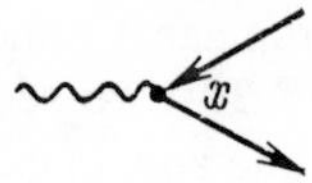 . (9)

The relations (6)–(9) form a complete set of rules of correspondence for constructing Feynman diagrams in spinor electrodynamics. These rules have been chosen so that the diagram corresponding to one of the terms occurring in S_n consists of n vertices and a certain number of internal and external photon and electron lines. Then one electron line enters and one electron line leaves each vertex. Thus, the electron lines of the complete diagram are continuous at the vertices and form either closed loops or open zigzag lines which begin and end at the edges of the diagram. The sequence of the arguments of the pairings

$$\overline{\psi}(x)\,\psi(x)\,\overline{\psi}(y)\,\psi(y)\,\overline{\psi}(z)\ldots\psi(u)\,\overline{\psi}(v)\,\psi(v)$$

determines the sequence of the vertices of the individual electron lines of the diagram, while the pairs of free operators $\overline{\psi}(x_j)\psi(x_i)$ within a normal product correspond to the beginning (x_i) and to the end (x_j) of individual electron lines which do not form loops.

As an example, we consider one of the second-order terms which occur in the expression for $S_2(x, y)$:

$$\begin{aligned} i^2e^2\overline{\psi}(x)\,\hat{A}(x)\,\psi(x)\,\overline{\psi}(y)\,\hat{A}(y)\,\psi(y) &= \\ &= e^2\,\mathrm{Sp}\,\{:iS^c(y-x)\,\hat{A}(x)\,iS^c(x-y)\,\hat{A}(y):\} = \\ &= -e^2\,\mathrm{Sp}\,\{:S^c(y-x)\,\hat{A}(x)\,S^c(x-y)\,\hat{A}(y):\}. \end{aligned} \quad (10)$$

Utilizing the rules of correspondence, we obtain the appropriate Feynman diagram, the so-called "photon self-energy" diagram presented in Figure 19.2.

As a second example, let us consider one of the third-order terms occurring in $S_3(x, y, z)$:

Fig. 19.2. Photon self-energy diagram in spinor electrodynamics.

$$(ie)^3 :\overline{\psi}(x)\,\hat{A}(x)\,\psi(x)\,\overline{\psi}(y)\,\hat{A}(y)\,\psi(y)\,\overline{\psi}(z)\,\hat{A}(z)\,\psi(z): =$$
$$= -e^3 D_0^c(x-z) :\overline{\psi}(x)\,\gamma^\nu S^c(x-y)\,\hat{A}(y)\,S^c(y-z)\,\gamma_\nu \psi(z):. \qquad (11)$$

In this case the rules of correspondence lead us to the diagram shown in Figure 19.3.

An examination of these simplest diagrams shows that motion along an electron line corresponds exactly to the order of the matrix elements from right to left in the corresponding term of the S-matrix. For example, in the case of the diagram shown in Figure 19.3 we obtain exactly the same order of noncommuting matrix factors,

$$:\overline{\psi}(x)\,\gamma^\nu S^c(x-y)\,\hat{A}(y)\,S^c(y-z)\,\gamma_\nu \psi(z):,$$

as in (11).

We note, further, that when Fermi fields are present, one must be very careful with the signs. Indeed, by definition the pairing $\overline{\psi}(x)\psi(y)$ differs in sign from the standard pairing $\psi(y)\overline{\psi}(x)$ which enters into the rules of correspondence:

$$\overline{\psi}(x)\,\psi(y) = -\,\psi(y)\,\overline{\psi}(x).$$

This change of sign is essential for diagrams containing closed electron loops. Thus, for example, according to the rules of correspondence the diagram in Figure 19.2 corresponds to the expression

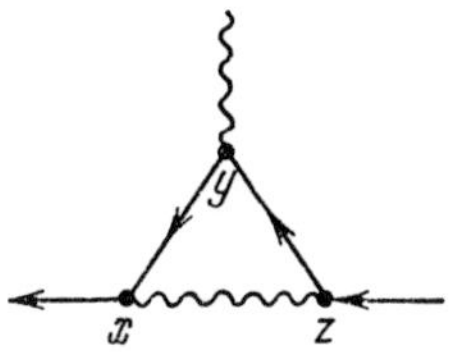

Fig. 19.3. Vertex diagram of the third order in spinor electrodynamics.

$$(ie)^2 \overline{\psi(x)\bar{\psi}}(y)\,\overline{\psi(y)\bar{\psi}}(x),$$

which differs in sign from (10). It is also evident that in the more general case such a difference will occur for any arbitrary group of pairings corresponding to each of the closed loops of the diagram, irrespective of the order of these loops. From this it follows that the expression which is obtained by means of the rules of correspondence must in addition be multiplied by

$$\eta=(-1)^c,$$

where c is the number of closed fermion loops in the diagram under consideration. We shall call this observation the *rule of signs.*

19.4. The Yang–Mills field. Of great interest are quantum-field models involving non-Abelian gauge fields and the implementation of their interaction with matter fields. As in the Abelian case (electrodynamics), such models turn out to be very economical in having structurally simple interaction terms and in the simplest cases involve only a single coupling constant g. This constant also enters into the Lagrangian of the "pure" Yang-Mills field (i.e., not interacting with other fields)

$$\mathscr{L}_{\mathrm{YM}}(B)=\mathscr{L}_0(B)+\mathscr{L}_1(g,\ B),$$

which together with the quadratic form $\mathscr{L}_0$ contains cubic and fourth-order terms as well:

$$\mathscr{L}_1(g,\ B)=2g\boldsymbol{F}_{\mu\nu}[\boldsymbol{B}_\mu\times\boldsymbol{B}_\nu]+g^2([\boldsymbol{B}_\mu\times\boldsymbol{B}_\nu])^2. \tag{12}$$

(Here and henceforth for definiteness an $SU(2)$ gauge theory will be considered.) The analogy with the example examined above of the nonlinear scalar field (4) suggests that we first quantize the Yang–Mills field in the linear approximation ($\mathscr{L}_{\mathrm{YM}}\to\mathscr{L}_0$) and then take into account the third- and fourth-order components from $\mathscr{L}_1$ with the help of perturbation theory in powers of the constant g.

The expression for $\mathscr{L}_0$ might then be chosen as a sum "over the components"

$$\mathscr{L}_0(B)=\sum_a\tilde{\mathscr{L}}_0(B^a) \tag{13}$$

of the quadratic nondegenerate forms similar to the Lagrangian of the electromagnetic field in the form of (4.13) or (4.19). However, as was pointed out in Section 11.1, the longitudinal components of the Yang-Mills field, contrary to the electromagnetic field, do not split away from the transverse ones. Owing to this the scattering matrix turns out not to be unitary in the space of physical states. This manifests itself in the following. The left-hand side of the condition of unitarity (15.9′) involves a summation

$$(S\overset{+}{S})_{\alpha\beta}=\sum_{\gamma}S_{\alpha\gamma}\overset{+}{S}_{\gamma\beta}=\sum_{\tilde{\gamma}}S_{\alpha\tilde{\gamma}}\overset{+}{S}_{\tilde{\gamma}\beta}+\sum_{\chi}S_{\alpha\chi}\overset{+}{S}_{\chi\beta} \tag{14}$$

over the intermediate states $|\gamma\rangle$ which include, besides the transverse three-dimensional states $|\tilde{\gamma}\rangle$ corresponding to massless transverse quanta of the Yang–Mills field (these states $|\tilde{\gamma}\rangle$ are referred to below as physical states), also the states $|\chi\rangle$ in which the scalar component $\chi(x)$ is present. In the Yang–Mills theory, in contrast to electrodynamics, owing to the fact that the function $\chi(x)$ satisfies equation (11.12) and therefore interacts with the transverse components, the matrix elements

$$S_{\chi\alpha}=\langle\chi|S|\alpha\rangle,$$

where $|\alpha\rangle$ represent physical states turn out to differ from zero. Therefore, to provide for unitarity in the space of physical states, it is necessary to modify the operators

$$S_n\to\tilde{S}_n$$

so that they obey the relations (16.4), in which matrix product on the right-hand side implies contraction over the physical states $|\tilde{\gamma}\rangle$ only:

$$\tilde{S}_n+\overset{+}{\tilde{S}}_n+\sum_k P(\ldots)\,\tilde{S}_k\overset{+}{\tilde{S}}_{n-k}=0,$$

$$(\tilde{S}_k\overset{+}{\tilde{S}}_{n-k})_{\alpha\beta}\equiv\sum_{\tilde{\gamma}}(\tilde{S}_k)_{\alpha\tilde{\gamma}}\,(\overset{+}{\tilde{S}}_{n-k})_{\tilde{\gamma}\beta}.$$

This results in the following: In all computations involving summation over intermediate states $|\gamma\rangle$ it is necessary to subtract from the sum the contribution of the states containing scalar quanta of the Yang–Mills field:

$$\sum_{\gamma}\to\sum_{\gamma}-\sum_{\chi}=\sum_{\tilde{\gamma}}. \tag{15}$$

A rigorous proof of the procedure under discussion was obtained by quantizing the Yang–Mills field with the aid of path integrals (see, for example, Chapter III of the book by Faddeev and Slavnov (1978)).

We shall now present elements of the diagram technique of the perturbation theory in powers of the interaction coupling constant g for the free Yang–Mills field.

In addition to the elements corresponding to the Yang–Mills fields: internal lines

$$\overline{B^a_\nu(x)\, B^b_\mu(y)} \sim$$ 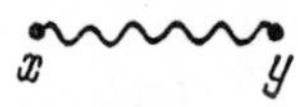 , (16)

external lines

$$B^a_\nu(x) \sim$$  , (17)

third-order vertices

$$gF_{\mu\nu}[B_\mu \times B_\nu]$$ 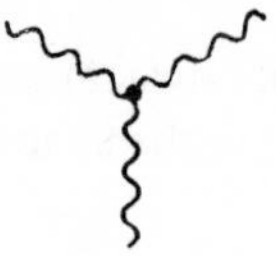 , (18)

and fourth-order vertices

$$g^2([B_\nu \times B_\mu])^2$$  , (19)

it is necessary to add elements corresponding to the subtracted longitudinal components. These components are described by the scalar massless isovector field ξ_a and are represented in the Feynman diagrams by internal lines of the field ξ:

$$\overline{\xi^a(x)\, \xi^b(x)}$$ 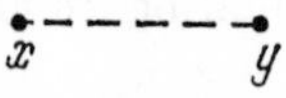

together with third-order vertices connecting the scalar field with the Yang-Mills field:

$$g\,\xi(x)[B_\nu \times \partial_\nu \xi]$$ 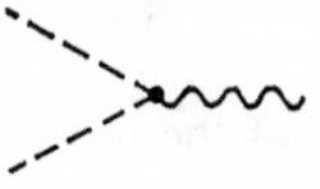 .

The auxiliary scalar field ξ is often called the "Faddeev–Popov ghost field".

The elements written out are conventionally related to the effective Lagrangian

$$\mathscr{L}_{\text{eff}} = -\frac{1}{4} F_{\mu\nu} F^{\mu\nu} + \frac{a}{2} (\partial B_\nu)^2 - \partial_\mu \bar{\xi}\, \partial^\mu \xi + g\bar{\xi}\, [B_\nu \times \partial^\nu \xi] \tag{20}$$

In this expression "complex" notation has been introduced for the ghost field $\xi \to (\xi, \bar{\xi})$. The point is that the closed loops of ghost lines correspond to contributions containing the "additional" factor (-1). Therefore, formally it is convenient to consider the scalar field ξ to be subjected to Fermi–Dirac quantization, and this is taken into account by the new notation

$$\overline{\xi(x)\bar{\xi}(y)} = -\overline{\bar{\xi}(y)\xi(x)}$$  , (21)

$$g\bar{\xi}(x)[B_\nu(x) \times \partial^\nu \xi(x)]$$  . (22)

The Feynman rules now include also the sign factor

$$\eta = (-1)^g, \tag{23}$$

where g is the number of closed ghost loops in the diagram.

For illustration we consider the terms of order g^2 in the scattering matrix $S_2(x, y)$, with two unpaired operators of the Yang–Mills field, similar to the expression (10). These terms appear, first, from various pairings of the operators occurring in the third-order vertices:

$$\begin{aligned} S_2(x, y) \sim i^2 g^2 \varepsilon_{abc} \varepsilon_{def} :\{ & F^a_{\nu\mu}(x) B^b_\nu(x) B^c_\mu(x) F^d_{\rho\sigma}(y) B^e_\rho(y) B^f_\sigma(y) + \\ & + F_{\nu\mu}(x) B^b_\nu(x) B^c_\mu(x) F^d_{\rho\sigma}(y) B^e_\rho(y) B^f_\sigma(y) + \\ & + F^a_{\nu\mu}(x) B^b_\nu(x) B^c_\mu(x) F^d_{\rho\sigma}(y) B^e_\rho(y) B^f_\sigma(y) \}:, \end{aligned} \tag{24}$$

and, second, from terms involving interaction of the ghost field ξ with the Yang–Mills field:

$$S_2(x, y) \sim ig^2 \varepsilon_{abc} \varepsilon_{def} :\bar{\xi}^a(x)\, \partial_\nu \xi^b(x)\, B^c_\nu(x)\, \bar{\xi}^d(y)\, \partial_\mu \xi^e(y)\, B^f_\mu(y):. \tag{25}$$

Up to obvious modifications of the pairings occurring under the sign of derivatives—for example,

$$\partial_\nu \overline{\xi^a(x)\,\xi^b}(y) = \frac{\delta_{ab}}{i(2\pi)^4}\frac{\partial}{\partial x^\nu}\int \frac{e^{-ik(x-y)}}{-k^2-i\varepsilon}\,dk = \frac{\delta_{ab}}{(2\pi)^4}\int \frac{e^{-ik(x-y)}k_\nu\,dk}{k^2+i\varepsilon} \tag{26}$$

—these terms correspond to the two Feynman diagrams shown in Figure 19.4.

We refer the reader to Appendix VIII where the Feynman rules for Yang—Mills fields interacting with matter fields may be found.

20. THE FEYNMAN RULES IN THE p-REPRESENTATION.

20.1. Transition to the momentum representation. We shall now proceed to consider the evaluation of matrix elements of the scattering matrix which are needed for obtaining transition probabilities. The evaluation of matrix elements is conveniently carried out in the momentum representation, since it is in this representation that state amplitudes are usually written down:

$$\Phi_{\dots k \dots} = a_1^+(\boldsymbol{k}_1)\dots a_s^+(\boldsymbol{k}_s)\,\Phi_0. \tag{1}$$

To obtain matrix elements of the nth-order terms of the S-matrix

$$S_{(n)} = \int dx_1 \dots dx_n\, K(x_1, \dots, x_n) : \dots u_\alpha(x_i)\dots u_\beta(x_j)\dots : \tag{2}$$

for transitions between states like (1), it is necessary to calculate the expressions

$$\overset{*}{\Phi}_{\dots k' \dots} : \dots u_\alpha \dots u_\beta \dots : \Phi_{\dots k \dots} = \overset{*}{\Phi}_f : \dots u_\alpha \dots u_\beta \dots : \Phi_{in}. \tag{3}$$

The creation operators $u^+(x)$ occurring in the normal product must then be commuted with the annihilation operators $a^-(\boldsymbol{p})$ from the amplitude $\overset{*}{\Phi}_f$, while the annihilation operators $u^-(x)$ are to be commuted with the creation operators $a^+(\boldsymbol{k})$ from Φ_{in} until one of the u^+ acts on $\overset{*}{\Phi}_0$ or one of the u^- happens to be next to Φ_0, which will yield zero. The matrix element (3) may turn out to

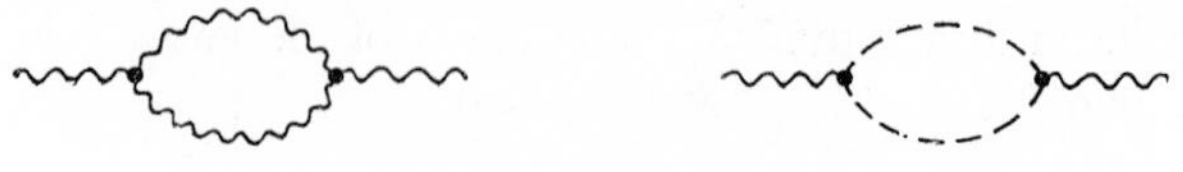

(a) (b)

Fig. 19.4. Diagrams of second order with two external vector lines for the Yang–Mills field in vacuum.

differ from zero, if for each operator $u(x)$ occurring in the normal product one can find an operator a^+ in Φ_{in} or a^- in $\overset{*}{\Phi}_f$ corresponding to the same field which will "cancel" the operator u as a result of being commuted with it. Thus (3) will differ from zero in the case when the sum of the number of particles of each field in the initial state Φ_{in} and in the final state $\overset{*}{\Phi}_f$ will be exactly equal to the number of operator functions of the given field in the normal product. The matrix element (3) turns out also to differ from zero in the case when in addition to the operators which "cancel" the normal product, $\overset{*}{\Phi}_f$ and Φ_{in} also contain operators which cancel one another. The total number of particles in the states $\overset{*}{\Phi}_f$ and Φ_{in} will then exceed the number of operators in the normal product by some even number. However, such matrix elements differ from zero only in the case when the momenta of the above "extra" particle are the same in states $\overset{*}{\Phi}_f$ and Φ_{in}, i.e., the "extra" particles do not experience interaction.

By restricting ourselves to the case when for no particles does the momentum in the initial state equal the momentum in the final state, we arrive at the conclusion that the matrix element (3) may be represented by a product of the results of commuting the operators

$$u^-(x_j) = (2\pi)^{-3/2} \int e^{-ipx_j} \sum_\sigma v_\sigma^-(\boldsymbol{p})\, a_\sigma^-(\boldsymbol{p})\, d\boldsymbol{p} \quad \text{with} \quad a^+(\boldsymbol{k}_j)$$

and

$$u^+(x_l) = (2\pi)^{-3/2} \int e^{ipx_l} \sum_\sigma v_\sigma^+(\boldsymbol{p})\, a_\sigma^+(\boldsymbol{p})\, d\boldsymbol{p} \quad \text{with} \quad a^-(k_f).$$

Thus, after the commutations have been carried out explicitly, the matrix element (3) is in fact expressed in the form of the product

$$\prod(\ldots x \ldots) = \prod_j \left\{(2\pi)^{-3/2} e^{-ik_j x_j} v^-(\boldsymbol{k}_j)\right\} \prod_f \left\{(2\pi)^{-3/2} e^{ik_f x_f} v^+(\boldsymbol{k}_f)\right\}, \tag{4}$$

where the factors occurring in $\prod_j$ correspond to particles in the initial state, while the factors occurring in $\prod_f$ correspond to particles in the final state.

Coming back to the initial matrix element of the S-matrix

$$\overset{*}{\Phi}_f S_{(n)} \Phi_{in} = \int dx_1 \ldots dx_n\, K(x_1, \ldots, x_n) \prod(\ldots x \ldots), \tag{5}$$

we may now perform integration over the variables $x_1, \ldots, x_n$. Since the dependence of the product $\prod$ is purely exponential, the matrix element, up to factors $v^-(\boldsymbol{k})$, $v^+(\boldsymbol{k})$ and powers of 2π, may be reduced to the Fourier transform of the coefficient function $K(x_1, \ldots, x_n)$. Since, in turn, this function may be expressed by the product of pairings, its Fourier transform is represented by the integral contraction of the Fourier transforms of the pairings.

Substituting the integral representations of the pairings

$$\frac{1}{i} D^c_{\alpha\beta}(x-y) = (2\pi)^{-4} \int e^{-iq(x-y)} \Delta_{\alpha\beta}(q)\, dq$$

into the right-hand side of (5), we see that the integration over the configuration variables now reduces to the evaluation of integrals of the form

$$\int dx_j e^{ix_j\left(\sum_\alpha p_\alpha\right)} = (2\pi)^4 \delta\left(\sum p_\alpha\right). \tag{6}$$

20.2. Feynman's rules for the evaluation of matrix elements. Summing up the above arguments, we arrive at the Feynman diagrams and rules in the momentum representation.

To evaluate the matrix element of the transition $s \to r$ (i.e. of the process in which s particles of the initial state transform into r particles of the final state) of the nth order, it is necessary first to draw the corresponding diagram according to the rules of the preceding section. On this diagram one must arrange the momentum variables, i.e., ascribe to each of the internal and external lines its four-momentum. One must distinguish between external lines describing incoming particles (the initial state) and outgoing particles (the final state) as well as take into account all topologically nonequivalent possibilities.

Thereafter the evaluation of the matrix element is carried out in accordance with the following rules:

(1)A factor $(2\pi)^{-3/2} v_\sigma^-(p_{in})$ is made to correspond to each external line of the initial state with the four-momentum p_{in};

(2) a factor $(2\pi)^{-3/2} v_\rho^+(\boldsymbol{k}_f)$ is made to correspond to each external line of the final state with the four-momentum k_f;

(3) to each vertex there corresponds a factor

$$igO_\alpha (2\pi)^4 \delta\left(\sum p_i\right),$$

where gO_α are the coupling constant and matrix from the corresponding interaction Lagrangian, the arguments of the δ-function contain the algebraic sum of the momenta of all the lines (both internal and external) meeting at the given vertex, and the factor i allows for the factor i^n of S_n in the expression for S;

(4) to each internal line with the four-momentum q_j we assign the factor

$$(2\pi)^{-4}\,\Delta_{\alpha\beta}\,(q_j),$$

where

(5) integration is carried out over all the arguments q_j;

(6) in case there are Fermi fields present, it is necessary to take into account the rule of signs and to multiply the whole expression by the factor η (see Section 19.3);

(7) the expression obtained in this manner must further be multiplied by a numerical factor—the *symmetry factor* c_n. At the end of the following section we shall discuss this factor.

20.3. An illustration for the ϕ^4 model. As an illustration, let us consider the contribution to the matrix element of the 2 → 2 scattering corresponding to the diagram in Figure 19.1b. We denote the four-momenta of the initial particles by the symbols p_1, p_2, and the momenta of the final particles by k_1, k_2.

There are three topologically nonequivalent ways of ascribing the momenta p_i, k_f to the external lines. They are presented in Figure 20.1a, b, c. The momenta of the internal lines are denoted by the symbols q_j, while the short arrows correspond to the algebraic signs of the respective momenta.

Using the above-formulated rules and taking into account that for the scalar field

$$v^-(p) = v^+(p) = (2p_0)^{-1/2}, \quad \Delta(q) = \frac{1}{i\,(m^2 - q^2 - i\varepsilon)},$$

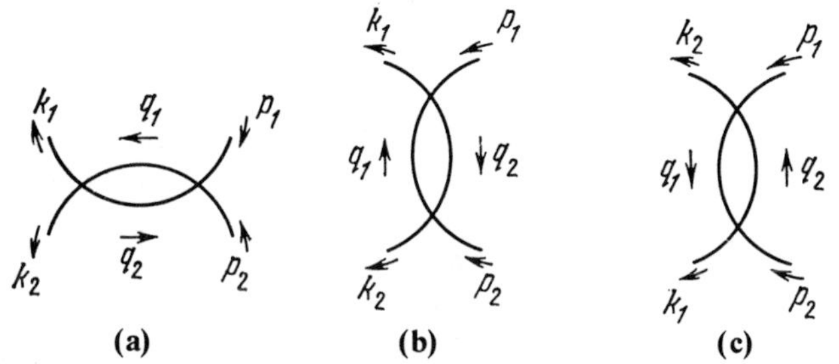

Fig. 20.1. Diagrams of second-order scattering in the φ^4 model in the momentum representation.

we obtain for the diagram in Figure 20.1a the following:

$$\mathcal{M}_a = [(2\pi)^{12}\, 2p_1^0 2p_2^0 2k_1^0 2k_2^0]^{-1/2} \int \frac{dq_1}{(2\pi)^4\, i\,(m^2 - q_1^2)} \times$$
$$\times \int \frac{dq_2}{(2\pi)^4\, i\,(m^2 - q_2^2)}\, i^2 h^2 (2\pi)^8\, \delta(p_1 + p_2 + q_2 - q_1)\, \delta(q_1 - k_1 - k_2 - q_2) =$$
$$= \frac{h^2 \delta(p_1 + p_2 - k_1 - k_2)}{(2\pi)^6 (16 p_1^0 p_2^0 k_1^0 k_2^0)^{1/2}} \int \frac{dq_1}{(m^2 - q_1^2)\,[m^2 - (q_1 - k_1 - k_2)^2]}.$$

For the diagrams in Figure 20.1b,c, respectively, we get

$$\mathcal{M}_b = \frac{h^2 \delta(p_1 + p_2 - k_1 - k_2)}{(2\pi)^6 (16 p_1^0 p_2^0 k_1^0 k_2^0)^{1/2}} \int \frac{dq_1}{(m^2 - q_1^2)\,[m^2 - (q_1 - k_1 + p_1)^2]},$$
$$\mathcal{M}_c = \frac{h^2 \delta(p_1 + p_2 - k_1 - k_2)}{(2\pi)^6 (16 p_1^0 p_2^0 k_1^0 k_2^0)^{1/2}} \int \frac{dq}{(m^2 - q^2)\,[m^2 - (q - k_1 + p_2)^2]}.$$

To obtain the final contribution to the scattering matrix element $\overset{*}{\Phi}_f S \Phi_i$, it is necessary to take into account also the symmetry factor. For its determination we consider possible variants of commutations and pairings of operators in the matrix element being investigated:

$$\overset{*}{\Phi}{}_2' S \Phi_2 = \frac{i^2 h^2}{2!} \int dx\, dy\, \langle a_1^- a_2^- T\, [\varphi^4(x)\, \varphi^4(y)]\, a_1^+ a_2^+ \rangle_0. \tag{7}$$

The following ordering of commutations corresponds to the diagram in Figure 20.1a:

$$\langle a_1^- a_2^- \varphi_x \varphi_x \varphi_x \varphi_x \varphi_y \varphi_y \varphi_y \varphi_y a_1^+ a_2^+ \rangle_0.$$

8 3 4 3 3

Here near each bracket denoting the type of commutation we include its combinatorial weight. We start by considering the annihilation operator a_1^-. It may act on any of the eight operators occurring in the product $\phi_x^4 \phi_y^4$. Therefore the "weight" of the first contraction $a_1^- \phi_x$ is equal to 8. Since the topology of Figure 20.1a requires the operator a_2^- to act on the same vertex as does a_1^-, the weight of the second contraction $a_2^- \phi_x$ is equal to 3, and so on.

It is also necessary to take into account the weight of the time-ordered pairings of the remaining operators. It is easy to see that

$$\langle T :\varphi_x \varphi_x: :\varphi_y \varphi_y: \rangle_0 = 2\, (\overline{\varphi_x \varphi_y})^2. \tag{8}$$

Collecting the factors, we obtain

$$c_{(2)}^{a} = \frac{1}{2!}\, 8\cdot 3\cdot 4\cdot 3\cdot 2 = \frac{(4!)^2}{2}.$$

Analogous arguments for the diagrams in Figure 20.1b,c give

$$c_{(2)}^{\delta} = c_{(2)}^{b} = c_{(2)}^{a} = (4!)^2/2.$$

Thus, the matrix element for 2 → 2 scattering in second order perturbation theory for the $h\phi^4$ model will be

$$\mathcal{M}_2 = c_{(2)}^{a}\,\mathcal{M}_a + c_{(2)}^{b}\,\mathcal{M}_b + c_{(2)}^{c}\,\mathcal{M}_c =$$

$$= \frac{\delta\,(p_1+p_2-k_1-k_2)\,18h^2}{(p_1^0 p_1^0 k_1^0 k_2^0)^{1/2}\,(2\pi)^4\, i}\,[I\,(k_1+k_2) + I\,(k_1-p_1) + I\,(k_1-p_2)], \tag{9}$$

where

$$I\,(k) = \frac{i}{\pi^2}\int \frac{dq}{(m^2-q^2-i\varepsilon)\,[m^2-(q-k)^2-i\varepsilon]}\,. \tag{10}$$

We shall now formulate the prescription for obtaining the symmetry factor:

(1) The nth term of the expansion of the S-matrix contains the factor $(n!)^{-1}$. This factor is always compensated by the factor $n!$, which takes into account permutations of the vertices $x_1, \ldots, x_n$, which due to integration over $dx_1, \ldots, dx_n$ are all equivalent.

(2) In the case when the interaction Lagrangian contains k identical operator factors (in the case under consideration $k = 4$), if the orders of contraction and of time-ordered pairings are established, since we go through the symmetric possibilities, we must ascribe to each vertex a factor equal to $k!$.

(3) In case there exist ν topologically equivalent variants of pairing of internal lines, they must be taken into account only once. this corresponds to the introduction of the factor $(\nu!)^{-1}$.

20.4. Spinor electrodynamics. We now present a summary of the Feynman rules for spinor electrodynamics (Table 2). These rules must be supplemented by the rule of signs and by taking into account the symmetry factor.

As was pointed out, the diagrams introduced actually describe several different processes simultaneously. Consider, for example, the diagram

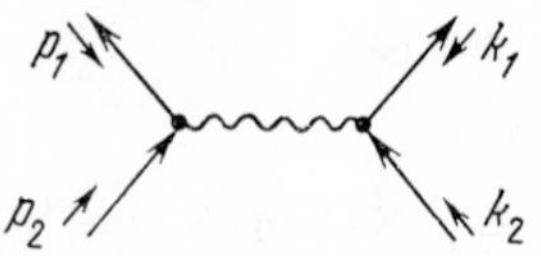

Fig. 20.2. Second-order diagram of spinor electrodynamics with four external fermion lines.

depicted in Figure 20.2. For reasons of symmetry all the incoming momenta are conventionally directed inwards.

If we assume that the zero components of the four-vectors p_1 and p_2 are positive, and that those of k_1 and k_2 are negative ($p_1^0, p_2^0 > 0$; $k_1^0, k_2^0 < 0$), then the diagram will correspond to the one-photon annihilation of an electron $e^-(p_2)$ and a positron $e^+(p_1)$ and the subsequent creation of a pair $e^-(K_2 = -k_2)$, $e^+(K_1 = -k_1)$. This process is represented in Figure 20.3a, where the time axis is directed from left to right.

If, on the other hand, we assume that $p_2^0, k_2^0 > 0$ and that $p_1^0, k_1^0 < 0$, we will then obtain the electron–electron scattering process involving the exchange of a single photon (the Möller scattering). In this case it is convenient to introduce the notation $K_1 = -k_1$, $P_1 = -p_1$ and to turn the diagram 90° clockwise. The time in Figure 20.3b thus obtained will flow from left to right.

The case of p_2^0, $k_1^0 > 0$, p_1^0, $k_2^0 < 0$ corresponds to the Möller electron–positron scattering. Denoting $P_1 = -p_1$ and $K_2 = -k_2$, we obtain Figure 20.3c. The matrix element of diagram (c) must be added to the matrix element of diagram (a).

Thus, two different diagrams, Figure 20.3a and c, contribute to the same physical process

$$e^+(p_1) + e^-(p_2) \to e^+(K_1) + e^-(K_2) \tag{11}$$

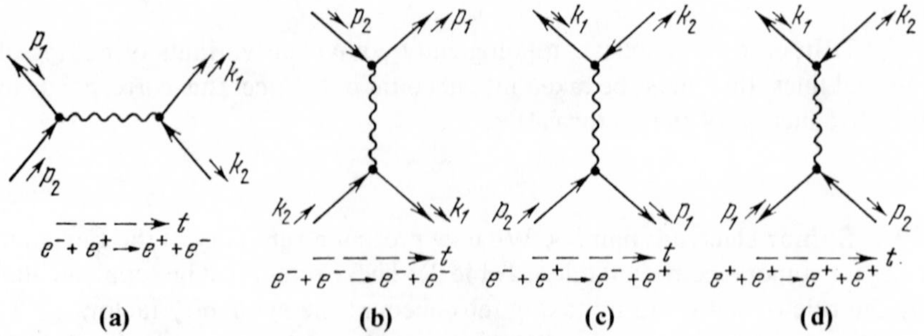

Fig. 20.3. Various physical processes described by the diagram in Figure 20.2.

Table 2. Feynman's rules for matrix elements in spinor electrodynamics.

	Particle and its state	Factor in the matrix element
1	Electron in the initial state with momentum $\boldsymbol{p}$ 	$(2\pi)^{-3/2}\, v^{s,-}(p)$
2	Positron in the initial state with momentum $\boldsymbol{p}$ 	$(2\pi)^{-3/2}\, \bar{v}^{s,-}(p)$
3	Electron in the final state with momentum $\boldsymbol{p}$ 	$(2\pi)^{-3/2}\, \bar{v}^{s,+}(p)$
4	Positron in the final state with momentum $\boldsymbol{p}$ 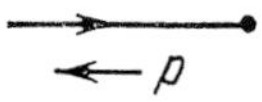	$(2\pi)^{-3/2}\, v^{s,+}(p)$
5	Photon in the initial or final state with polarization e_ν and momentum $\boldsymbol{k}$. 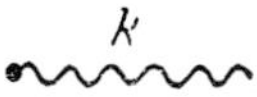	$\dfrac{e^{\nu}_{\mu}}{(2\pi)^{3/2}\sqrt{2k_0}}$ $(\nu \neq 0)$
6	Motion of an electron from 1 to 2 (or of a positron from 2 to 1). 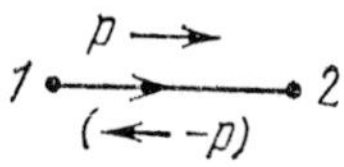	$\dfrac{1}{(2\pi)^4 i} \displaystyle\int \frac{m+\hat{p}}{m^2-p^2-i\varepsilon}\, dp$

Table 2 *(continued)*

	Particle and its state	Factor in the matrix element
7	Motion of a photon between vertices with summation indices μ and ν. 	$\frac{g^{\mu\nu}}{(2\pi)^4 i} \int \frac{dk}{k^2+i\varepsilon}$
8	Vertex with summation index ν with electron ling p_1 and photon line k incoming, and electron line p_2 outgoing. 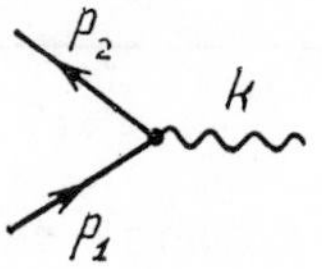	$ie\gamma^{\nu} (2\pi)^4 \delta (p_2 - p_1 - k)$

(it is only necessary to take into account that in Figure 20.3c the four-vectors p_1 and K_1 are denoted by k_1 and P_1). The total matrix element of this process in the approximation considered (second-order perturtabion theory) is equal to the sum of two terms

$$\mathcal{M}\,(11) = \mathcal{M}\ (20.3\text{a}) + \mathcal{M}\ (20.3\text{c}), \tag{12}$$

corresponding to these two diagrams.

Finally, the case $k_1^0, p_1^0 > 0; p_2^0, k_2^0 < 0$ corresponds to the Möller positron–positron scattering (see Figure 20.3d). To obtain the scattering matrix element for the latter, it is necessary to take into account the antisymmetry of physical states under permutation of the two identical Fermi particles, the positrons.

In the evaluation of the quantity

$$\int dx\, dy \,\langle a^- (k_2)\, a^- (q_2)\, \bar{\psi}\, (x)\, \hat{A}\, (x)\, \psi\, (x)\, \bar{\psi}\, (y)\, \hat{A}\, (y)\, \psi\, (y)\, a^+ (k_1)\, a^+ (q_1)\rangle$$

there arise two alternatives for the "contraction" of the Fermi operators represented in Figure 20.4. The matrix element of the process

$$e^+ (k_1) + e^+ (q_1) \to e^+ (k_2) + e^+ (q_2) \tag{13}$$

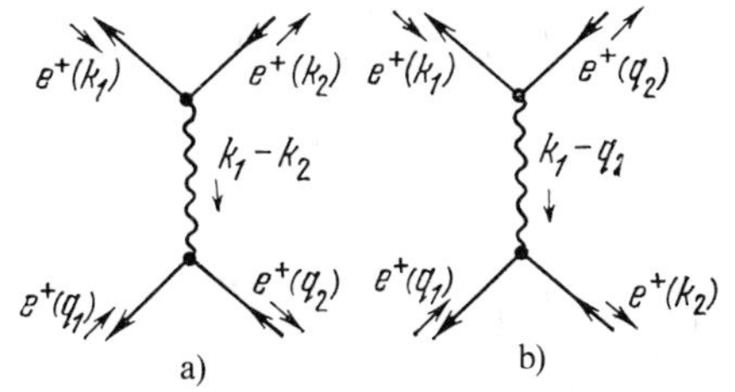

Fig. 20.4. The direct and exchange diagrams of the Möller positron scattering.

will now be represented by the difference

$$\mathcal{M}\,(13) = \mathcal{M}\;(20.4\mathrm{a})\;-\mathcal{M}\;(20.4\mathrm{b}) \tag{14}$$

between the matrix elements corresponding to the diagrams in Figure 20.4, which provides for the aforementioned antisymmetry and for fulfillment of the Pauli principle.

In conclusion we point out that in the examples presented above there occur propagators of the scalar, spinor, and electromagnetic fields which possess to internal symmetry. In these examples, also, no vector field with mass is present. To complete the picture, we therefore present also the expressions in the momentum representation for the propagator of the pion field

$$\int \overline{\varphi_a(x)\,\varphi_b}(y)\,e^{ik(x-y)}\,dx = \frac{\delta_{ab}}{i\,(m^2-k^2-i\varepsilon)} \tag{15}$$

and for the propagator of the massive vector field (compare with (18.14)):

$$\int \overline{U_\nu(x)\,U_\mu}(y)\,e^{ik(x-y)}\,dx = i\,\frac{g_{\nu\mu}-k_\nu k_\mu/m^2}{m^2-k^2-i\varepsilon}, \tag{16}$$

21. TRANSITION PROBABILITIES

21.1. The general structure of matrix elements. As was established in the example of Section 20.3, the transition matrix element contains a singular factor in the form of a δ-function corresponding to the conservation law of the total energy-momentum four-vector of the initial and final particles. It is not difficult to verify that from the Feynman rules for the matrix elements (see

Section 20.2, item 3) it follows that the matrix element of the transition $s \to r$ is always proportional to the four-dimensional δ-function

$$\delta\left(\sum_i^s p_i - \sum_f^r p_f'\right). \tag{1}$$

This fact reflects the general property of invariance under translations of the scattering matrix. If the interaction process takes place in the external fields, then the corresponding homogeneity is violated and the factor (1) partly or completely vanishes. For example, in the case of scattering in an external field depending on the space coordinates and independent of the time, we obtain, instead of (1), the one-dimensional δ-function

$$\delta\left(\sum_i p_i^0 - \sum_f p_f'^0\right)$$

representing the energy conservation law.

Returning to the general case of interaction in vacuum, we obtain

$$\overset{*}{\Phi}_{p'} S \Phi_p = \delta\left(\sum p - \sum p'\right) F(p', p), \tag{2}$$

The function F is described by a set of Feynman diagrams with a number of external lines equal to $s + r$.

We now introduce some useful definitions. We shall call a diagram G *connected* if it cannot be decomposed into parts which are not joined together by lines. If, on the other hand, the diagram G may be decomposed into connected "parts" $G_1, G_2, \ldots,$ we then call G *disconnected*, while its "pieces" G_i are called its connected components. A connected diagram is called *weakly connected* (or one-particle reducible) if by opening one line it may be changed into an disconnected one, and we shall say that it is *strongly connected* (one-particle irreducible) if this is not possible.

The contribution of each of the connected components of a disconnected diagram to the matrix element (1) involves a δ-function. The contributions of connected diagrams contain one δ-function each. Therefore in the case of connected diagrams the function F on the right-hand side of (2) is smooth, i.e., does not contain as factors any other δ-functions.

To determine the transition probabilities, it is necessary to evaluate the squares of matrix elements and to deal with the square of a δ-function. We must here take into account that, as in ordinary quantum mechanics, we are interested in transition probabilities per unit time interval and unit space volume. Taking further into account that the four-dimensional momentum δ-

function appears as a result of integration over the four-dimensional volume VT ($V \to \infty$, $T \to \infty$), we find that

$$[\delta(p)]^2 = \delta(p)\,\delta(0) = \delta(p)\,(2\pi)^{-4}\int dx = \delta(p)\,(2\pi)^{-4}\,VT.$$

We therefore have

$$|\overset{*}{\Phi}_{p'} S \Phi_p|^2 = (2\pi)^{-4}\,VT\delta(\Sigma p - \Sigma p')\,|F(p', p)|^2. \tag{3}$$

Analogous reasoning for the transition probability in a static external field A_{ext} yields

$$|\overset{*}{\Phi}_{p'} S(A_{\text{ext}})\,\Phi_p|^2 = (2\pi)^{-1}\,T\delta(\Sigma p_0 - \Sigma p_0')\,|F(p', p;\; A_{\text{ext}})|^2. \tag{4}$$

This formula is entirely equivalent to the corresponding expressions for transition probabilities in nonrelativistic quantum mechanics (see, for example, Section 84 of the textbook by Blokhintsev (1976) and Section 42 in Landau and Lifshitz (1974)). The singular factor T vanishes when one deals with the transition probability per unit time interval. The existence of another singular factor V in (3) is a consequence of the fact that we are dealing with plane waves. this factor disappears from the expression for the transition probability normalized per unit volume.

21.2. Normalization of the state amplitude. Let us now investigate the problem of normalizing state amplitudes of the form

$$\Phi_k = a_1^+(\boldsymbol{k}_1) \ldots a_s^+(\boldsymbol{k}_s)\,\Phi_0.$$

Since such an amplitude corresponds to particles with strictly fixed momenta, it corresponds to plane waves, which do not possess a finite norm in coordinate space. However, a plane wave, and, consequently the amplitude Φ_k, can be normalized per unit volume.

In order to specify precisely the limiting transition $V \to \infty$, we shall consider a single-particle state, and we shall introduce a packet of plane waves

$$\Phi_1 = \int \chi_\sigma(\boldsymbol{k})\,a_\sigma^+(\boldsymbol{k})\,d\boldsymbol{k}\,\Phi_0. \tag{5}$$

The norm of this state is equal to

$$\overset{*}{\Phi}_1 \Phi_1 = \int |\chi_\sigma(\boldsymbol{k})|^2\,d\boldsymbol{k} = N.$$

By setting $N = 1$ we determine that the expression

$$|\chi_\sigma(\boldsymbol{k})|^2 \, d\boldsymbol{k}$$

gives the probability that the particle characterized by the internal quantum number σ has its momentum within the interval $d\boldsymbol{k}$ about the average value $\boldsymbol{k}$. The function $\chi(\boldsymbol{k})$ itself is thus the wave function of the particle in the momentum representation. Therefore its Fourier transform

$$f_\sigma(\boldsymbol{x}) = (2\pi)^{-3/2} \int e^{i\boldsymbol{k}\boldsymbol{x}} \chi_\sigma(\boldsymbol{k}) \, d\boldsymbol{k} \tag{6}$$

is the wave function in the configuration representation with a norm equal to

$$\int |f_\sigma(\boldsymbol{x})|^2 \, d\boldsymbol{x} = \int |\chi_\sigma(\boldsymbol{k})|^2 \, d\boldsymbol{k} = N.$$

In the case of $N = 1$ one may interpret the quantity

$$|f(\boldsymbol{x})|^2 \, d\boldsymbol{x} \tag{7}$$

as the probability in configuration space. By setting N equal to the number of particles (much larger than unity) we find that the expression (7) is the average number of particles within the infinitesimal volume element $d\boldsymbol{x}$.

Here it is appropriate to recall that in accordance with the general statistical interpretation of quantum mechanics, we are dealing with a quantum ensemble, i.e., with a large number of identical systems of microparticles or with a set of identical copies of one and the same system. In the given concrete case this image reflects quite adequately the situation which is experimentally realized. Thus, for example, a beam of particles from an accelerator is a rarefied beam of identical particles which do not interact with each other, and which all collide "in turn" with a target in the same way.

Now by letting the normalization N increase indefinitely, so that χ_σ tends to $(2\pi)^{3/2}\delta(\boldsymbol{k} - \boldsymbol{k}_0)$, we shall obtain a wave packet of increasingly better defined momentum. From (6) we find that

$$|f_\sigma(\boldsymbol{x})| \to 1,$$

and that therefore also

$$|f_\sigma(\boldsymbol{x})|^2 \to 1$$

and we obtain in the limit the situation in which there is one particle per unit volume.

By going to the limit we obtain the expression for the amplitude of the single-particle state normalized per unit volume:

$$\Phi_1 = (2\pi)^{3/2}\, a_\sigma^+ (\boldsymbol{k})\, \Phi_0.$$

In the case of several different particles it is necessary to consider, instead of (5), the expression

$$\Phi_s = \prod_{1 \leqslant i \leqslant s} \left\{ \int \chi_{\sigma_i} (\boldsymbol{k}_i)\, \overset{+}{a}_{\sigma_i} (\boldsymbol{k}_i)\, d\boldsymbol{k}_i \right\} \Phi_0, \tag{8}$$

where all the sets $\{\sigma_i, \boldsymbol{k}_i\}$ are different. The norm of such an amplitude will obviously be given by the product of the norms of S single-particle states:

$$\overset{*}{\Phi}_s \Phi_s = N_1 N_2 \ldots N_s. \tag{9}$$

By repeating the argument given above separately for each of the factors in (8) and (9), we conclude that the state amplitude normalized per unit volume of each of the particles present in this state has the form

$$\Phi_s = \prod_i \left\{ (2\pi)^{3/2}\, \overset{+}{a}_{\sigma_i} (\boldsymbol{k}_i) \right\} \Phi_0. \tag{10}$$

21.3. The general formula for the transition probability. We shall now examine the problem of determining the probability of such a process when in the initial state there are s particles with precisely defined momenta $\boldsymbol{k}_1, \ldots, \boldsymbol{k}_s$ and internal quantum numbers $\sigma_1, \ldots, \sigma_s$ charaterizing the mass, charge, and spin, and one is required to determine the average number of scattered particles having momenta within the infinitesimal volumes $d\boldsymbol{p}_1, \ldots, d\boldsymbol{p}_r$ and possessing the internal quantum numbers $\rho_1, \ldots, \rho_r$.

As a result of interaction the system will transform from the state (10) into the state described by the amplitude

$$S(1)\,(2\pi)^{3s/2}\, \Phi_k,$$

where Φ_k is defined by (20.1). Therefore the average number of particles which will turn out to be in the final state described by the normalized amplitude Φ_α, according to the general quantum-mechanical rule, will be equal to

$$(2\pi)^{3s} \frac{|\overset{*}{\Phi}_\alpha S(1)\, \Phi_k|^2}{\overset{*}{\Phi}_\alpha \Phi_\alpha}. \tag{11}$$

The choice of normalization is such that in the initial state the average numbers of particles per unit volume are equal to unity. If, on the other hand, these numbers are respectively equal to $n_1, n_2, \ldots, n_r$, then (11) must be multiplied by $n_1 n_2 \ldots n_r$.

We are now interested in the number of particles scattered into the momentum intervals $\Delta \boldsymbol{p}_1, \ldots, \Delta \boldsymbol{p}_r$ centered about the average values $\boldsymbol{p}_1, \ldots, \boldsymbol{p}_r$. In carrying out this calculation we shall assume, as was mentioned in Section 20.1, that the momentum of each particle has different values in the initial and final states. Experimentally, this requirement corresponds to the fact that particles with unchanged momentum are always included in the primary beam, while those particles whose momentum has been altered are considered as scattered particles.

We shall therefore take the amplitude of the final state in the form

$$\Phi_\alpha = (2\pi)^{3r/2} \int_\Delta \Phi_{\ldots p \ldots} \, d\boldsymbol{p}_1 \ldots d\boldsymbol{p}_r,$$

where the region Δ is the product of the volumes Δp_f. Thus,

$$\overset{*}{\Phi}_\alpha \Phi_\alpha = (2\pi)^{3r} \, \Delta \boldsymbol{p}_1 \ldots \Delta \boldsymbol{p}_r,$$

while in place of (11) we obtain the expression

$$\frac{n_1 \ldots n_s (2\pi)^{3s}}{\Delta \boldsymbol{p}_1 \ldots \Delta \boldsymbol{p}_r} \left| \int_\Delta d\boldsymbol{p}'_1 \ldots d\boldsymbol{p}'_r \overset{*}{\Phi}_{p'} S \Phi_k) \right|^2,$$

which by virtue of the definition of the region Δ is equal to

$$n_1 \ldots n_s (2\pi)^{3s} \, | \overset{*}{\Phi}_p S \Phi_k |^2 \, \Delta \boldsymbol{p}_1 \ldots \Delta \boldsymbol{p}_r. \tag{12}$$

The structure of the squares of matrix elements of the type $| \overset{*}{\Phi} \ldots S\Phi \ldots |^2$ has been investigated at the beginning of this section. By substituting the expression (3) into (12) and dividing it by VT we obtain the number of particles scattered into the interval $d\boldsymbol{p}_1, \ldots, d\boldsymbol{p}_r$ per unit time and per unit volume:

$$(2\pi)^{3s-4} n_1 \ldots n_s \, | F(p, k) |^2 \, \delta\left(\textstyle\sum p - \sum k\right) d\boldsymbol{p}_1 \ldots d\boldsymbol{p}_r. \tag{13}$$

Making use of (4), we also find the corresponding expression for scattering per unit time by a stationary classical field in the form

$$(2\pi)^{3s-1} n_1 \ldots n_s \, | F(p, k, A_{\text{ext}}) |^2 \, \delta\left(\textstyle\sum p^0 - \sum k^0\right) d\boldsymbol{p}_1 \ldots d\boldsymbol{p}_r. \tag{14}$$

Thus, in order to calculate the transition probability, it is first of all necessary to calculate the matrix element (2) and to substitute the resulting function F into (13) (or into (14)).

Notice that in connection with the transition from (20.1) to (10) it turns out to be convenient to change the normalization of the matrix element. Instead of (2) we assume

$$\overset{*}{\Phi}_{\dots p \dots} S \Phi_{\dots k \dots} = \frac{i\,(2\pi)^{4-\frac{3}{2}(s+r)}}{\left\{\prod\limits_i (2k_i^0) \prod\limits_f (2p_f^0)\right\}^{1/2}} \delta^{(4)}\left(\sum p - \sum k\right) \mathcal{M}(p,\ k). \quad (15)$$

Equation (13) then takes on the form

$$2\pi\, n_1 \dots n_s \frac{|\mathcal{M}(p,\ k)|^2}{\prod\limits_i (2k_i^0)} R_r, \quad (16)$$

where

$$R_r = \prod_{1 \leqslant f \leqslant r} \left\{\frac{dp_f}{(2\pi)^3\, 2p_f^0}\right\} (2\pi)^3\, \delta\left(\sum p - \sum k\right) \quad (17)$$

is the density of final states in momentum space, the dimension of which is $3r$. If in the final state there turn out to be two particles ($r = 2$), it is then convenient to make use of the density normalized with respect to the two-dimensional volume instead of the six-dimensional one (i.e., by eliminating four integrations with the aid of δ-functions). This density may be represented in the form

$$R_2 = \frac{d\Omega_1}{4\pi \cdot 8\pi^2} \cdot \frac{1}{[(p_1)^2 Q_0 - p_1^0\,(p_1 Q)]}, \quad (18a)$$

where $Q = \sum_i k_i$; $p_1 = |\mathbf{p}_1|$ is a solution of the equation $(Q - p_1)^2 - m_2^2 = 0$.

In the limit when one of the particles is much heavier than the other one ($m_2 \gg m_1, p_1$), we have

$$R_2 = \frac{p_1\, d\Omega_1}{32\pi^3 m_2}. \quad (18b)$$

In the center-of-mass system we have

$$R_2 = \frac{p_1\, d\Omega_1}{32\pi^3 Q^0}. \quad (18c)$$

21.4. Scattering of two particles. *Let us consider an important case: the scattering of two particles* ($2 \to 2$). The transition matrix element occurring in the left-hand side of (15) may differ from zero even in the case when there is no interaction present, if $S = 1$ and the momenta and quantum numbers of the particles in the initial state are equal to those in the final state.

To deal purely with the effect of interaction, the "diagonal" matrix element is subtracted from the total one, i.e., the following expression is considered:

$$\overset{*}{\Phi}_{\ldots p \ldots}(S-1)\,\Phi_{\ldots k \ldots},$$

which does not differ from the initial one when $p_i \neq k_i$. In accordance with (15) we shall assume (at the same time changing the normalization) the following:

$$\overset{*}{\Phi}_0 a_1^-(\boldsymbol{p}_1)\, a_2^-(\boldsymbol{p}_2)\,(S-1)\, a_1^+(\boldsymbol{k}_1)\, a_2^+(\boldsymbol{k}_2)\,\Phi_0 = \\ = \frac{i\delta^{(4)}(p_1+p_2-k_1-k_2)}{2\pi\,(p_1^0 p_2^0 k_1^0 k_2^0)^{1/2}} f(p_1,\ p_2;\ k_1,\ k_2). \tag{19}$$

The function f introduced here represents the scattering amplitude in the relativistic case. The norm of the right-hand side corresponds to the case of scattering of spinless particles. It is not difficult to verify that the relativistic amplitude f is a dimensionless function. It may be shown that it is also a relativistically invariant function.

For the scattering of a boson (with creation operator $a_n^+(\boldsymbol{k})$, by a fermion (with creation operator $b_\nu^+(\boldsymbol{q})$) the scattering amplitude is usually introduced in a somewhat different manner:

$$\overset{*}{\Phi}_0 a_m^-(\boldsymbol{k}')\, b_\mu^-(\boldsymbol{q}')\,(S-1)\, a_n^+(\boldsymbol{k})\, b_\nu^+(\boldsymbol{q})\,\Phi_0 = \\ = \frac{i\delta(k'+q'-k-q)}{2\pi\sqrt{k_0 k_0'}} f_{\nu n,\,\mu m}(\boldsymbol{k}',\ \boldsymbol{q}';\ \boldsymbol{k},\ \boldsymbol{q}). \tag{20}$$

Here n and m are the isotopic indices of the bosons, while ν and μ are the spin and isotopic indices of the fermion. The representation (20) may be applied, for example, to the scattering of light on electrons (Compton scattering) or to pion–nucleon scattering. The matrix scattering amplitude $f_{\nu n,\mu m}$ has the dimensionality $[m^{-1}] = \text{cm}^1$ and can be made dimensionless by the introduction of an appropriate mass. This, however, is usually not done, in order to retain the conventional correspondence between the nonrelativistic limit and the dimensional scattering amplitude in quantum mechanics.

Taking the norm of (20), we obtain instead of (16) the following:

$$n_1 n_2 (2\pi)^3\, 4\, \frac{q_0'}{k_0}\, |f|^2 R_2. \tag{21}$$

This expression is equal to the number of particles scattered within R_2 per unit volume and per unit time. It is usually represented as the product $n_1 n_2 v(\boldsymbol{k})\, d\sigma$, where $v(\boldsymbol{k})$ is the modulus of the velocity of the incident particle in the laboratory coordinate system (lab system) (equal to unity for the photon, and to $|\boldsymbol{k}|/k_0$ for particles with finite rest mass). The factor $d\sigma$ has the dimensions of an area (cm^2), is proportional to the element of solid angle containing the particle after scattering, and is called the *differential effective cross section*. In accordance with (21) and (18a) it may be represented in the form

$$\frac{d\sigma}{d\Omega_{k'}} = \frac{q_0'\,(k')^3}{k\left[(k')^2 Q_0 - k_0'\,(k'Q)\right]}\,|f|^2, \qquad k = |\boldsymbol{k}|. \tag{22}$$

In the static limit of pion–nucleon scattering

$$\frac{d\sigma}{d\Omega} = |f|^2. \tag{23}$$

By integrating the differential cross section over the total solid angle we obtain the *total* effective cross section in the lab system:

$$\sigma = \int_{4\pi} d\Omega\,\frac{d\sigma}{d\Omega}. \tag{24}$$

We recall also that with the help of the so-called optical theorem (see Blokhintsev (1976), Section 86; Landau and Lifshitz (1974), Section 125; Messiah (1962), Chapter XIX), the imaginary part of the forward elastic-scattering amplitude, which is defined as the limit of the scattering amplitude in going to small angles,

$$f(E,\ \theta = 0) = \lim_{\theta \to 0} f(E,\ \theta), \tag{25}$$

can be expressed in terms of the total effective cross section

$$\sigma_{\text{tot}} = \sigma_{\text{elast}} + \sigma_{\text{inelast}} \tag{26}$$

by means of a linear relation.

For the pion–nucleon scattering amplitude introduced in (20), the optical theorem is usually written as follows:

$$\operatorname{Im} f_{\nu n,\,\mu m}(k,\ q;\ k,\ q) = \frac{|\boldsymbol{k}|_{\text{l.c.s.}}}{4\pi}\,\sigma_{\text{tot}}^{\nu n}, \tag{27}$$

whereas in the case of the dimensionless pion–nucleon scattering amplitude normalized in accordance with (19), this theorem may be conveniently expressed in terms of invariant variables:

$$\operatorname{Im} f(\boldsymbol{k}_1, \boldsymbol{k}_2; \boldsymbol{k}_1, \boldsymbol{k}_2) = \frac{\sqrt{S(S-4m^2)}}{4\pi} \sigma_{\text{tot}}(S), \tag{28}$$
$$S = (k_1^0 + k_2^0)^2 - (\boldsymbol{k}_1 + \boldsymbol{k}_2)^2.$$

21.5. The two-particle decay. Let us now assume that in the initial state there is one particle ($s = 1$) of mass M, while in the final state there are two particles ($r = 2$) with masses m_1 and m_2. This case corresponds to the decay of particle M into two particles, which is kinematically possible if $M > m_1 + m_2$.

According to (16) the average number of particles to be found in the final state within the intervals $d\boldsymbol{p}_1$, $d\boldsymbol{p}_2$ per unit time and per unit volume will be

$$2\pi n \frac{|\mathcal{M}(p, k)|^2}{2k^0} R_2.$$

Passing over to the rest frame of the initial particle, i.e., to the center-of-mass system of the decay products, we obtain with the help of (18c), when $n = 1$, the following:

$$\frac{p_1}{32\pi^2 M^2} |\mathcal{M}(p, k)|^2 \, d\Omega.$$

By integrating this expression over the solid angle, we find the total probability of the two-particle decay per unit time:

$$w = \frac{p}{32\pi^2 M^2} \int d\Omega \, |\mathcal{M}(p, \Omega)|^2. \tag{29}$$

The quantity w has the dimension of mass (in the usual system of units equal to c^{-1}) and coincides with the energy width: $w = \Gamma$. In case there exist several decay modes, the total width is represented by a sum of expressions of the form (29):

$$\Gamma_{\text{tot}} = \sum_i \Gamma_i,$$

while the lifetime of the initial particle, τ, is determined by the relation

$$\tau = \frac{1}{\Gamma_{\text{tot}}} \left(= \frac{\hbar}{\Gamma_{\text{tot}}} \right). \tag{30}$$

CHAPTER VI

EVALUATION OF INTEGRALS AND DIVERGENCES

22. THE METHOD FOR EVALUATING INTEGRALS

22.1. Integrals over virtual momenta. In the general case the number of vertices in a diagram is smaller than the number of internal lines, as a consequence of which the matrix element is represented by an expression involving integration over virtual four-momenta:

$$\mathcal{M}(\ldots p \ldots) = \int dk_1 \ldots dk_c F(p, k).$$

The number of integrations, c, is equal to the number of "loops" in the diagram, i.e., to the number of topologically nonequivalent closed contours occurring in it.

The integrand F is the product of propagators (free Green's functions) having the structure

$$\Delta_{\alpha\beta}(k) = \frac{P_{\alpha\beta}(k)}{m^2 - k^2 - i\varepsilon}$$

which correspond to the internal lines having momenta over which the integration is performed, together with factors corresponding to the vertices. The latter consist of coupling constants, of matrices, and possibly of components of the four-momenta "passing" through the vertex (in case the interaction Lagrangian contains derivatives of the field functions).

Thus, the most general expression for an integral over the four-momentum has the form

$$I(p, L) = \frac{i}{\pi^2} \int dk \prod_l^L \{m_l^2 - (k + p_l)^2 - i\varepsilon\}^{-1} P(k, p), \tag{1}$$

where P is a polynomial in the components of the four-vectors $k, \ldots$, $p, \ldots$ with matrix coefficients.

Our immediate aim will be to develop a technique for evaluating integrals of the type (1).

22.2. The α-representation and Gaussian quadratures. In order to evaluate the integral (1), we shall make use of the following auxiliary device. We represent the factors in the denominator of the integrand in the form of an integral over a parameter:

$$\frac{1}{m^2 - k^2 - i\varepsilon} = i \int_0^\infty e^{i\alpha\,(k^2 - m^2 + i\varepsilon)}\, d\alpha. \tag{2}$$

We shall call this transformation the transition to the α-representation. After this transition, integration in (1) will reduce to the calculation of Gaussian integrals.

The fundamental four-dimensional Gaussian integral has the form

$$\frac{i}{\pi^2} \int e^{i\,(ak^2 + 2bk)}\, dk = \frac{1}{a^2}\, e^{-ib^2/a}. \tag{3}$$

Here b is some four-vector. Since both the left- and right-hand sides may be factored into their respective component integrations, (3) is equivalent to the product of four one-dimensional Gaussian quadratures of the form

$$I\,(a,\ \beta) = \int_{-\infty}^{\infty} \exp i\,(-at^2 + 2\beta t)\, dt.$$

For definiteness we shall consider the integral I to be the limit of the expression

$$I(a - i\eta,\, b) = \int_{-\infty}^{\infty} \exp\left[i\,(-at^2 + 2bt) - \eta t^2\right] dt, \quad a > 0, \quad \eta \to +0,$$

for the calculation of which we introduce a change of variable $t \to x$:

$$t = \frac{1-i}{\sqrt{2}}\, x + \frac{b}{a}, \quad i\,(-at^2 + 2bt) = -ax^2 + ib^2/a,$$

which consists of a displacement of the origin of the coordinate system by b/a and of a subsequent rotation of the contour of integration through an angle of 45° in the complex plane. Then in the limit $\eta = 0$ we obtain

$$I\,(a,\ b) = \frac{1-i}{\sqrt{2}}\, e^{ib^2/a} \int_{-\infty}^{\infty} e^{-ax^2}\, dx = (\pi/ai)^{1/2} \exp\,(+\,ib^2/a), \quad a > 0.$$

The value of the analogous integral for $a < 0$ may be obtained from this by taking the complex conjugate

$$I\,(-|\,a\,|,\ b) = \overset{*}{I}\,(a,\ b) = (\pi i/|\,a\,|)^{1/2} \exp\,(-\,ib^2/|\,a\,|).$$

Returning to the initial expression on the left-hand side of (3), we represent it by the product

$$\int e^{i(ak_0^2+2b_0k_0)}\,dk_0 \prod_{1\leqslant n\leqslant 3}\int e^{-i(ak_n^2+2b_nk_n)}\,dk_n =$$
$$= I(-a,\ b_0)\,I(a,\ b_1)\,I(a,\ b_2)\,I(a,\ b_3) = (\pi^2/ia^2)\exp\frac{b_0^2-b_1^2-b_2^2-b_3^2}{ia},$$

which completes the proof of (3).

Notice that in performing the integration over k_0, instead of taking the complex conjugate of the integral $I(a, b)$, it would be possible to carry out the integration over the purely imaginary argument $k_0 = ik_4$, which is equivalent to rotating the contour of integration over k_0 through 90°. Such an operation, leading to a four-dimensional Euclidean integral

$$\int d^4k = \int dk^0 \int d\mathbf{k} \to i\int dk_4 \int d\mathbf{k} \equiv i\int (d_4k)_E,$$

is sometimes called the "Wick rotation".

The other necessary integrals containing the factors k_μ in the numerator (and corresponding to the polynomial $P(k, p)$ in (1)) can be obtained from the fundamental Gaussian integral (3) by means of repeated differentiation with respect to the components b_ν:

$$\frac{i}{\pi^2}\int e^{i(ak^2+2bk)}\,[k^\nu]\,dk = \frac{1}{a^2}\,e^{-ib^2/a}\left[-\frac{b^\nu}{a}\right], \tag{4a}$$

$$\frac{i}{\pi^2}\int e^{i(ak^2+2bk)}\,[k^\nu k^\mu]\,dk = \frac{1}{a^2}\,e^{-ib^2/a}\left[\frac{2b^\nu b^\mu + iag^{\nu\mu}}{2a^2}\right], \tag{4b}$$

$$\frac{i}{\pi^2}\int e^{i(ak^2+2bk)}\,[k^2]\,dk = \frac{1}{a^2}\,e^{-ib^2/a}\left[\frac{b^2+2ia}{a^2}\right]. \tag{4c}$$

Making use of (2), we transform the right-hand side of (1) into the form

$$\frac{i^{L+1}}{\pi^2}\int_0^\infty d\alpha_1\ldots d\alpha_L \int dk\,P(k,\ p)\exp i\,(Ak^2+2Bk-M^2+i\varepsilon A).$$

Here $A = \alpha_1 + \ldots + \alpha_L$, $B = \Sigma_l\alpha_l p_l$, $M^2 = \Sigma\alpha_l(m_l^2 - p_l^2)$. Representing, further, P in the form

$$P(k,\ p) = P(p) + k^\nu P_\nu(p) + k^\nu k^\mu P_{\nu\mu}(p) + \ldots,$$

with the aid of (3) and (4) we obtain

$$I(p) = i^L \int_0^\infty \frac{d\alpha_1 \ldots d\alpha_L}{A^2} \exp i\left(-\frac{B^2}{A} - M^2 + i\varepsilon A\right) \times$$
$$\times \left[P(p) - \frac{B^\nu}{A} P_\nu(p) + \frac{2B^\nu B^\mu + iAg^{\nu\mu}}{2A^2} P_{\nu\mu}(p) + \ldots\right].$$

To carry out singular integration we introduce a change of variables $(\alpha_1, \ldots, \alpha_L) \to (x_1, \ldots, x_{L-1}, A)$:

$$\alpha_l = x_l A, \quad l = 1, 2, \ldots, L-1, \quad \alpha_L = (1 - x_1 - \ldots - x_{L-1})A,$$
$$\left|\frac{\partial\alpha_1 \ldots \partial\alpha_L}{\partial x_1 \ldots \partial x_{L-1} \partial A}\right| = A^{L-1}. \tag{5}$$

Taking into account that

$$B^2/A + M^2 = AD(x, p); \quad D(x, p) = (\sum x_l p_l)^2 + \sum x_l (m_l^2 - p_l^2), \tag{6}$$

we obtain

$$I(p; L) = i^L \int_0^\infty dA A^{L-3} \int_0^1 \{dx\}_L e^{iA(-D+i\varepsilon)} \left\{Q_0 + \frac{1}{A} Q_1 + \ldots\right\}, \tag{7}$$

where

$$\int_0^1 \{dx\}_L = \int_0^1 dx_1 \ldots \int_0^1 dx_L \delta\left(1 - \sum_l x_l\right) =$$
$$= \int_0^1 dx_1 \int_0^{1-x_1} dx_2 \int_0^{1-x_1-x_2} dx_3 \ldots \int^{1-x_1-x_2-\ldots-x_{L-2}} dx_{L-1},$$
$$Q_0(p, x) = P(p) - b^\nu P_\nu + b^\nu b^\mu P_{\nu\mu},$$
$$Q_1(p, x) = (i/2) P_\nu^\nu + \ldots, \quad b = \sum_l x_l p_l.$$

It is not difficult to now carry out the singular integration over A with the aid of formulas like (2):

$$\int_0^\infty e^{iA(-D+i\varepsilon)} A^{L-3} dA = \frac{-\Gamma(L-2)}{i^L (D-i\varepsilon)^{L-2}}, \quad L > 2. \tag{8}$$

In integrating over A on the right-hand side of (7) when $L \leq 2$ and/or the particles have spin ($P_{\nu\mu} \neq 0$), the formula (8) may turn out to be inapplicable, because the index L will turn out to be zero or negative. Such integrals contain a nonintegrable singularity at $A = 0$ due to the ultraviolet divergences of the initial integrals (1), and we postpone considering them until Section 23.

The procedure presented above is quite sufficient for calculating Feynman integrals not involving ultraviolet divergences. As a result of performing integrations over A, $I(p)$ will be represented in the form of a multiple integral over the parameters x_l:

$$I_L(p) = i^2 \int_0^1 \frac{\{dx\}}{(D - i\varepsilon)^{L-2}} \left\{ (L-3)!\, Q_0 + \frac{(L-2)!}{iD} Q_1 + \ldots \right\}. \tag{9}$$

The remaining quadratures over the parameters x_l are, generally speaking, not singular and when $l \leq 3$ may be carried out to the end.

22.3. Feynman's parametrization. We present one more—frequently used—method of calculating the integral (1). The initial step consists in taking the formula used by Feynman in his early work:

$$\frac{1}{a_1 a_2 \ldots a_L} = (L-1)! \int_0^1 \frac{\{dx\}_L}{\left(\sum\limits_{1 \leqslant l \leqslant L} a_l x_l\right)^L}. \tag{10}$$

Assuming $(k + p_l)^2 - m_l^2 + i\varepsilon = a_l$, we represent (1) by an integral over the parameters x_l:

$$I_L(p;\ L) = (-1)^L (L-1)! \int_0^1 \{dx\}_L\, J_L(p,\ x), \tag{11}$$

where

$$J_L(p,\ x) = \frac{i}{\pi^2} \int \frac{dk\, P(k,\ p)}{[Z(p,\ k,\ x)]^L};$$

here Z is a quadratic form in the components of k:

$$Z(p,\ k,\ x) = \sum_l x_l [(k + p_l)^2 - m_l^2 + i\varepsilon] = k^2 + 2bk + b^2 - D + i\varepsilon.$$

After shifting the integration variable we have

$$J_L(p,\ x) = \frac{i}{\pi^2} \int \frac{dq\, P(q - b,\ p)}{(q^2 - D + i\varepsilon)^L}, \tag{12}$$

where D is the form introduced earlier by means of (6).

This integral in a number of cases can be reduced to the simpler one

$$\tilde{J}_3(p, x) = \frac{i}{\pi^2} \int \frac{dq}{(q^2 - D + i\varepsilon)^3}. \tag{13}$$

To verify this, we first of all note that all the integrals of the form $\int q^\nu F(q^2)\, dq$, $\int q^\nu q^\mu q^\rho F(q^2)\, dq, \ldots$, containing an odd number of components of the integration vector q^ν, for symmetry reasons turn out to be zero. By the same argument, integrals with an even number of components are simplified in the following way:

$$\int q^\nu q^\mu F(q^2)\, dq = (^1/_4)\, g^{\nu\mu} \int q^2 F(q^2)\, dq,$$

$$\int q^\nu q^\mu q^\rho q^\sigma F(q^2)\, dq = (^1/_{24})\,(g^{\nu\mu} g^{\rho\sigma} + g^{\nu\rho} g^{\mu\sigma} + g^{\nu\sigma} g^{\mu\rho}) \int (q^2)^2 F(q^2)\, dq.$$

Therefore the polynomial $P(q - b, p)$ can be replace by a polynomial in powers of q^2 and, consequently, by a polynomial in powers of the argument $q^2 - D$. Because of this the initial integral (12) will be represented in terms of a linear combination of the integrals

$$\tilde{J}_l(p, x) = \frac{i}{\pi^2} \int \frac{dq}{(q^2 - D + i\varepsilon)^l}, \quad l \leqslant L. \tag{14}$$

These latter, in turn, as a result of applying the operation of integration over D for $l \geq 3$, reduce to $\tilde{J}_3$. For $l \leq 2$ the integral (14) does not exist. The integral (13) may be calculated by combining (2), (3), and (8). In this way we obtain

$$\frac{i}{\pi^2} \int \frac{dq_0\, d\mathbf{q}}{(q_0^2 - \mathbf{q}^2 - D + i\varepsilon)^3} = \frac{i}{\pi^2} \int \frac{dq_0\, d\mathbf{q}}{(-q_4^2 - \mathbf{q}^2 - D + i\varepsilon)^3} =$$

$$= \frac{1}{\pi^2} \int \frac{d_E q}{(q_E^2 + D)^3} = \frac{1}{2D}. \tag{15}$$

Differentiating with respect to D, we have also

$$\tilde{J}_L(p, x) = \frac{(-1)^{L+1}}{(L-1)(L-2)} \frac{1}{D^{L-2}} \quad (L \geqslant 3). \tag{16}$$

By substitution into (11) we obtain

$$\tilde{I}_L(p) = -(L-3)! \int_0^1 \frac{\{dx\}_L}{[D(x, p) + i\varepsilon]^{L-2}} \tag{17}$$

which coincides with (9) in the simplest case ($P(k, p) = 1$).

Thus, both ways of integrating lead to identical expressions.

22.4. Ultraviolet divergences. As was mentioned above repeatedly, matrix elements defined formally in accordance with Feynman's rules in the form of integrals (1) over virtual momenta may turn out to be nonexistent owing to the fact that the integrand does not fall off sufficiently fast in the region of large k.

The divergences of integrals over virtual momenta in the region of large k are called ultraviolet. Ultraviolet divergences are typical of perturbation-theory matrix elements in relativistic quantum field theory. They do not represent an exception but are rather a rule.

The simplest divergence corresponds to the one-loop integral with two spinless internal lines (Figure 22.1),

$$I(k) = \frac{i}{\pi^2} \int \frac{dp}{(m^2 - p^2 - i\varepsilon)[m^2 - (k-p)^2 - i\varepsilon]}$$

which diverges logarithmically for large p.

The integral $I(p)$ is the Fourier transform of the square of the causal Green's function:

$$\int e^{ip(x-y)} I(p)\, dp \sim i 2^8 \pi^6 [D^c(x-y)]^2.$$

The expression on the right-hand side is indefinite owing to the singular nature of the function D^c on the light cone. It arises in the evaluation with the aid of Wick's theorem of the product $T(\mathscr{L}(x)\,\mathscr{L}(y))$ occurring in $S_2(x, y)$. In Section 14 it was established that this expression is not defined when $x = y$, and in Section 16 it was shown that its definition may be completed with the help of the quasilocal operator $\Lambda_2(x, y)$.

It turns out to be possible to choose the operator Λ_2 so that the completely defined expressions of the type (16), (17) are finite. Such a procedure is known as the removal of ultraviolet divergences by renormalization. We shall examine it in greater detail in Sections 27 and 28.

To analyse the nature of the ultraviolet divergences, as well as for an appropriate choice of Λ_2, $\Lambda_3, \ldots$, it turns out to be convenient to use an intermediate auxiliary regularization of divergent integrals. We shall now proceed to examine different versions of this procedure.

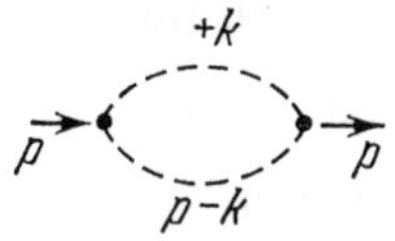

Fig. 22.1. Second-order scalar diagram with one loop.

23. AUXILIARY REGULARIZATIONS

23.1. The necessity of regularization. As was established in Section 18.3, the causal Green's functions occurring in Feynman's rules are singular functions with singularities on the light cone. From a mathematical point of view they represent distributions. The integrals over virtual momenta which were introduced in the preceding section and which correspond to closed loops of diagrams are the Fourier transforms of products of the causal Δ_c-functions whose arguments correspond to elements of these loops.

As was pointed out at the end of the preceding section, the simplest integral

$$I(k) \sim \frac{i}{\pi^2} \int \frac{dp}{(m^2 - p^2 - i\varepsilon)\,[m^2 - (p-k)^2 - i\varepsilon]}, \qquad (1)$$

which corresponds to a one-loop diagram with two scalar internal lines (see Figure 22.1) is the Fourier transform of two propagators with identical arguments:

$$I(k) = 16\pi^2 i \int e^{ikx}\,[D_c(x)]^2\,dx. \qquad (2)$$

Such products, in accordance with (18.22), contain squares of singular distributions of the form $\delta(\lambda)$, λ^{-1} (where $\lambda = x^2$) and are not, therefore, well-defined quantities. In other words, from a mathematical point of view the definitions of the products of propagators and of their Fourier transforms need to be completed.

We emphasize that the necessity of a special definition of products is, on the whole, typical for distributions. The point is that a distribution is determined by fixing the rules for integrating its product with sufficiently regular functions, while from such rules the prescription for the integration of the product of several distributions does not follow.

To construct the necessary definitions, one may make use of an improper transition to a limit and try to define the respective products $\Pi_{r \neq s}\{\Delta_c(x_r - x_s)\}$ and their Fourier transforms of the form (22.1) with the help of a sequence of regular expressions. Here two approaches are possible.

The first is based on obtaining regular approximations of the causal functions themselves, i.e., on the substitution for the distributions

$$\Delta_c(x) \rightarrow \operatorname{reg} \Delta_c(x)$$

of certain regularized approximations which are continuous on the light cone together with the necessary number of their derivatives. The products of the regularized propagators $\Pi\{\operatorname{reg} \Delta_c(x_r - x_s)\}$ then also turn out to be regular.

A very frequently used procedure of this type is the Pauli–Villars regularization.

The second method takes the initial expressions for the propagators and instead consists in the construction of regular approximations of their products and of the integrals over the virtual four-momenta. Regularization by means of a cutoff as well as dimensional regularization are examples of this method.

23.2. Pauli–Villars regularization. Pauli–Villars regularization consists in replacing the singular Green's functions of the massive free field with the linear combination

$$\Delta(m) \to \mathrm{reg}_M \Delta(m) = \Delta(m) + \sum_i c_i \Delta(M_i). \tag{3}$$

Here the symbol $\Delta(m)$ stands for the Green's functions of the field of mass m. Henceforth we shall always assume that the symbol Δ denotes the causal Green's function, although the Pauli–Villars procedure is formulated in the same manner both for Δ_c and for Δ_{ret}, Δ_{adv}, Δ^+, Δ^-. On the left-hand side $\Delta(M_i)$ are auxiliary quantities representing Green's functions of fictitious fields with masses M_i, while c_i are certain coefficients satisfying special conditions. These conditions are chosen so that the regularized function $\mathrm{reg}\Delta(x; m)$ considered in the configuration representation turns out to be sufficiently regular in the vicinity of the light cone, or (what is equivalent) such that the function $\mathrm{reg}\,\tilde{\Delta}(p; m)$ in the momentum representation falls off sufficiently fast in the region of large $|p^2|$.

Thus, for fields of integral spin we see on the basis of the concluding formulas of Section 18.3 that the strongest singularities on the light cone, $\delta(\lambda)$ and λ^{-1}, enter into $D(x; m)$ with coefficients independent of the mass m. Therefore, imposing on c_i the condition

$$1 + \sum_i c_i = 0, \tag{4a}$$

we obtain for $\mathrm{reg}\,\Delta(x; m)$ an expression which does not contain the singularities $\delta(\lambda)$ and λ^{-1}. Its Fourier transform $\mathrm{reg}\,\tilde{\Delta}(p; m)$ in the region of large $|p^2|$ will fall off like $|p^2|^{-2}$. In a similar manner the condition

$$m^2 + \sum_i c_i M_i^2 = 0 \tag{4b}$$

provides for the absence in $\mathrm{reg}\,\Delta(x)$ of singularities of the form of $\theta(\lambda)$, $\ln\lambda$ and leads to $\mathrm{reg}\,\tilde{\Delta}(p)$ falling off in the ultraviolet region as $|p^2|^{-6}$.

The removal of the Pauli-Villars regularization is achieved by passing to the limit $M_i \to \infty$.

The simplest case of the Pauli–Villars regularization reduces to the introduction of a single auxiliary mass M with the coefficient $c = -1$ satisfying the condition (4a). In the momentum representation

$$D_c(p) = \frac{1}{m^2 - p^2} \to \text{reg}_M D_c(p) = \frac{1}{m^2 - p^2} - \frac{1}{M^2 - p^2}. \tag{5}$$

By applying the regularization (5) to the integrand of (1) we obtain

$$I(k) \to \text{reg}_M I(k) =$$
$$= \frac{i}{\pi^2} \int dp \left[\frac{1}{m^2 - p^2} - \frac{1}{M^2 - p^2}\right]\left[\frac{1}{m^2 - (p-k)^2} - \frac{1}{M^2 - (p-k)^2}\right].$$

The right-hand side of this equation is a finite expression and can be calculated with the help of the technique presented in Section 22. In this way we obtain

$$\text{reg}_M I(k) = i^2 \int_0^\infty \frac{d\alpha\, d\beta}{(\alpha+\beta)^2} e^{ik^2 \frac{\alpha\beta}{\alpha+\beta}} [e^{-i\alpha m^2} - e^{-i\alpha M^2}][e^{-i\beta m^2} - e^{-i\beta M^2}] =$$
$$= \int_0^1 dx \ln\left[\frac{m^2 - x(1-x)k^2}{xM^2 + (1-x)m^2 - x(1-x)k^2} \cdot \frac{M^2 - x(1-x)k^2}{(1-x)M^2 + xm^2 - x(1-x)k^2}\right].$$

Here we have made use of formulas of the form

$$\int_0^\infty \frac{da}{a} (e^{iAa} - e^{iBa})\, e^{-\varepsilon a} = \ln \frac{B + i\varepsilon}{A + i\varepsilon}$$

(see Appendix VI).

We may now try to remove the regulator. By letting M^2 tend to infinity we have

$$\text{reg}_M I(k) \approx -\ln \frac{M^2}{\mu^2} + \int_0^1 dx \ln\left[\frac{m^2 - x(1-x)k^2 - i\varepsilon}{\mu^2}\right] + O\left(\frac{1}{M^2}\right), \tag{6}$$

where μ^2 is an arbitrary positive parameter. Thus, it turns out to be impossible to remove the regularization directly. One may only explicitly isolate from $I(k)$ the ultraviolet infinity

$$\text{reg}_M I(k) = -\ln \frac{M^2}{\mu^2} + I_{\text{fin}\,(M)}(k^2, \mu^2), \tag{7}$$

where the finite part of I_{fin} can be calculated explicitly (see Section 24.1 below).

We shall present also without computations the result of evaluating, with the help of the Pauli–Villars regularization, the quadratically divergent integral

$$J_{\mu\nu}(k)=\frac{i}{\pi^2}\int\frac{p_\mu p_\nu\,dp}{[m^2-(p-k/2)^2]\,[m^2-(p+k/2)^2]}, \tag{8}$$

which enters as an essential part into the contribution of the photon self-energy diagram (see Section 24.2 below). In the limit $M^2\to\infty$ we obtain

$$=\frac{1}{4}g_{\mu\nu}M^2+\left[g_{\mu\nu}\left(\frac{k^2}{12}-\frac{m^2}{2}\right)+\frac{k_\mu k_\nu}{12}\right]\ln\frac{M^2}{\mu^2}+J^{\text{fin}(M)}_{\mu\nu}(k), \tag{9}$$

where the finite part

$$J^{\text{fin}(M)}_{\mu\nu}(k)=\frac{3}{4}g_{\mu\nu}m^2+$$

$$+\frac{1}{4}\int_0^1 dx\,[2g_{\mu\nu}Z(x,\ k^2)-k_\mu k_\nu(1-2x)^2]\ln\frac{Z(x,\ k^2)}{\mu^2},$$

$$Z(x,\ k^2)=m^2-x(1-x)\,k^2-i\varepsilon$$

can be integrated to the end.

A number of other formulas for frequently encountered divergent integrals are presented in Appendix VI.

23.3. Dimensional regularization. Dimensional regularization consists in substituting for the integral over a four-dimensional manifold of virtual momenta a symbol formally corresponding to an integral over a space of $n=4-2\varepsilon$ dimensions. Here ε is considered to be a small positive quantity, while the regularized integration multiplicity n turns out to be noninteger.

Here we denote, in accordance with established usage, the noninteger part of the dimensionality by the same symbol which is used for the definition of a causal function. In the present section we shall make use of this notation only in this new sense, tacitly implying that $m^2=m^2-i\delta\ (\delta\to+0)$.

In the Euclidean case the operation of introducing dimensional regularization looks as follows:

$$\int(d^4p)_E=\int_{\Omega(4)}d\Omega\int_0^\infty p^3\,dp\to\int d^np\equiv\mu^{2\varepsilon}\int_{\Omega(n)}d\Omega\int_0^\infty p^{n-1}\,dp, \tag{10}$$

where the volume $\Omega(n)$ of a unit sphere in the n-dimensional space is interpolated with the aid of the Euler Γ-function:

$$\Omega(n)=2\pi^{n/2}/\Gamma(n/2),$$

while the dependence on the parameter μ with the dimensions of mass has been introduced in order to preserve the overall dimensionality.

Instead of this one may also directly interpolate the formulae of the preceding section. The fundamental Gaussian quadrature takes the form

$$\frac{i}{\pi^2}\int d^n p e^{i(ap^2+2bp)} = \left(\frac{ia\mu^2}{\pi}\right)^{\varepsilon}\frac{1}{a^2}e^{-ib^2/a}. \tag{11}$$

The fundamental integral within the framework of the Feynman parametrization may now be calculated by combining (11) and (22.8). It turns out to be equal to

$$\operatorname{reg}_\varepsilon J_l(D) = \frac{i}{\pi^2}\int\frac{d^n p}{(p^2-D)^l} = \left(\frac{\mu^2}{\pi}\right)^{\varepsilon}\frac{(-1)^{l+1}\Gamma(l+\varepsilon-2)}{D^{l+\varepsilon-2}\Gamma(l)}. \tag{12}$$

We shall now calculate the integral (1) by the method of dimensional regularization with the help of the α-representation:

$$I(k)\to \operatorname{reg}_\varepsilon I(k) = \frac{i\mu^{2\varepsilon}}{\pi^2}\int\frac{d^n p}{(m^2-p^2)[m^2-(p+k)^2]} =$$

$$= \frac{i^3\mu^{2\varepsilon}}{\pi^2}\int_0^\infty d\alpha\, d\beta\, e^{i\beta k^2 - i(\alpha+\beta)m^2}\int d^n p e^{i(\alpha+\beta)p^2+2i\beta pk} =$$

$$= -\left(\frac{i\mu^2}{\pi}\right)^{\varepsilon}\int_0^\infty\frac{d\alpha\, d\beta}{(\alpha+\beta)^{2-\varepsilon}}e^{i\frac{\alpha\beta}{\alpha+\beta}k^2 - i(\alpha+\beta)m^2} =$$

$$= -\left(\frac{i\mu^2}{\pi}\right)^{\varepsilon}\int_0^1 dx\int_0^\infty\frac{da}{a^{1-\varepsilon}}e^{iaZ(x,\,k^2)}.$$

To evaluate the integral that arises, we make use of the quadrature (22.8). We find

$$\operatorname{reg}_\varepsilon I(k) = -\left(\frac{\mu^2}{\pi}\right)^{\varepsilon}\Gamma(\varepsilon)\int_0^1\frac{dx}{[m^2-x(1-x)k^2]^{\varepsilon}}. \tag{13}$$

Now it is necessary to remove the regularization, i.e., to make ε tend to zero. By utilizing the relation

$$\Gamma(\varepsilon)\to\frac{1}{\varepsilon}-C+O(\varepsilon)\qquad (C=0{,}5772\ldots),\quad \varepsilon\to+0,$$

we obtain

$$\operatorname{reg}_\varepsilon I(k)\to -1/\varepsilon \quad I^{\text{fin}(e)}(k), \tag{14}$$

where

$$I^{\text{fin}(\varepsilon)}(k) = \int_0^1 dx\ln\left[\frac{m^2-x(1-x)k^2}{\mu^2}\right]+\ln\pi+C.$$

Thus, in the framework of the method of dimensional regularization the singular part of an integral logarithmically divergent in the ultraviolet region becomes the pole ε^{-1}.

We shall present also, without calculations, the result of the evaluation of the quadratically diverging integral (8):

$$\text{reg}_{\varepsilon}\, J_{\mu\nu}(k) = \left(\frac{\mu^2}{\pi}\right)^{\varepsilon} \frac{\Gamma(\varepsilon)}{4} \int_0^1 \frac{dx}{(Z)^{\varepsilon}} \left[2g_{\mu\nu}\frac{Z}{1-\varepsilon} - k_\mu k_\nu (1-2x)^2\right],$$

where

$$Z = Z(x,\ k^2) = m^2 - x(1-x)k^2.$$

In the limit of small ε we have

$$\text{reg}_{\varepsilon}\, J_{\mu\nu}(k) = \frac{1}{2\varepsilon}\left[g_{\mu\nu}m^2 - \frac{g_{\mu\nu}m^2 + k_\mu k_\nu}{6}\right] + J_{\mu\nu}^{\text{fin}(\varepsilon)}(k), \tag{15}$$

where

$$J_{\mu\nu}^{\text{fin}(\varepsilon)}(k) = \frac{g_{\mu\nu}}{2}\left(m^2 - \frac{k^2}{6}\right) + \frac{1}{4}\int_0^1 dx\,[2g_{\mu\nu}Z - k_\mu k_\nu(1-2x)^2]\ln Z. \tag{16}$$

Comparing (6) and (14), we conclude that in the method of dimensional regularization the logarithmical ultraviolet divergences appear in the form of poles in ε. Comparison of (9) and (15) confirms this conclusion and shows that quadratic ultraviolet divergences also lead, in this method, to the pole ε^{-1}.

23.4. Regularization by means of a cutoff. Regularization by means of a cutoff represents the simplest regularization from an intuitive point of view and consists in cutting off integrals over momenta at a certain upper limit. After the reduction of an integral over the four-momentum to its Euclidean form, this regularization may be formulated in the following way:

$$\int dp = i\int (d^4p)_E = i\int dp_4 \int d\mathbf{p} \to \text{reg}_\Lambda \int dp = \int_{\Omega_\Lambda} dp = i\int_{\Omega(4)} d\Omega \int_0^\Lambda p^3\, dp.$$

It is not difficult to verify that for divergent integrals over momenta represented in the Feynman parametrization, regularization by means of a cutoff gives

$$\text{reg}_\Lambda\, J_2(D) = \frac{i}{\pi^2}\int_{\Omega_\Lambda} \frac{dq}{(q^2 - D + i\varepsilon)^2} = \ln\frac{D - i\varepsilon}{\Lambda^2} + 1, \tag{17}$$

$$\text{reg}_\Lambda\, J_{\mu\nu}(D) = \frac{i}{\pi^2}\int_{\Omega_\Lambda} \frac{q_\mu q_\nu\, dq}{(q^2 - D + i\varepsilon)^2} = \frac{g_{\mu\nu}}{4}\left(\Lambda^2 - 2D\ln\frac{\Lambda^2}{D} - D\right). \tag{18}$$

In utilizing the α-representation it is more convenient to introduce the smooth cutoff (Feynman's regularization)

$$\int dp \rightarrow \mathrm{reg}_{\Lambda\,(F)} \int dp = \int dp \frac{\Lambda^2}{\Lambda^2 - p^2}, \tag{19}$$

which turns out to be sufficient for logarithmically divergent integrals. In the simplest cases it reduces to a version of the Pauli–Villars regularization, and allows one, by modifying the initial integral, to make use of the formulae of the preceding section without having to change them. Thus, for the integral (1) we obtain

$$\begin{aligned}
\mathrm{reg}_{\Lambda\,(F)}\, I(k) &= \frac{i\Lambda^2}{\pi^2} \int \frac{dp}{(m^2 - p^2)(\Lambda^2 - p^2)[m^2 - (p-k)^2]} = \\
&= \frac{i}{\pi^2} \int \frac{dp}{m^2 - (p-k)^2} \left(\frac{1}{m^2 - p^2} - \frac{1}{\Lambda^2 - p^2} \right) \frac{\Lambda^2}{\Lambda^2 - m^2} = \\
&= i^2 \int_0^\infty \frac{d\alpha\, d\beta}{(\alpha+\beta)^2} e^{i\frac{\alpha\beta}{\alpha+\beta}k^2 - i\alpha m^2} \left(e^{-i\beta m^2} - e^{-i\beta\Lambda^2} \right) \frac{\Lambda^2}{\Lambda^2 - m^2} = \\
&= \frac{\Lambda^2}{\Lambda^2 - m^2} \int_0^1 dx \ln \left[\frac{m^2 - x(1-x)k^2}{xm^2 + (1-x)\Lambda^2 - x(1-x)k^2} \right] \rightarrow -\ln \frac{\Lambda^2}{\mu^2} + \\
&\quad + \left\{ 1 + \int_0^1 dx \ln \left[\frac{m^2 - x(1-x)k^2}{\mu^2} \right] \right\}.
\end{aligned} \tag{20}$$

24. ONE-LOOP DIAGRAMS

We shall now consider in greater detail a number of the simplest Feynman integrals corresponding to one-loop diagrams, and thus involving one four-dimensional integration.

24.1. The scalar "fish". The simplest second-order diagram with two internal lines (see Figure 22.1) is referred to as the "fish". In the case when both internal lines correspond to a scalar field of mass m, the matrix element I has the form of the integral

$$I(k) \sim \frac{i}{\pi^2} \int \frac{dp}{(m^2 - p^2 - i\varepsilon)[m^2 - (p-k)^2 - i\varepsilon]}. \tag{23.1}$$

This integral diverges logarithmically in the region of large p. In the preceding section it was examined with the aid of regularizations of various types. Comparison of the formulas (23.6), (7), (14), (20) shows that the divergent expression (23.1) reduces, after isolation of the singular additive part independent of k, to the finite expression

$$I_{\text{fin}}(k) = \int_0^1 dx \ln\left[\frac{m^2 - x(1-x)k^2 - i\varepsilon}{\mu^2}\right]. \tag{1}$$

In other words, we may write

$$I(k) = I_{\text{sing}} + I_{\text{fin}}(k). \tag{2}$$

The divergent constant I_{sing} is equal under the Pauli–Villars regularization to $\ln(\mu^2/M^2)$, under the dimensional regularization to $\ln \pi + \boldsymbol{C} - 1/\varepsilon$, and under regularization by means of a cutoff to $1 + \ln(\mu^2/\Lambda^2)$.

The finite part I_{fin} contains the dependence on the arbitrary parameter μ^2 and is defined up to an additive constant. The definition of the constant that we chose in (1) is absolutely arbitrary. It is convenient to break up the right-hand side of (1) into two terms

$$J\left(\frac{k^2}{4m^2}\right) + \left(\ln\frac{m^2}{\mu^2} - 2\right),$$

where the integral

$$J(z) = \int_0^1 dx \ln[1 - 4x(1-x)z] + 2 = \int_0^1 dx \ln\frac{1-4x(1-x)z}{1-4x(1-x)}$$

may be represented in the form

$$J(z) = (1-z)\int_1^\infty \frac{d\sigma}{\sqrt{\sigma(\sigma-1)}\,(\sigma - z + i\varepsilon)}. \tag{3}$$

The expression (3) obtained has the same structure as the Cauchy integral. We shall call such formulae *spectral representations.* The spectral representation (3) defines a function of the complex variable z which has a singularity on the cut $\operatorname{Im} z = 0$, $\operatorname{Re} z \geq 1$. On the remaining part of the real axis it is real:

$$\begin{aligned} J(x) &= 2\left(\frac{1-x}{x}\right)^{1/2} \operatorname{arctg}\left(\frac{x}{1-x}\right)^{1/2} \quad \text{for} \quad 0 \leqslant x \leqslant 1, \\ &\left.\begin{aligned} &= 2\left(\frac{x-1}{x}\right)^{1/2} \operatorname{Arth}\left(\frac{x}{x-1}\right)^{1/2} = \\ &= 2\left(\frac{x-1}{x}\right)^{1/2} \ln\left(\sqrt{1-x} + \sqrt{-x}\right) \end{aligned}\right\} \quad \text{for} \quad x \leqslant 0. \end{aligned} \tag{4}$$

For $x > 1$ the integrand in (3) contains a pole. The rules for going round it are determined by the infinitesimal additional term $+i\varepsilon$. Using the symbolic formula (AV.6), we obtain

$$J(x) = -i\pi\left(\frac{x-1}{x}\right)^{1/2} + 2\left(\frac{x-1}{x}\right)^{1/2}\ln\left(\sqrt{x}+\sqrt{x-1}\right) \quad \text{for} \quad x \geqslant 1. \tag{5}$$

Thus, in the ultraviolet region, when $|p^2| \gg m^2$,

$$\begin{aligned} I_{\text{fin}}(p^2, \mu^2) &= -\ln(|p^2|/\mu^2) - i\pi\theta(p^2) + \text{const} = \\ &= -\ln(-p^2/\mu^2) + \text{const}. \end{aligned} \tag{6}$$

24.2. Self-energies of the photon and of the electron. In spinor electrodynamics diagrams topologically equivalent to the one considered may differ in types of external and internal lines (Figure 24.1a, b). When the external lines are photon lines while the internal lines are electron lines, we have the diagram of the self-energy of the photon. The corresponding integral

$$I^{\mu\nu}(k) = \frac{1}{i\pi^2}\int dq\, \text{Sp}\left(\frac{1}{m+\hat{q}}\gamma^{\mu}\frac{1}{m+\hat{q}+\hat{k}}\gamma^{\nu}\right), \tag{7}$$

multiplied by $e^2/16\pi^2 = \alpha/4\pi$, forms the so-called *polarization operator*

$$\alpha\Pi^{\mu\nu}(k) = \frac{\alpha}{4\pi}I^{\mu\nu}(k). \tag{8}$$

The Feynman integral (7) diverges quadratically in the region of large q. For its regularization we use a version of the Pauli–Villars procedure:

$$S_c(q) = \frac{1}{i}\frac{1}{m+\hat{q}} \to \text{reg}\, S_c(q) = \frac{m-\hat{q}}{i}\left(\frac{1}{m^2-q^2} - \frac{1}{M^2+m^2-q^2}\right).$$

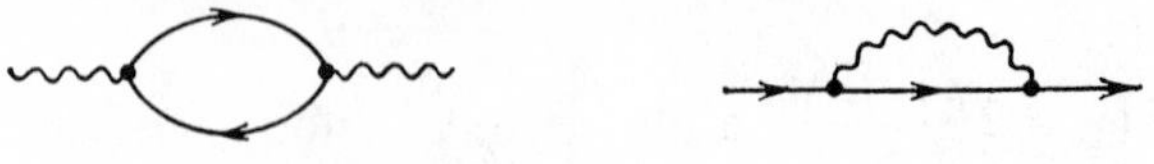

(a) (b)

Fig. 24.1. One-loop diagrams of the second order in spinor electrodynamics: (a) photon self-energy; (b) electron self-energy.

The regularized expression

$$I_M^{\nu\sigma}(k) = \frac{1}{i\pi^2} \int \frac{dq \, \mathrm{Sp}\,(\ldots)}{(m^2-q^2)(M^2+m^2-q^2)[m^2-(k+q)^2][M^2+m^2-(k+q)^2]}, \quad (9)$$

where

$$\mathrm{Sp}\,(\ldots) = \mathrm{Sp}\,[(m-\hat{q})\gamma^\nu(m-\hat{q}-\hat{k})\gamma^\sigma] =$$
$$= 4g^{\nu\sigma}m^2 + 8q^\nu q^\sigma + 4(q^\nu k^\sigma + k^\nu q^\sigma) - 4g^{\nu\sigma}(q^2+qk),$$

reduces with the aid of standard transformations to the form

$$I_M^{\nu\sigma}(k) = 4i \int_0^1 dx \int_0^\infty \frac{d\lambda}{\lambda} \exp i\lambda[x(1-x)k^2 - m^2 - i\varepsilon] \times$$
$$\times (1-e^{-ix\lambda M^2})(1-e^{-i(1-x)\lambda M^2}) \times$$
$$\times \left[ix(1-x)(2k^\nu k^\sigma - g^{\nu\sigma}k^2) + g^{\nu\sigma}\left(\frac{1}{\lambda} + im^2\right)\right].$$

To perform integration over λ, we transform the terms containing λ and λ^2 in the denominator according to the formulae

$$\int_0^\infty \frac{d\lambda}{\lambda} e^{-\varepsilon\lambda}(e^{ia\lambda} - e^{iA\lambda}) = \ln \frac{A+i\varepsilon}{a+i\varepsilon},$$

$$\int_0^\infty \frac{d\lambda}{\lambda^2} e^{-\varepsilon\lambda} f(\lambda) = \int_0^\infty \frac{d\lambda}{\lambda} e^{-\varepsilon\lambda} f'(\lambda).$$

The final result of the calculations in the limit $M^2 \to \infty$ will be

$$\Pi_M^{\nu\sigma}(k) = \frac{1}{2\pi} M^2 g^{\nu\sigma} + \frac{1}{3\pi} \ln \frac{M^2}{m^2}(g^{\nu\sigma}k^2 - k^\nu k^\sigma) + \Pi_{\mathrm{fin}}^{\nu\sigma}(k), \quad (10)$$

where

$$\Pi_{\mathrm{fin}}^{\nu\sigma}(k) = \frac{1}{\pi}(k^2 g^{\nu\sigma} - k^\nu k^\sigma)\, I\left(\frac{k^2}{4m^2}\right), \quad (11)$$

while

$$I(z) = 2\int_0^1 dx\,(1-x)\,x \ln[1-4x(1-x)z].$$

Explicit integration over x allows us to obtain an expression akin to the function $J(z)$ introduced in (3):

$$I(z) = \frac{1}{4} \int_0^z [J'(t) - J'(0)] \frac{dt}{t}. \quad (12)$$

The spectral representation for I has the form

$$I(z) = -\frac{z}{6}\int_1^\infty \frac{1+2\sigma}{\sigma^2}\sqrt{\frac{\sigma-1}{\sigma}}\,\frac{d\sigma}{\sigma-z}. \tag{13}$$

To the electron self-energy diagram (Fig. 24.1b) there corresponds the linearly divergent expression (for its physical meaning see equation (27.4) below):

$$\alpha\Sigma(p) \sim \frac{\alpha i}{4\pi^2}\int \frac{dq}{q^2+i\varepsilon}\gamma^\nu \frac{1}{m-\hat{q}-\hat{p}}\gamma_\nu, \tag{14}$$

which after regularization of the photon propagator

$$\frac{1}{q^2} \to \frac{1}{q^2} + \frac{1}{M^2-q^2} = \frac{M^2}{q^2(M^2-q^2)} \tag{15}$$

reduces to the following:

$$\Sigma_M(p) = \frac{1}{2}\int_0^1 dx\,(2m - x\hat{p})\ln\frac{p^2x(1-x)-(1-x)m^2-xM^2}{xp^2(1-x)-(1-x)m^2}.$$

In the limit $M^2 \to \infty$ we obtain

$$\Sigma_M(p) \simeq \left(m - \frac{\hat{p}}{4}\right)\ln\frac{M^2}{m^2} + \Sigma_{\text{fin}}(p), \tag{16}$$

where

$$\Sigma_{\text{fin}}(p) = \frac{1}{2}\int_0^1 dx\,(2m - x\hat{p})\left(\ln\frac{m^2}{m^2-xp^2} + \ln\frac{x}{1-x}\right) =$$

$$= \left[\hat{p}\left(\frac{p^2+m^2}{4p^2}\right) - m\right]\left(\frac{p^2-m^2}{p^2}\right)\ln\left(\frac{m^2-p^2}{m^2}\right) + m - \hat{p}\left(\frac{3}{8}+\frac{m^2}{4p^2}\right). \tag{17}$$

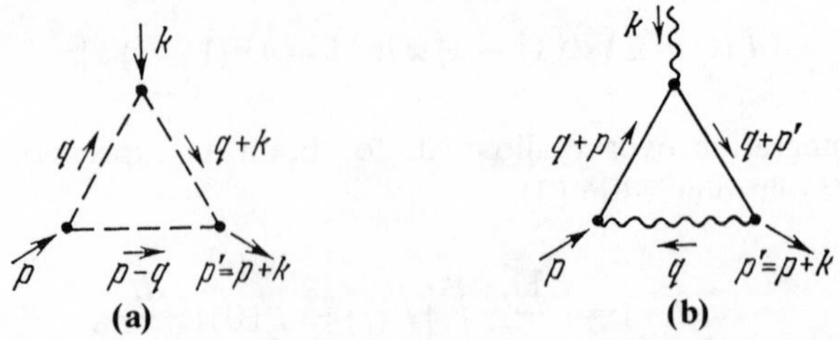

Fig. 24.2. One-loop vertex diagrams of the third order: (a) in the scalar model, (b) in spinor electrodynamics.

24.3. Triangular vertex diagrams. Let us now turn to one-loop diagrams consisting of three lines. In the simplest case, when all these lines are scalar (Figure 24.2a), the corresponding Feynman integral

$$\Gamma(p, k)=\frac{1}{i\pi^2}\int\frac{dq}{m^2-q^2}\cdot\frac{1}{m^2-(q+k)^2}\cdot\frac{1}{m^2-(q-p)^2} \tag{18}$$

converges. As a result of integration over q, it reduces to the form

$$\Gamma(p^2, k^2, p'^2)=\int_0^1\frac{\{dx\}_3}{m^2-x_1x_2k^2-x_1x_3p^2-x_2x_3p'^2}, \quad p'=p+k. \tag{19}$$

Here the abbreviated notation (AVI.3) introduced after (22.7) is used.

Let us examine the physically important diagram of spinor electrodynamics depicted in Fig. 24.2b. The corresponding integral

$$\frac{i}{\pi^2}\int\frac{dq}{-q^2}\gamma^\mu\frac{1}{m-\hat{q}-\hat{p}}\gamma^\nu\frac{1}{m-\hat{q}-\hat{p}'}\gamma_\mu \tag{20}$$

diverges logarithmically for large q. Performing regularization of the photon propagator in accordance with formula (15), we have

$$\Gamma_M^\nu(p, p')=\frac{M^2i}{\pi^2}\int\frac{dq}{q^2(q^2-M^2)}\frac{\gamma^\mu(m+\hat{q}+\hat{p})\gamma^\nu(m+\hat{q}+\hat{p}')\gamma_\mu}{[m^2-(q+p)^2][m^2-(q+p')^2]}, \tag{21}$$

and after necessary transformations, in the limit of large M we obtain

$$\mathrm{reg}_M\Gamma^\nu(p', p, k)=\gamma^\nu\ln\frac{M^2}{\mu^2}+\Gamma^\nu_{\mathrm{fin}} \tag{22}$$

$$\Gamma^\nu_{\mathrm{fin}}=2\int_0^1\{dx\}_3\left(\gamma^\nu\ln\frac{\mu^2}{Z}+\frac{N^\nu}{Z}\right), \tag{23}$$

where

$$\begin{aligned}Z&=m^2(1-x_1)-x_1x_2p'^2-x_1x_3p^2-x_2x_3k^2,\\ N^\nu&=\gamma^\nu\{m^2+(1-x_3)(1-x_2)k^2-x_1(1-x_2)p'^2-x_1(1-x_3)p^2\}+\\ &+k^\nu\{(1-x_2)(1-x_3)(\hat{p}-\hat{p}')+2m(x_2-x_3)\}+\\ &+(p^\nu+p'^\nu)\{(1-x_2)(1-x_3)(\hat{p}+\hat{p}')-2mx_1\}-\\ &-x_1\hat{p}'\gamma^\nu\hat{p}-2x_2(1-x_2)p'^\nu\hat{p}'-2x_3(1-x_3)p^\nu\hat{p}.\end{aligned}$$

On the right-hand side of (23) certain simplifications may be carried out and two integrations may be performed explicitly. As these calculations are cumbersome, we shall not present them here. We point out only that in the important particular case when the electron before and after the act of interaction is on the mass shell, the corresponding projection of the vertex function may be represented in the form

$$\bar{u}(p')\,\Gamma^{\nu}_{\text{fin}}\,u(p) = f_1(k^2)\,\bar{u}(p')\,\gamma^{\nu}u(p) + f_2(k^2)\,\bar{u}(p')\left\{\frac{p^{\nu}+p'^{\nu}}{2m} - \gamma^{\nu}\right\}u(p), \tag{24}$$

i.e., it may be expressed in terms of the two form factors

$$f_1(k^2) = 2\int_0^1 \{dx\}_3 \ln\frac{\mu^2}{Z_0} + \int_0^1 \{dx\}_3 \frac{N_1}{Z_0}, \tag{25}$$

$$f_2(k^2) = \int_0^1 \{dx\}_3 \frac{N_2}{Z_0}, \tag{26}$$

where

$$Z_0 = m^2(1-x_1)^2 - x_2^2x_3^2k^2,$$
$$N_1 = 2m^2(1-4x_1+x_1^2) - 2k^2(1-x_2)(1-x_3), \quad N_2 = 4m^2x_1(x_1-1).$$

The integrals in (25) and (26) may be expressed in terms of elementary functions. After the substitution $x_2 = t(1 - x_1)$ we obtain for f_2

$$f_2(k^2) = 2\int_0^1 \frac{m^2\,dt}{t(1-t)k^2 - m^2 + i\varepsilon} = -8\frac{m^2}{k^2Q}\operatorname{arctg}\frac{1}{Q}; \qquad Q = \left(\frac{4m^2}{k^2} - 1\right)^{1/2} \quad \text{for} \quad k^2 \leqslant 4m^2 \tag{27}$$

and, specifically,

$$f_2(0) = -2. \tag{28}$$

At the same time the substitution $x_2 = t(1 - x_1)$ in the second term of the form factor f_1 exposes the nonintegrable singularity at $x_1 = 1$. This singularity is a manifestation of the *infrared* catastrophe and corresponds to the singularity of the integrand of the initial Feynman integral (20) at small values of q, i.e., when integration is carried out over the momenta of the *soft* virtual photons. A similar singularity is encountered also in the derivative of the operator for the electron self-energy (17) evaluated on the mass shell. We shall not, however, discuss here the problem of infrared divergences in quantum electrodynamics, and we refer the reader to a brief exposition of the physical reasons for their appearance presented below in Section 29.4 as well as to the literature cited therein. We note only that if a small mass λ_0 is ascribed to the virtual photon in the diagram in Figure 24.2b then the aforementioned singularity may be expressed in terms of the logarithm of this mass:

$$\int_0^1 \frac{dx_1}{1-x_1} \to \ln \frac{\lambda_0}{m} \tag{29}$$

(see on this topic Part 5 of Appendix VI). At the same time the form factor $f_1(k^2)$ can be "displaced by a constant":

$$f_1(k^2) \to f_1(k^2) + c$$

through the renormalization of the coupling constant (i.e., of the electric charge e; see Sections 25 and 28.1 below) as a result of which the divergent additive components in (25) independent of the momenta have no direct physical meaning.

The second term in (24), with the aid of the relation

$$2\bar{u}(\boldsymbol{p}')[p^{\nu} + p'^{\nu} - 2m\gamma^{\nu}]u(\boldsymbol{p}) = \bar{u}(\boldsymbol{p}')[\gamma^{\nu}\hat{k} - \hat{k}\gamma^{\nu}]u(\boldsymbol{p}); \qquad k = p' - p, \tag{30}$$

following from the Dirac equations, may be transformed into a form convenient for evaluating the radiative correction to the magnetic moment of the electron (see Section 30.2 below).

24.4. Ultraviolet divergences of higher orders. The examples examined above of the simplest one-loop diagrams lead to the conclusion that ultraviolet divergences represent a typical, frequently encountered phenomenon (a "rule" rather than an "exception"). In passing over to more complicated diagrams (higher orders of perturbation theory, greater numbers of loops) the picture remains essentially the same.

As an illustration we shall schematically consider the process of Compton scattering. In the lowest (second) order of perturbation theory this process is described by two Feynman diagrams depicted in Figure 24.3. These diagrams have no closed loops, and the corresponding matrix elements do not contain any integrations over the four-momenta.

Fig. 24.3. Second-order diagrams for Compton scattering.

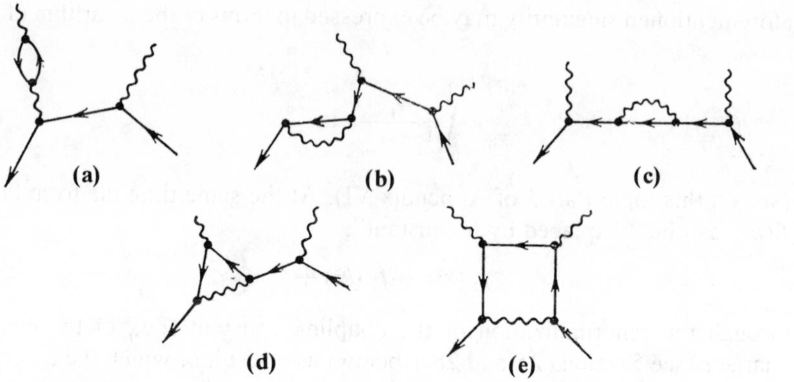

Fig. 24.4. Some fourth-order diagrams for Compton scattering.

In the fourth order the process under consideration is described by eight one-loop diagrams, some characteristic representatives of which are presented in Figure 24.4. Two of the diagrams contain polarization insertions into the external photon lines (see Figure 24.4a); two diagrams contain self-energy insertions (subdiagrams) into the external electron-positron lines (Figure 24.4b); one diagram involves a self-energy subdiagram on the internal electron–positron line (Figure 24.4c), and two diagrams contain triangular vertex subdiagrams (Figure 24.4d). In accordance with the results of Section 24.3, all these diagrams contain ultraviolet divergences. Finally, the last diagram presented, in Figure 24.4e, has a closed loop with four vertices. The corresponding integral over the momenta converges.

The sixth order of perturbation theory corresponds to two-loop

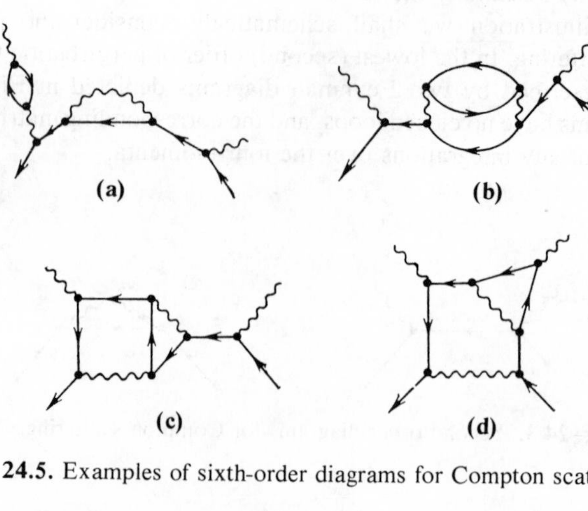

Fig. 24.5. Examples of sixth-order diagrams for Compton scattering.

diagrams; some of them are depicted in Figure 24.5. Here are encountered diagrams with two independent one-loop subdiagrams (similar to those which are depicted in Figure 24.5a) as well as diagrams in which the divergent one-loop subdiagram enters into the element involving the second integration over the momentum. Examples of such cases are presented in Figure 24.5b, c, d. Then, as a rule, both integrations contain ultraviolet divergences, which thus overlap each other.

Our immediate aim is to establish the general structure of the divergent contributions to the transition matrix elements and to construct a procedure for their removal (the renormalization procedure).

25. ISOLATION OF THE DIVERGENCES.

25.1. The structure of one-loop divergences. As we have seen in the preceding section, there is a remarkable property characteristic of all divergent one-loop integrals: ultraviolet infinities may be isolated in the form of additive components (see (24.2), (24.10), (24.16), (24.22)). The latter have the form of polynomials in powers of the components of the external momenta with divergent (in the case when the regularization is removed) coefficients.

If we now pass from the momentum to the configuration representation, then the divergent contributions will be represented by a polynomial in the derivatives of the δ-functions. This, for example, (23.1), (23.2), and (24.2) give

$$\operatorname{reg}[D^c(x)]^2 = \frac{\pi^2}{i(2\pi)^8}\int e^{-ipx} I_\Lambda(p^2)\,dp = \frac{1}{16\pi^2 i} I_{\text{sing}}\,\delta(x) + \widetilde{D_c^2}(x), \tag{1}$$

where the singular constant in the Pauli–Villars regularization is equal to $I_{\text{sing}} = \ln(\mu^2/M^2)$, while the function

$$\widetilde{D_c^2}(x) = \frac{\pi^2}{i(2\pi)^8}\int e^{-ipx}\tilde{I}(p^2,\ \mu^2)\,dp \tag{2}$$

is free from ultraviolet divergences.

In a similar manner

$$\begin{aligned}\operatorname{reg}[\Pi^{\mu\nu}(x)] &= -ie^2\,\mathrm{Sp}[\gamma^\mu \operatorname{reg} S_c(x)\,\gamma^\nu \operatorname{reg} S_c(-x)] = \\ &= \frac{\alpha}{2\pi} g^{\mu\nu} M^2 \delta(x) + \frac{\alpha}{3\pi}\ln\frac{M^2}{m^2}(g^{\mu\nu}\Box + \partial^\mu\partial^\nu)\,\delta(x) + \tilde{\Pi}^{\mu\nu}(x), \end{aligned}\tag{3}$$

$$\begin{aligned}\operatorname{reg}[\Sigma(x)] &= -ie^2\gamma^\nu S^c(x)\,\gamma_\nu \operatorname{reg} D_0^c(x) = \\ &= \frac{\alpha}{4\pi}\ln\frac{M^2}{m^2}(4m - i\hat{\partial})\,\delta(x) + \tilde{\Sigma}(x), \end{aligned}\tag{4}$$

$$\begin{aligned}\operatorname{reg}\Gamma^\nu(x,\ z;\ y) &= ie^2 \operatorname{reg} D_0^c(x-z)\,\gamma^\mu S^c(x-y)\,\gamma^\nu S^c(y-z)\,\gamma_\mu = \\ &= \frac{\alpha}{4\pi}\gamma^\nu \ln\frac{M^2}{m^2}\,\delta(x-y)\,\delta(x-z) + \tilde{\Gamma}^\nu(x,\ z;\ y), \end{aligned}\tag{5}$$

where the functions $\tilde{\Pi}$, $\tilde{\Sigma}$, $\tilde{\Gamma}$ are Fourier transforms of the finite expressions (24.11), (24.17), and (24.33).

25.2. The contribution to the *S*-matrix. Let us turn to the structure of the contributions to the scattering matrix from the divergent terms of the right-hand sides (1), (3), (4), and (5).

We shall start with the scalar "fish". In the model $\mathcal{L} = g\varphi^3$ the second-order term in the S-matrix contains a piece corresponding to the self-energy of the particle (Figure 25.1):

$$S_2(x, y) = i^2 g^2 T(\varphi^3(x)\varphi^3(y)) = 18g^2 :\varphi(x)[D^c(x-y)]^2\varphi(y): + \ldots,$$

which contains an ultraviolet divergence. Upon regularization on the basis of (1) we obtain

$$S_2(x, y) = ig^2 C :\varphi(x)\,\delta(x-y)\,\varphi(y): + 18g^2 :\varphi(x)\widetilde{[D^c(x-y)]^2}\varphi(y): + \\ + \ldots = ig^2 C :\varphi(x)\,\delta(x-y)\,\varphi(y): + \tilde{S}_2(x, y), \quad (6)$$

where $C = (9/8\pi^2)\ln(m^2/\mu^2)$, and $\tilde{S}_2$ is free from ultraviolet divergences. In the evaluation of matrix elements of $\tilde{S}_2$ the finite expression (2) corresponds to the scalar "fish".

Subsituting (6) into the expansion for the S-matrix, we find

$$S = 1 + \int S_1(x)\,dx + \frac{i^2}{2}\int S_2(x, y)\,dx\,dy + \ldots = \\ = 1 + i\int [\mathcal{L}(x) + \Delta\mathcal{L}(x)]\,dx + \frac{i^2}{2}\int \tilde{S}_2(x, y)\,dx\,dy + \ldots, \quad (7)$$

where

$$\Delta\mathcal{L}(x) = \frac{g^2}{2} C :\varphi^2(x):. \quad (8)$$

Thus, isolation from S_2 of the divergent quasilocal contribution $\phi(x)\delta(x-y)\phi(y)$ is equivalent to adding the local term $\Delta\mathcal{L}(x)$ to S_1. Such terms are called *counterterms*.

The scalar "fish" enters also into the second-order term in the model $\mathcal{L} = h\varphi^4$:

$$S_2(x, y) = i^2 h^2 T(\varphi^4(x)\varphi^4(y)) = 72h^2 :\varphi^2(x)\,D_c^2(x-y)\,\varphi^2(y): - \\ - i96h^2 :\varphi(x)\,D_c^3(x-y)\,\varphi(y): + \ldots, \quad (9)$$

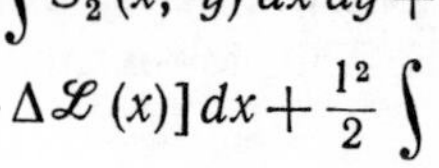

Fig. 25.1. Second-order diagram of self-energy in the scalar φ^3 model.

Fig. 25.2. Second-order diagram for scattering in the φ^4 model.

which contributes to the scattering amplitude 2 ↔ 2 (Figure 25.2).

The second term of the right-hand side of (9) corresponds to the "walnut" diagram depicted in Figure 25.3. The corresponding marix element diverges quadratically. Calculations give

$$\text{reg}\,[D_c(x-y)]^3 = -\frac{A_2}{48}\,\delta(x) - \frac{A_0}{48}\,\square_x\,\delta(x) + \widetilde{D_c^3}(x), \tag{10}$$

where for the Pauli–Villars regularization

$$A_0 = a_0 \ln\frac{M^2}{m^2}, \quad A_2 = a_2 M^2,$$

while $\widetilde{D}_c^3$ is free from ultraviolet divergences. Substituting (9), (10), and (1) into the expansion of the S-matrix, we obtain a formula of the form of (7) where $\Delta\mathscr{L}$ consists of three counterterms

$$\Delta\mathscr{L}(x) = Ch^2 :\varphi^4(x): + A_2 h^2 :\varphi^2(x): - A_0 h^2 :\partial_\mu\varphi\,\partial^\mu\varphi:. \tag{11}$$

In the S-matrix of spinor electrodynamics the second-order term contains divergent components corresponding to (3) and (4):

$$S_2(x, y) = -i :\overline{\psi}(x)\,\Sigma(x-y)\,\psi(y): - \\ -i :A_\mu(x)\,\Pi^{\mu\nu}(x-y)\,A_\nu(y): + (x \leftrightarrow y) + \ldots,$$

while the third-order term includes components (Figure 25.4) containing as factors the self-energies (3), (4) as well as the triangular vertex diagram (5):

$$S_3(x, y, z) = -ie :\overline{\psi}(z)\,A_\nu(x)\,\Gamma^\nu(z, y; x)\,\psi(y): + \\ -ie :\overline{\psi}(z)\,\Sigma(z-y)\,S^c(y-x)\,\gamma^\nu A_\nu(x)\,\psi(x): + \\ -ie :\overline{\psi}(x)\,\gamma^\nu A_\nu(x)\,S^c(x-z)\,\Sigma(z-y)\,\psi(y): + \\ +ie :A_\nu(x)\,\Pi^{\nu\mu}(x-z)\,D^c_{\nu\rho}(z-y)\,\overline{\psi}(y)\,\gamma^\rho\,\psi(y): + \\ + (x \leftrightarrow y) + (x \leftrightarrow z) + (y \leftrightarrow z) + \ldots$$

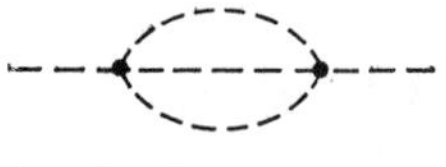

Fig. 25.3. Second-order "walnut" diagram representing self-energy in the φ^4 model.

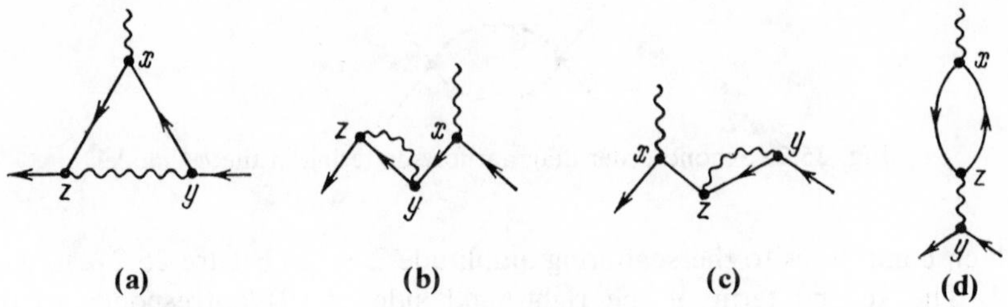

Fig. 25.4. Third-order divergent contributions to the scattering matrix of spinor electrodynamics.

Isolation from Σ, Π, and Γ of the local terms

$$\begin{aligned}\Sigma(x-y) &= (am+ib\hat{\partial})\,\delta(x-y)+\Sigma'(x-y),\\ \Pi^{\mu\nu}(x-y) &= cg^{\mu\nu}\delta(x-y)+d\,(g^{\mu\nu}\,\Box+\partial^{\mu}\partial^{\nu})\,\delta(x-y)+\Pi'^{\mu\nu},\\ \Gamma^{\nu}(x,\,z;\,y) &= f\gamma^{\nu}\delta(x-y)\,\delta(x-z)+\Gamma'^{\nu}(x,\,z;\,y)\end{aligned} \tag{12}$$

leads to the following reconstruction of the second-order contribution:

$$\begin{aligned}\frac{1}{2}\int S_2(x,\,y)\,dx\,dy = &-i\int\bar{\psi}(x)\,(am+ib\hat{\partial})\,\psi(x)\,dx+\\ &-i\int A_{\mu}(x)\,[cg^{\mu\nu}+d\,(g^{\mu\nu}\,\Box+\partial^{\mu}\partial^{\nu})]\,A_{\nu}(x)\,dx+\\ &+\frac{1}{2}\int S_2'(x,\,y)\,dx\,dy\end{aligned} \tag{13}$$

as well as of the third-order contribution:

$$\begin{aligned}\frac{1}{3!}\int S_3(x,\,y,\,z)\,dx\,dy\,dz &=\\ = ie:\int A_{\mu}(x)\,&[cg^{\mu\nu}+d\,(g^{\mu\nu}\,\Box_x+\partial^{\mu}_x\,\partial^{\nu}_x)]\,D_{\rho\nu}(x-y)\times\\ &\times\bar{\psi}(y)\,\gamma^{\rho}\psi(y):dx\,dy-ie\int:\bar{\psi}(x)\,(am+ib\hat{\partial}_x)\times\\ &\times S^c(x-y)\,\gamma^{\nu}A_{\nu}(y)\,\psi(y):\,dx\,dy+\\ -ie\int:\bar{\psi}(x)\,&\gamma^{\nu}A_{\nu}(x)\,S^c(x-y)\,(am+b\hat{\partial}_y)\,\psi(y):\,dx\,dy+\\ -ief\int:\bar{\psi}(x)\,&A_{\nu}(x)\,\gamma^{\nu}\psi(x):\,dx+\frac{1}{3!}\int S_3'(x,\,y,\,z)\,dx\,dy\,dz.\end{aligned} \tag{14}$$

It may now be seen that the first two terms on the right-hand side of (13) and the fourth term on the right-hand side of (14) may be included in the first-order term

$$\begin{aligned}\int S_1(x)\,dx &\to i\int[\mathscr{L}(x)+\Delta\mathscr{L}(x)]\,dx,\\ \Delta\mathscr{L}(x) &= -:\bar{\psi}(x)\,(am+ib\hat{\partial})\,\psi(x):-c:A_{\nu}(x)\,A^{\nu}(x):+\\ -d:A_{\mu}(x)\,\Box\,A^{\mu}(x)&:+d:(\partial^{\mu}A_{\mu})\,(\partial^{\nu}A_{\nu}):-ef:\bar{\psi}(x)\,\hat{A}(x)\,\psi(x):.\end{aligned} \tag{15}$$

At the same time the first three terms of the right-hand side of (14) may be represented in the form

$$\frac{i^2}{2}\int T\{\mathscr{L}(x)+\Delta\mathscr{L}(x),\ \mathscr{L}(y)+\Delta\mathscr{L}(y)\}\,dx\,dy-$$
$$-\frac{i^2}{2}\int T(\mathscr{L}(x)\,\mathscr{L}(y))\,dx\,dy=i^2\int T(\mathscr{L}(x),\ \Delta\mathscr{L}(y))\,dx\,dy+\ldots,$$

where the symbol $\Delta\mathscr{L}$ stands for the right-hand side of (15) taken without the last term.

Thus, we have verified that isolation from S_2 and S_3 of the singular contributions corresponding to the quasilocal terms in (12) is equivalent to adding the local counterterms (15) to the interaction Lagrangian $\mathscr{L}(x)$.

This equivalence forms the basis of the method for removing ultraviolet divergences known as the *renormalization method.*

25.3. Counterterms and renormalization. The idea of renormalization is based on the fact that the counterterms (9), as a rule, have the same structure as separate components of the initial total Lagrangian.

Let us consider the model of a real scalar field with a quartic interaction

$$\mathscr{L}_{\text{tot}}(x)=\frac{1}{2}\,\partial_\mu\varphi\,\partial^\mu\varphi-\frac{m^2}{2}\,\varphi^2(x)+h\varphi^4(x).$$

Comparison with (11) shows that the divergent counterterms have the structure of separate terms of the total Lagrangian.

To establish the physical meaning of adding counterterms of such a type to the Lagrangian, we shall divert our attention from the divergences for a while, and consider the counterterms

$$\Delta\mathscr{L}(x)=\frac{a_0}{2}\,\partial_\mu\varphi\,\partial^\mu\varphi-\frac{a_2}{2}\,\varphi^2(x)+c\varphi^4(x) \tag{11'}$$

with finite coefficients a_0, a_2, c. Summing $\mathscr{L}_{\text{tot}}$ with $\Delta\mathscr{L}$, we obtain

$$\mathscr{L}'=\frac{z_1}{2}\,(\partial_\mu\varphi)^2-\frac{z_1 m'^2}{2}\,\varphi^2+z_2 h\varphi^4, \tag{16}$$

where $z_1 = 1 + a_0$, $z_1 m'^2 = m^2 + a_2$, $z_2 h = h + c$.

Let us first discuss the meaning of the coefficient z_1. The first two terms on the right-hand side of (16) represent the Lagrangian of the free scalar field with mass m' multipled by z_1. The absolute normalization of the free Lagrangian manifests itself only when the fields are quantized, i.e., in equations of the type of (7.1). It is not difficult to verify that by performing the substitution

$$\varphi\to\varphi'=\sqrt{z_1}\,\varphi, \tag{17}$$

we arrive at the renormalized field $\varphi'(x)$ which satisfies the normal commutation relations (8.4). Substituting (17) into (16), we transform $\mathscr{L}'$ into the form

$$\mathscr{L}' = \frac{1}{2}(\partial_\mu \varphi')^2 - \frac{m'^2}{2}\varphi'^2 + h'\varphi'^4, \tag{18}$$

where

$$h' = z_1^{-2} z_2 h = \frac{h+c}{(1+a_0)^2}, \quad m'^2 = \frac{m^2+a_2}{1+a_0}. \tag{19}$$

Thus, we see that the joint effect of the three counterterms (11), (11′) reduces to the renormalization of two physical parameters of the model: the mass m and the coupling constant h.

In reality the coefficients of the counterterms (of the types C, A_0, A in (11), (8)) are singular, i.e., tend to infinity when the regularization is removed. As a result, relations of the type (19) connecting the initial "bare" values of m and h with the physical "renormalized" values m' and h' also become singular.

Such renormalization relations for the mass and norm of the operator wave function have already been obtained here in the heavy-nucleon model (Section 13.4). We recall that these relations became singular upon transition to the pointlike nucleon. In relativistic quantum field theory the nature of ultraviolet divergences is also due to the local character of the interaction. This fact is particularly well illustrated by the examples, considered in Section 24, of the electron self-energy diagrams as well as the vertex diagram. the regularization utilized therein of the photon propagator (24.15) is equivalent to a "smearing out" of a pointlike electron, i.e., to the transition to the nonlocal interaction

$$j_\mu(x)\, A^\mu(x) \to j_\mu(x) \int K(x-y)\, A^\mu(y)\, dy,$$

where the form factor

$$K(x) = \frac{1}{(2\pi)^4} \int e^{iqx} \frac{M^2}{M^2 - q^2}\, dq$$

in the limit $M^2 \to \infty$ tends to the δ-function.

In the renormalization method, which is treated in greater detail below (see Sections 28, 29), singular renormalization relations of the type (19) are considered purely formally. It must be recognized that the bare values of constants of the type of m and h occurring in the initial Lagrangian and connected with free fields (more precisely, with fields with infinitesimal interaction) have no physical meaning, because such idealized fields do not exist in nature and it is not possible to connect the bare parameters with any experiments, even with idealized mental experiments.

At the same time, allowing for radiative corrections of higher orders leads to the conclusion that not only S_1, but also the higher S_n ($n \geq 2$) depend only on the renormalized masses and coupling constants. Therefore it turns out to be possible to connect the renormalized parameters with experimentally measured quantities and consider them as physical masses and charges.

The fact that the matrix elements free from divergences, obtained as a result of renormalization, depend only on the renormalized physical masses and coupling constants

$$g' = f_g(m, g, \Lambda), \quad m' = f_m(m, g, \Lambda) \tag{20}$$

and that there do not exist any additifonal dependences on the bare m, g and the singular parameter Λ (or separately on any other parameters) represents an important property of a certain class of models which are called *renormalizable*.

25.4. Divergences and distributions. Another possible point of view is that the removal of ultraviolet divergences in effect reduces to the subtraction from the diverging matrix elements of the corresponding polynomials. We draw attention to the fact that to divergent Feynman integrals over virtual momenta in the configuration representation there always correspond products of a certain number of propagators $D_c(x)$, $C_0^c(x)$, $S^c(x)$, . . . As was established in Section 18.3, these latter contain components of the form $\theta(x^0)$, $\delta(x^2)$ and from a mathematical point of view are distributions. Unlike ordinary functions encountered in mathematical analysis which establish the correspondence number → number, distributions are defined as kernels of linear functionals, and they establish the correspondence function → number. Therefore the operation of multiplication of distributions by each other requires a special definition. Without going into details of this quite complicated problem (see the *Introduction*, Sections 18, 19), we point out here that a transition such as

$$[D^c(x)]^2 \to \text{reg}\,[D^c(x)]^2 \to \widetilde{D_c^2}(x)$$

may be considered as the *definition* of the product D_c^2. In other words, one may assume

$$D_c(x) \cdot D_c(x) \underset{\text{def}}{=} \widetilde{D_c^2}(x), \tag{21}$$

where the right-hand side is defined in accordance with (2).

When one reasons in such a manner, the ultraviolet divergences do not occur at all in the S-matrix and there is no need for infinite renormalizations of masses and charges. It is necessary only to fix uniquely the arbitrariness in the finite quantities related to parameters of the type μ^2 from the formula (24.1).

It then turns out that in renormalizable models of the quantum field theory, matrix elements are expressed in terms of certain functions of the parameters m, g of the Lagrangian and of arbitrary constants like μ:

$$\tilde{g} = \varphi_g (m, g, \mu), \quad \tilde{m} = \varphi_m (m, g, \mu) \tag{22}$$

and they do not depend separately on m, g, and μ. These subjects will be considered in greater detail in the following chapter.

CHAPTER VII

REMOVAL OF DIVERGENCES

26. THE GENERAL STRUCTURE OF DIVERGENCES

26.1. Higher-order divergences. In the preceding section we examined ultraviolet divergences in the two lowest orders of perturbation theory in the scalar nonlinear model, and in spinor electrodynamics. There it was established, first of all, that in the case of the simplest one-loop Feynman diagrams the ultraviolet divergences may be isolated into quasilocal terms. Second, it was shown that, owing to their quasilocal nature, such terms give a contribution to the scattering matrix which has the same structure as the lower-order terms of perturbation theory, and because of this they ultimately reduce to local contributions to the Lagrangian, i.e., to counterterms.

We shall now carry out a similar investigation of higher-order diagrams, and show that, although in the general case divergent contributions are not of a quasilocal nature, they nevertheless also reduce to counterterms. For simplicity we shall restrict ourselves to the scalar nonlinear model with the quartic self-interaction

$$\mathscr{L} = -\frac{16\pi^2}{4!} h\varphi^4(x). \tag{1}$$

Second-order diagrams depicted in Figure 25.2 and 25.3 were considered in Section 25.2. In the third order the scattering process is described by the two-loop diagrams presented in Figure 26.1.

The diagram of Figure 26.1a represents an iteration of the "fish" diagram, and the corresponding matrix element in the configuration representation is proportional to the expression

$$D_c^2(x-y)\, D_c^2(y-z). \tag{2}$$

Upon regularization with the aid of (25.1) we have

$$\mathrm{reg}_M[D_c^2(x-y)\,D_c^2(y-z)]=\left(\frac{i}{16\pi^2}\ln\frac{M^2}{\mu^2}\right)^2\delta(x-y)\,\delta(y-z)+$$
$$+\frac{i}{16\pi^2}\ln\frac{M^2}{\mu^2}\left[\delta(x-y)\,\widetilde{D_c^2}(y-z)+\delta(y-z)\,\widetilde{D_c^2}(x-y)\right]+$$
$$+\widetilde{D_c^2}(x-y)\,\widetilde{D_c^2}(y-z). \quad (3)$$

Correspondingly, the Fourier transform in the momentum representation reduces to the square of the integral (23.7), i.e., to the expression

$$[\mathrm{reg}_M I(k^2)]^2=\left(\ln\frac{M^2}{\mu^2}\right)^2-2\ln\frac{M^2}{\mu^2}I_{\mathrm{fin}}(k^2)+(I_{\mathrm{fin}}(k^2))^2, \quad (4)$$

where I_{fin} is described by (24.1).

The structure of the singular components of the diagram in Fig. 26.1b is most simply established in the momentum representation. We parametrize the momenta as in Figure 26.2, and consider the following two-loop Feynman integral:

$$K(k_1,\ldots,k_4)\sim\int\frac{i\,dq}{\pi^2}\,\frac{1}{m^2-q^2}\cdot\frac{1}{m^2-(q-k)^2}\times$$
$$\times\int\frac{i\,dp}{\pi^2}\cdot\frac{1}{m^2-p^2}\cdot\frac{1}{m^2-(q+p-k_4)^2}. \quad (5)$$

As it is not difficult to verify, this integral diverges in the region of large momenta. We now introduce a "smooth cutoff" with respect to each integration variable:

$$\int dq\to\mathrm{reg}_{M(F)}\int dq=\int\frac{dq\,M^2}{M^2-q^2}.$$

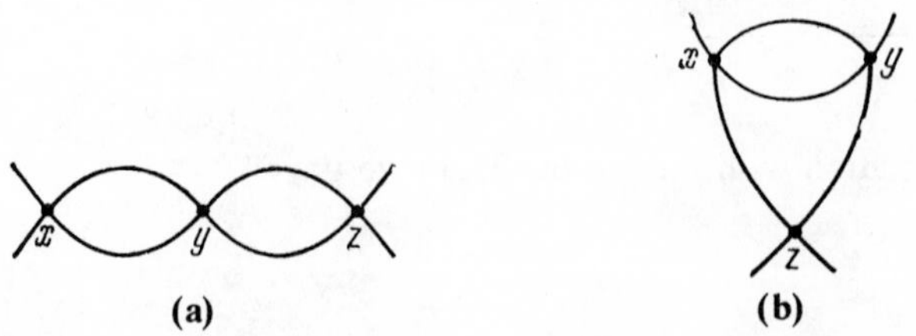

Fig. 26.1. Two-loop scattering diagrams of third order in the ϕ^4 model.

We obtain

$$\mathrm{reg}_{M\,(F)}\,K=\int \frac{i\,dq}{\pi^2}\cdot\frac{M^2}{M^2-q^2}\cdot\frac{1}{m^2-q^2}\cdot\frac{1}{m^2-(q-k)^2}\,\mathrm{reg}_{M\,(F)}\,I\,(q-k_4). \quad (6)$$

In this convergent expression we have established a definite sequence for the integrations, and have utilized the notation of (23.1). Applying (23.7) in the integrand, we find

$$\mathrm{reg}_{M\,(F)}\,K=-\ln\frac{M^2}{\mu^2}\,\mathrm{reg}_{M\,(F)}\,I\,(k)+ \\ +\int \frac{i\,dq}{\pi^2}\,\frac{M^2}{M^2-q^2}\cdot\frac{1}{m^2-q^2}\cdot\frac{1}{m^2-(q-k)^2}\,I_{\mathrm{fin}}\,(q-k_4). \quad (7)$$

The structure of the first term is evident from (23.7). The second term also contains a singularity in the limit $M^2 \to \infty$, as follows from the asymptotic formula (24.6) for I_{fin}. It is not difficult to verify that this singular contribution is additive. Calculations show that it is proprotional to the square of the "large" logarithm $\ln M^2$. Omitting the calculations, we present the final result:

$$\mathrm{reg}_{M\,(F)}\,K=\left(\ln\frac{M^2}{\mu^2}\right)^2-\ln\frac{M^2}{\mu^2}\,I_{\mathrm{fin}}\,(k)-\frac{1}{2}\left(\ln\frac{M^2}{\mu^2}\right)^2+K_{\mathrm{fin}}\,(\ldots k\ldots). \quad (8)$$

Here the first two terms of the right-hand side correspond to the term of the right-hand side of (7).

The resulting expressions (4) and (8) have the same structure. In the limit of large M they are polynomials in $\ln M$. The coefficients of $(\ln M)^2$ are numbers, while the coefficients of $\ln M$ are functions of the momentum variables.

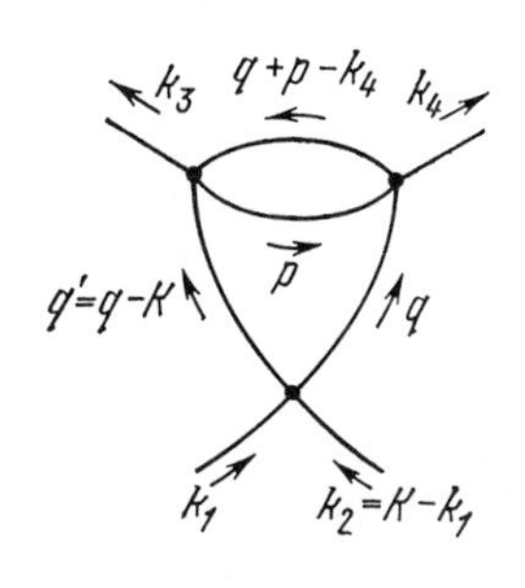

Fig. 26.2. Notation for momenta adopted in the integral (5).

In the case of higher orders the scattering diagrams (four-legged diagrams) also diverge logarithmically. The complete expression for each of them in the limit of large M is represented by a polynomial in powers of ln m. The highest power of the divergent logarithm is equal to the number of loops, l (i.e., to the number of independent 4-momentum integrations) of the diagram. the coefficient of $(\ln M)^l$ is a number, while the coefficients of the lower powers of ln M are functions of the momentum variables.

26.2. The connection with counterterms and renormalizations. We shall now show that the terms involving lower powers of divergent logarithms, which do not have a quasilocal structure in the x-representation, also reduce to counterterms and renormalizations.

To this end we introduce a graphical representation for the process of isolation of divergences. We shall take the following picture to correspond to (24.2):

(9)

where the shaded fish corresponds to the finite function I_{fin}, while the small cross within the circle corresponds to the divergent logarithm. Then the contribution of the "fish" diagram to the second-order scattering matrix element (25.9) will be represented in the form

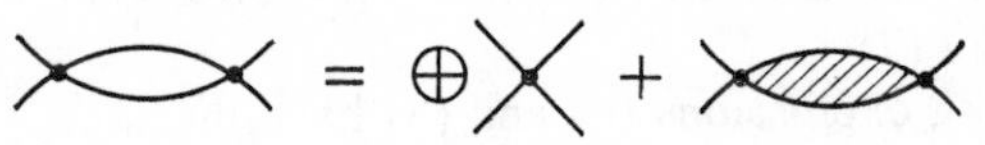

(10)

The contribution of the diagram of Fig. 26.1a, in conformity with (4), will be

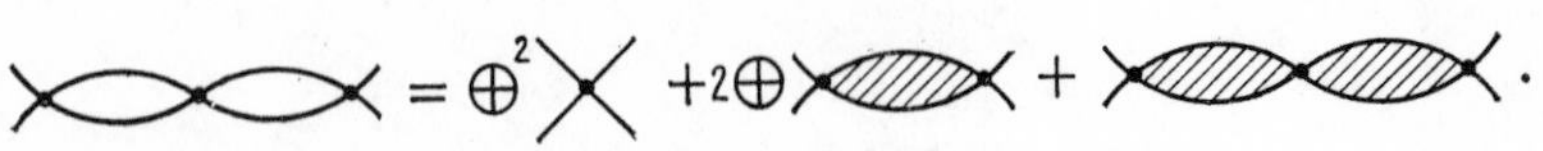

(11)

For the diagram of Fig. 26.2b, similarly,

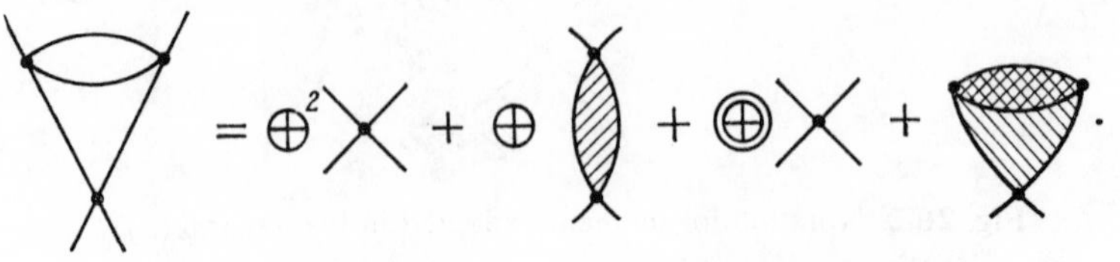

(12)

In the latter picture, the small cross within the double circle represents the third term of (8), i.e., the divergent part of the second term of the right-hand side of (7). The meaning of all the remaining symbols is evident.

By collecting the contributions to the scattering matrix of the terms (10), (11), (12) we obtain

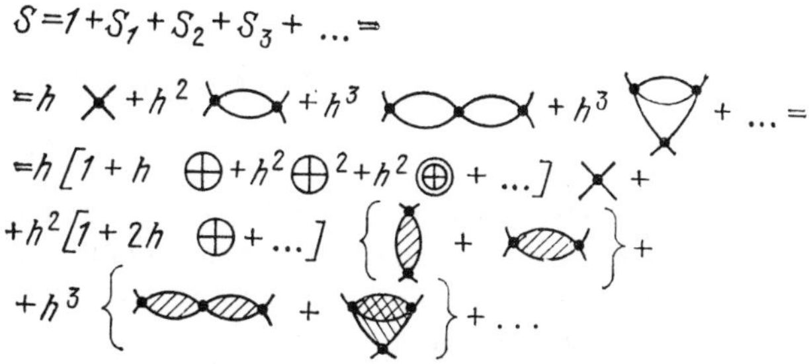

(13)

We see that the divergent contributions from the higher-order multiloop diagrams, which have a quasilocal structure, contribute to the first-order term S_1, i.e., to the counterterms of the Lagrangian, which in the example under consideration leads to the renormalization of the coupling constant:

$$h \to h_* = h\,(1 + h \oplus + h^2 \oplus^2 + h^2 \textcircled{$\oplus$} + \ldots).$$

The singular components of the second type, which are not of a quasilocal character and which in the momentum representation depend on the momenta (and are similar to the second terms on the right-hand sides of (4) and (8)), reduce in our example to structures similar to the second-order terms of the scattering matrix, and effectively renormalize the coupling constant in S_2:

$$h^2 \to h^2 (1 + 2h \oplus + \ldots) \simeq h^2 (1 + h \oplus + \ldots)^2 = h_*^2.$$

The illustration considered above is typical of a whole class of quantum-field models of interacting fields, which includes self-interaction of the scalar field $h\phi^4$ as well as the physically important case of spinor electrodynamics. For models belonging to this class, all the divergent components of the matrix elements of arbitrarily high order reduce to the counterterms added to the interaction Lagrangians occurring in the components of the S-matrix of lower orders, and thus turn out to be equivalent to the renormalization of the physical parameters of the model, i.e., of the mass, the coupling constants, and the wave functions.

Such models of interacting fields are called *renormalizable.*

26.3. The degree of divergence of diagrams. Let us now consider in greater detail the above described property of renormalizability, and formulate the criterion to be satisfied by renormalizable models of quantum field theory.

Take a typical Feynman integral in the momentum representation:

$$J(\ldots k \ldots) = \int \prod_{1 \leqslant \nu \leqslant n} \{\delta(\Sigma p_\nu - k_\nu)\} \prod_{1 \leqslant l \leqslant L} \{\Delta_l^c(p_l)\, dp_l\}, \tag{14}$$

corresponding to a connected diagram of the $n$the order. Here the arguments of the δ-functions contain algebraic sums of the moment of the initial lines of the diagram as well as the external momenta k_ν for the vertex ν in question. We represent the propagators in the form

$$\Delta_l^c(p) = \frac{P_l(p)}{m_l^2 - p^2 - i\varepsilon}, \tag{15}$$

where P_l is a certain polynomial of degree r_l which is characterized by the spin of the respective field.

As has already been established, integrals of the type of (14), owing to the existence of ultraviolet divergences, do not exist in general and must be regularized. We shall not, however, introduce regularization here, but we shall examine the properties of the integrands in the region of large $|p_l|$ in order to determine the nature of the corresponding ultraviolet divergences.

Since we are investigating connected diagrams, $4(n - 1)$ integrations may be carried out with the aid of δ-functions (the one remaining δ-function expresses the law of conservation of the total four-momentum) and we are left with $4(L - n + 1)$ independent variables of integration, where L denotes the total number of internal lines.

Just as in the case of integration over three-dimensional space it is convenient to use the radius as a variable of integration, we shall introduce the corresponding "radial" momentum P in carrying out the integration over the $4(L - n + 1)$-dimensional space. Then the product of the independent differentials dp_l gives the factor $P^{4(L-n+1)}dP/P$. Taking into account only the terms of the highest degree in the functions Δ^c, we obtain the factor

$$P^{\sum_l r_l - 2L} = P^{\sum_l (r_l - 2)},$$

and therefore, in carrying out the integration over P, the factor multiplying dP/P will behave for large P as P^{ω}, where

$$\omega(G) = \sum_{l} (r_l + 2) - 4(n - 1). \tag{16}$$

The number ω is called the *index of the diagram.*

When $\omega(G) < 0$, the integral over P turns out to be convergent; otherwise it diverges.

Naturally, the convergence of the integral over P does not yet imply that an integral of type (14) converges as a whole. Here a situation may arise similar to that occurring when in the course of evaluating the integral

$$\int_0^\infty dx \int_0^\infty dy \frac{x}{(y^2+1)^2} ,$$

the integral

$$\int_0^\infty \frac{\rho^2\, d\rho}{(1+\rho^2 \sin^2 \varphi)^2}$$

over the radial variable $\rho = x/\cos\varphi = y/\sin\varphi$ converges, while the remaining integral over φ turns out to be divergent because of a singularity at the point $\varphi = 0$. Examples of this type are given by integrals corresponding to the diagrams depicted in Figure 26.3. Both diagrams have a negative index $\omega = -2$; however, each of them possesses a subdiagram which diverges either logarithmically or quadratically.

Thus the condition

$$\omega(G) < 0 \tag{17}$$

is necessary for convergence to take place, but is not sufficient.

As is seen from the examples of divergent diagrams considered in Sections 24, 25, the degree of divergence ω of an integral determines, when $\omega \geq 0$, the structure of the divergent contributions of the corresponding regularized integral in the limit of the removal of auxiliary regularization. To an integral with an index $\omega > 0$ there corresponds the expression

$$I_{\omega}(p) \sim \operatorname{reg} I_{\omega}(p,\ M),$$

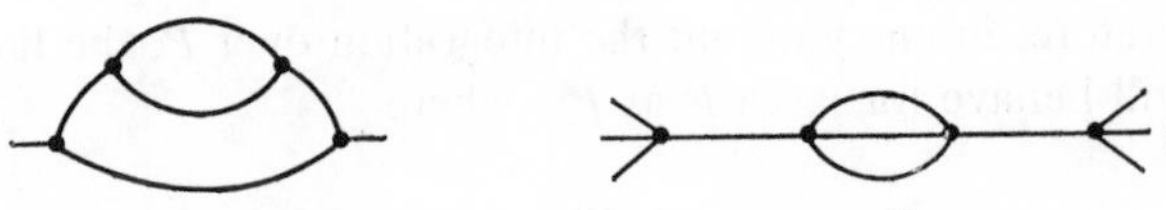

Fig. 26.3. Examples of scalar diagrams which have negative indices but contain divergent subdiagrams.

where

$$\text{reg}\, I_\omega(p,\ M) \to a_0 M^\omega + a_1 p M^{\omega-1} + \ldots\ a_{\omega-1} p^{\omega-1} M + a_\omega p^\omega + I^{\omega}_{\text{fin}}(p).$$

Here I_{fin} is the "finite part" of the integral I, which is independent of the regularization parameter, and a_α are coefficients which may depend on $\ln M$.

We see that the diagram index determines the degree of the polynomial in momentum variables with singular coefficients, the removal of which makes the integral finite. This polynomial, in the configuration representation, is a polynomial in derivatives of δ-functions.

Thus, if the diagram G under consideration combines n vertices and has s external lines, then the corresponding quasilocal operator will be represented by an expression of the type

$$:u_1(x_1)\ \ldots\ u_s(x_s)\colon Z_\omega(\partial/\partial x)\,\delta(x_1 - x_2)\ \ldots\ \delta(x_{n-1} - x_n),$$

while the degree of the polynomial Z_ω is equal to the index of the diagram ω. By integrating this expression over all the variables x_i but one, we obtain the counterterm of the Lagrangian. In the course of carrying out these trivial integrations the derivatives will pass from the δ-functions over to the field operators u_σ, and the result will be represented in the form of a normal product of a certain number of operator field functions and their derivatives. The maximum total degree of the derivatives will then turn out to be equal to the index of the diagram $\omega(G)$, while the degree of the whole expression in the operator functions will be equal to the number of external lines s.

26.4. The property of renormalizability. Therefore, if a theory leads to strongly connected diagrams with a nonnegative index, and with numbers $\omega(G)$, s of which turn out to have limited values, then to completely remove all the divergences, such a theory would require the introduction of counterterms of a finite number of types. By the type of counterterm we mean here its operator type and the degree of the derivatives occurring in each operator. Otherwise, the number of types of counteterms turns out to be infinite.

Let us now analyse the dependence of the indices $\omega(G)$ upon the number of external and internal lines of the diagram. For this purpose we introduce the concept of the index of a vertex, defined by the equation

$$\omega_i = \frac{1}{2} \sum_{l_{\text{int}}} (r_l + 2) - 4, \tag{18}$$

where the summation runs over all the internal lines which enter the ith vertex. It is not difficult to verify that the diagram indices are expressed in terms of the indices of the vertices entering into the diagram in the following manner:

$$\omega(G) = \sum_{1 \leqslant i \leqslant n} \omega_i + 4, \tag{19}$$

since each internal line arrives simultaneously at two vertices. For a given type of vertices the index ω_i takes on its maximum value Ω_i when all the lines arriving at the vertex are internal.

The maximum index of a vertex, Ω_i, is obviously a characteristic of the interaction Lagrangian, or, to be more precise, of its individual terms. It may be called the index of the Lagrangian. If all

$$\Omega_i \leqslant 0, \tag{20}$$

then from (19) it follows that $\omega(G) \leq 4$. On the other hand, if for at least certain types of vertices (certain terms of the interaction Lagrangian)

$$\Omega_i > 0, \tag{21}$$

then one may always construct such a diagram G, containing a sufficient number of vertices of this type, that $\omega(G)$ turns out to be larger than any given number. Thus, either the diagram index does not exceed 4 or it may be made to be arbitrary large.

Taking into account that

$$\omega_i = \Omega_i - \frac{1}{2} \sum_{l_{\text{ext}}} (r_l + 2),$$

where l_{ext} are the indices of the external lines occurring in the given vertex, we may write the dependence of $\omega(G)$ on the number of external lines in the form

$$\omega(G) = \sum_i \Omega_i + 4 - \frac{1}{2} \sum_{L_{ext}} (r_l + 2), \tag{22}$$

the summation in the last term being carried out over all the external lines of the diagram under consideration. Therefore, in the case of (20) the number of lines of a diagram with a positive index does not exceed 4. In this case both quantities $\omega(G)$ and s are less than or equal to 4, the number of types of the respective counterterms turns out to be finite and a detailed classification of them may be performed. On the other hand, in the case of (21) both sums may be made arbitrarily large for a nonnegative $\omega(G)$. Both characteristics, $\omega(G)$ and s, turn out to be unbounded, and in order to compensate the divergences of increasing orders, it is necessary to introduce counterterms of an increasing degree of "linearity" and an increasing number of derivatives. It turns out to be impossible to obtain a closed expression for the total effective Lagrangian.

In accordance with the above, all interactions may be divided into two classes:

(a) interactions of the renormalizable type (all $\Omega_i \leq 0$);

(b) interactions of the nonrenormalizable type (certain $\Omega_i > 0$).

Quantum-mechanical models involving interactions from the first type are renormalizable. Here, however, it is appropriate to point out that the condition $\Omega_i \leq 0$ is sufficient but not necessary. This observation is illustrated by the Weinberg–Salam model, to be considered below in Section 32.

27. DRESSED GREEN'S FUNCTIONS

27.1. Propagators of physical fields. For a more systematic investigation of the physical meaning of the various counterterms and of the renormalizations connected with them we shall have to make use of the so-called *dressed Green's functions* and *vertex parts*, i.e., of the objects allowing for radiative corrections, appearing as a result of the interaction of quantum fields, and corresponding, therefore, to real particles.

We shall start by considering one-particle (two-legged) Green's functions, i.e., propagators. The radiative corrections for free-field propagators are determined by diagrams of the self-energy type. Figure 27.1 represents self-energy insertions of lower orders to the propagator of a scalar

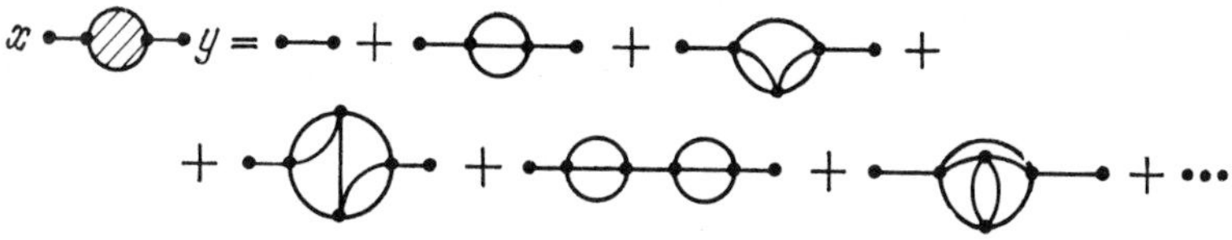

Fig. 27.1. Propagator of a scalar particle with lower-order radiative corrections in the φ^4 model.

particle in the $h\varphi^4$ model. By summing them we arrive at the dressed Green's function for a scalar field.

From diagrams of the self-energy type one may isolate a class of strongly connected or one-particle irreducible (1PI) diagrams (i.e., diagrams which cannot be reduced to two unconnected ones by means of the removal of one internal line). The sum of such 1PI diagrams is represented in Figure 27.2. It corresponds to the so-called mass operator $M(p^2)$.

The dressed Green's function of the scalar field, $\Delta(p;\, g)$ (i.e. the propagator of a scalar particle allowing for virtual processes of creation and absorption of additional particles in the course of the motion of the initial particle through vacuum) is connected with the mass operator $M(p^2;\, g)$ by the following relation corresponding to Figure 27.3:

$$\Delta(p^2;\, g) = \Delta_0(p^2) + \Delta_0 \frac{M(p^2;\, g)}{i} \Delta_0 + \Delta_0 \frac{M}{i} \Delta_0 \frac{M}{i} \Delta_0 + \dots \tag{1}$$

Here

$$\Delta_0(p^2) = \frac{1}{i(m^2 - p^2)},$$

while the normalization of the mass operator (note the factor i) corresponds to the simplest one-loop insertions introduced in Section 24. In the lowest order of perturbation theory, corresponding to the "walnut" diagram (see Figure 25.3), we have

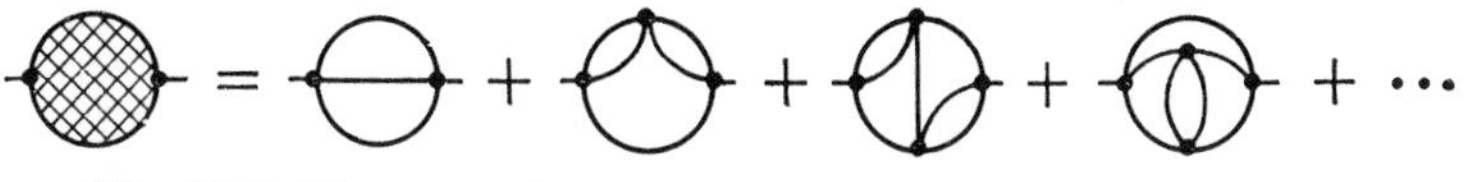

Fig. 27.2. The sum of strongly connected diagrams representing the mass operator.

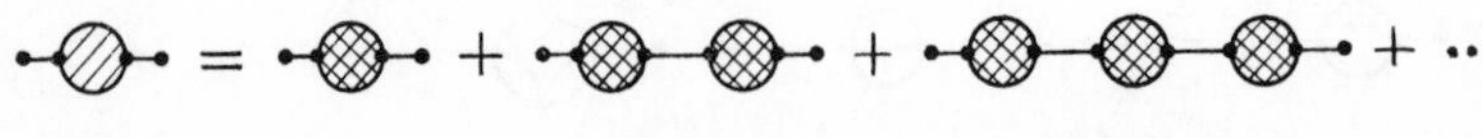

Fig. 27.3. Representation of the dressed Green's function in terms of diagrams containing iterations of the mass operator.

$$M(p^2;\ g)=g^2M_2(p^2)=\frac{96g^2}{(2\pi)^8\,i}\int dk\,dq\Delta(q)\,\Delta(k)\,\Delta(p+q+k). \qquad (2)$$

It is easy to see that the right-hand side of (1) has the same structure as a geometric progression and may be represented in a compact form. We obtain

$$\Delta=\Delta_0[1-iM\Delta_0+(-iM\Delta_0)^2+\ldots]=\Delta_0(1+iM\Delta_0)^{-1},$$

i.e.,

$$\Delta(p^2;\ g)=\frac{1}{i\,[m^2-p^2+M(p^2;\ g)]}. \qquad (3)$$

One may represent in a similar manner the dressed Green's functions of particles with spin, for example, of electrons and photons in spinor electrodynamics. Thus, by introducing the mass operator of the electron, Σ, in accordance with Figure 27.4, we obtain

$$G(p)=\frac{-i}{m-\hat{p}+\Sigma(p,\ \alpha)}. \qquad (4)$$

The contributions from the 1PI self-energy diagrams of a photon form the polarization operator $\Pi_{\mu\nu}(k;\ \alpha)$, etc. In the lowest order of perturbation theory these operators reduce to the expressions introduced in Section 24

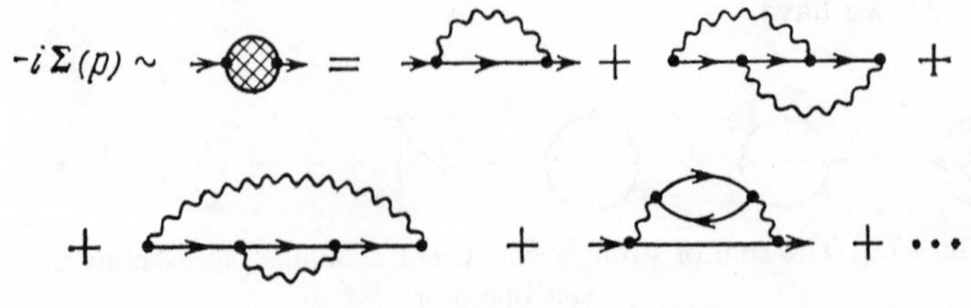

Fig. 27.4. One-particle irreducible diagrams forming the electron mass operator.

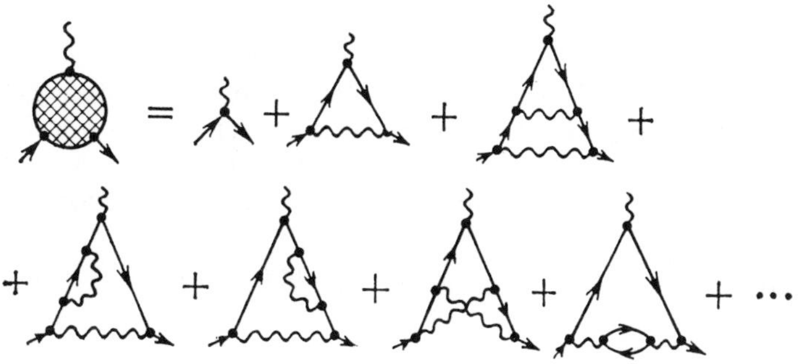

Fig. 27.5. One-particle irreducible diagrams of third and fourth orders representing the electron–photon vertex in spinor electrodynamics.

[see (24.8) and (24.14)]:

$$\Sigma(p;\alpha) = \alpha\Sigma(p) + \alpha^2 \ldots,$$
$$\Pi_{\mu\nu}(k;\alpha) = \alpha\Pi_{\mu\nu}(k) + \alpha^2 \ldots \tag{5}$$

We shall postpone considering the matrix structure of the polarization operator until Section 29, where the condition of gauge invariance will be introduced and utilized.

In a similar way one may introduce the sums of multilegged digrams, including the sums of one-particle irreducible diagrams with $l \geq 3$ external lines which form the so-called vertex functions or, simply, *vertices* (see, for example, Figure 27.5). Henceforth we shall use the collective term *dressed Green's functions* for expressions corresponding to sets of connected diagrams with a given number of external lines greater than or equal to two. Propagators of physical interacting fields correspond, in the framework of this terminology, to two-legged dressed Green's functions.

27.2. Higher Green's functions. The concepts introduced above were based on semi-intuitive reasoning utilizing the summation of terms corresponding to certain types of diagrams. We shall now present analytic formulae for the complete Green's functions.

The formal expression corresponding to the function with n legs will be constructed on the basis of the vacuum expectation value

$$\langle T u_1(x_1) \ldots u_n(x_n) S \rangle_0, \tag{6}$$

containing within the T-product sign the scattering matrix

$$S = T\,(\exp i\mathcal{A}), \tag{7}$$

where $\mathcal{A} = \int \mathcal{L}(x)\,dx$. On the one hand, expressions of the type (6) are akin (see Section 27.4 below) to the S-matrix elements for transitions $n_{in} \to n_f$, $n_{in} + n_f = n$. On the other hand, for $n = 2$ such expressions are natural generalizations of the one-particle free-field Green's functions to the case where interaction is present.

By expanding the T-exponential we obtain from (6), when $n = 2$, the series

$$\begin{aligned}\langle Tu_1(x)\,u_2(y)\,S\rangle_0 &= \langle Tu_1(x)\,u_2(y)\rangle_0 + i\,\langle Tu_1(x)\,u_2(y)\,\mathcal{A}\rangle + \\ &+ \frac{i^2}{2!}\langle Tu_1(x)\,u_2(y)\,\mathcal{A}^2\rangle_0 + \frac{i^3}{3!}\langle Tu_1(x)\,u_2(y)\,\mathcal{A}^3\rangle_0 + \ldots,\end{aligned} \tag{8}$$

the subsequent terms of which contain radiative corrections to the first term, $\langle Tu_1u_2\rangle_0$, which represents elementary (free) pairing.

It seems appropriate at this point to introduce an important technical remark concerning expressions where the operation $\langle T(\ldots)\rangle_0$ contains operator factors that do not explicitly depend on time which may appear in the argument of the exponential in (7) or, for example, as field operators in the momentum representation. One might be tempted to take a "time-independent" factor of this kind outside the T-product sign. However, the result of this operation after the calculation of the vacuum expectation value is an expression in which time-ordered pairing may turn out to be replaced by ordinary pairing, i.e., may lead to an incorrect result.

The correct procedure in expressions of this kind is always to begin with the operation $\langle T(\ldots)\rangle_0$, and only after this has been done integrate over the configuration space. For example, the third term on the right-hand side of (8) is, by definition,

$$\langle Tu_1(x)\,u_2(y)\,\mathcal{A}^2\rangle_0 \equiv \int dz_1\,dz_2\,G(x,\ y;\ z_1,\ z_2),$$

where

$$G(x,\ y;\ z_1,\ z_2) = \langle Tu_1(x)\,u_2(y)\,\mathcal{L}(z_1)\,\mathcal{L}(z_2)\rangle_0.$$

Therefore, the operations $\langle T\ldots\rangle_0$ and $\int dx$ do not commute, and an expression of the type of (8) must be considered as purely symbolic.

The terms on the right-hand side of (8) include terms corresponding to diagrams with two external lines leaving the points x and y, similar to the ones presented in Figure 27.1. However, besides the diagrams of Figure

27.1, to the third term on the right-hand side of (8) there corresponds the disconnected diagram depicted in Fig. 27.6a; the diagram in Fig. 27.6b corresponds to the fourth term, and so on. Such disconnected diagrams appear in all the higher-order terms. They correspond to a particular method of pairing the operators in which the operators $u_1(x)$ and $u_2(y)$ are paired through a part of the intermediate vertices $z_1, \ldots z_m$. Operators of the remaining vertices $z_{m+1}, \ldots, z_n$ in diagrams of a given order n are paired with each other only. The corresponding contributions can be factorized:

$$\langle T\{u_1 u_2 (i\mathcal{A})^m\}\rangle_0^{\mathrm{con}} \cdot \langle T(i\mathcal{A})^{n-m}\rangle_0.$$

The superscript introduced here corresponds to connected contributions. The vertices $z_1, \ldots, z_m$ may be chosen out of the n vertices in $n!/m!(n-m)!$ ways. Therefore, we have

$$\langle Tu_1u_2S\rangle_0 = \sum_n \frac{i^n}{n!} \sum_{1\leqslant m\leqslant n} \frac{n!}{m!\,(n-m)!} \langle Tu_1u_2\mathcal{A}^m\rangle_0^{\mathrm{con}} \cdot \langle T\mathcal{A}^{n-m}\rangle_0;$$

$$\sum_n \equiv \sum_{0\leqslant n\leqslant\infty}.$$

Rearranging the order of summation, we obtain

$$\langle Tu_1u_2S\rangle_0 = \sum_m \frac{i^m}{m!} \langle Tu_1u_2\mathcal{A}^m\rangle_0^{\mathrm{con}} \sum_l \frac{i^l}{l!} \langle T\mathcal{A}^l\rangle_0 =$$
$$= \langle Tu_1u_2S\rangle_0^{\mathrm{CB}} S_0; \quad S_0 = \langle S\rangle_0. \tag{9}$$

The vacuum contributions S_0 are thus seen to be factored out in the complete expression as well.

We therefore define the complete one-particle Green's function as follows:

$$\Delta_{12}(x, y; g) = i\,\frac{\langle Tu_1(x)\,u_2(y)\,S\rangle_0}{S_0}. \tag{10}$$

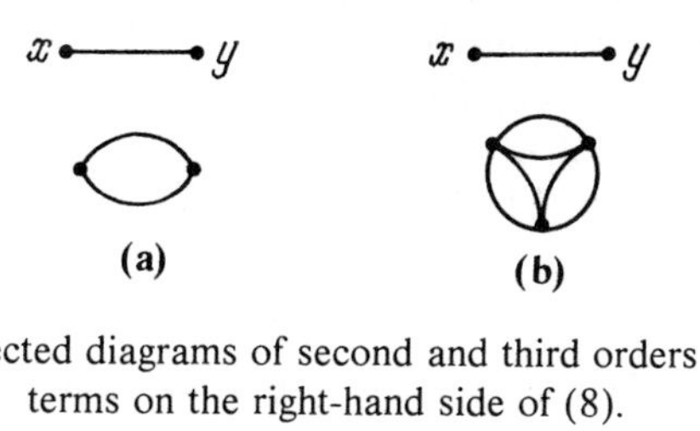

Fig. 27.6. Disconnected diagrams of second and third orders corresponding to terms on the right-hand side of (8).

The above discussion, which leads to the factorization of the vacuum contributions, may be repeated for the vacuum expectations for $k > 2$. Here, however, after the removal of the vacuum contribution the expressions

$$S_0^{-1} \langle Tu_1 \dots u_k S\rangle_0, \tag{11}$$

in general, may contain terms in which some of the operators $u_1, \dots, u_l$ appear in one of connected components of a disconnected diagram and the others, $u_{l+1}, \dots, u_k$, in another (or others). Such terms correspond to the product of diagrams occurring in the complete Green's functions with a lesser number of external lines.

Therefore, expressions of the type of (11) for $k > 2$ cannot be identified with complete k-legged Green's functions. The only exception is provided by the three-vertex

$$S_0^{-1} \langle T u_1 (x)\, u_2 (y)\, u_3 (z)\, S\rangle_0, \tag{12}$$

which is the sum of connected contributions.

For the connected four-vertex we present the following formula without giving the derivation:

$$G_{1234} (x, y, z, t) = S_0^{-1} \langle Tu_1 (x)\, u_2 (y)\, u_3 (z)\, u_4 (t)\, S\rangle_0 - \\ - i\Delta_{12} (x, y)\, \Delta_{34} (z, t) - i\Delta_{13} (x, z)\, \Delta_{24} (y, t) - i\Delta_{14} (x, t)\, \Delta_{23} (y, z). \tag{13}$$

27.3. Strongly connected multilegged functions (vertices). The Green's functions with several legs introduced above include all the weakly connected (i.e., one-particle reducible) contributions. This property is illustrated by Figures 27.3 and 27.7, in which the strongly connected components are shown-cross hatched.

The formulas which express connected multilegged objects (complete Green's functions) in terms of strongly connected vertices may be derived by use of the *generating functional* together with the representation of complete Green's functions and vertices in terms of corresponding variational derivatives. We refer the inquisitive reader to more special literature (see, for example, Section 38 of the *Introduction*), and present here the final formulas corresponding to Figure 27.7 for the scalar field:

$$G_3 (x, y, z) = \\ = \int \Delta (x - x')\, dx' \Gamma_3 (x', y', z')\, dy'\, \Delta (y' - y)\, dz'\, \Delta (z' - z), \tag{14}$$

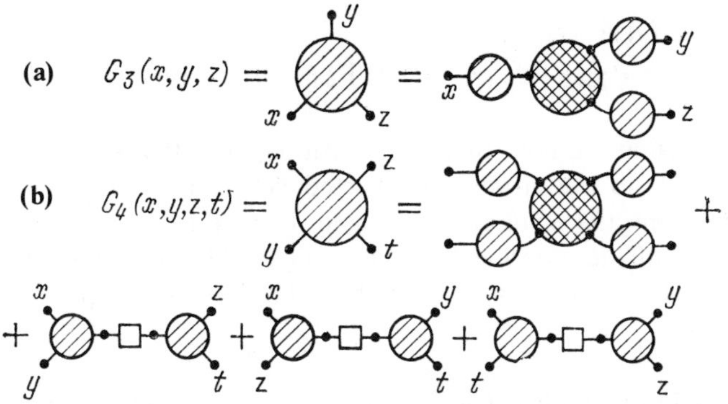

Fig. 27.7. Representation of the three- and four-legged Green's functions through the two-legged Green's function Δ and strongly connected vertices, which are shown cross-hatched. The square $\square$ stands for the "inverse propagator" Δ^{-1}.

$$
\begin{aligned}
&G_4(x, y, z, t) = \\
&= \int \Delta(x - x')\, dx'\, \Delta(y - y')\, dy' \Gamma_4(x', y', z', t')\, dz'\, \Delta(z' - z)\, dt'\, \Delta(t' - t) + \\
&\qquad + \int G_3(x, y, \tau)\, d\tau\, \Delta^{-1}(\tau - \tau')\, d\tau' G_3(\tau', z, t) + \\
&\qquad + \int G_3(x, z, \tau)\, d\tau\, \Delta^{-1}(\tau - \tau')\, d\tau' G_3(\tau', y, t) + \\
&\qquad + \int G_3(x, t, \tau)\, d\tau\, \Delta^{-1}(\tau - \tau')\, d\tau' G_3(\tau', y, t).
\end{aligned} \tag{15}
$$

Here the function Δ^{-1} is the inverse of Δ in the sense of the integral contraction, i.e., by definition

$$
\int \Delta(x - \tau)\, d\tau\, \Delta^{-1}(\tau - y) = \int \Delta^{-1}(x - \tau)\, d\tau\, \Delta(\tau - y) = \delta(x - y). \tag{16}
$$

We note that in the momentum representation the Fourier transforms of functions Δ and Δ^{-1} are mutually inverse in the algebraic sense:

$$
\tilde{\Delta}(p)\, \tilde{\Delta}^{-1}(p) = 1.
$$

With the aid of the function Δ^{-1} it is not difficult to write down formulas inverse to (14) and (15). We have, for example,

$$
\begin{aligned}
&\Gamma_3(x, y, z) = \\
&= \int \Delta^{-1}(x - x')\, dx'\, \Delta^{-1}(y - y')\, dy' G_3(x', y', z')\, dz'\, \Delta^{-1}(z' - z).
\end{aligned} \tag{17}
$$

We present one more formula of the type of (14) for spinor electrodynamics:

$$S_0^{-1}\langle T\psi(x)\overline{\psi}(y)A^{\mu}(z)S\rangle_0 = \\ = \int S^c(x-x')\,dx'D_c^{\mu\nu}(z-z')\,dz'\Gamma_\nu(x',y',z')\,dy'S^c(y'-y). \quad (18)$$

In conclusion we note that the transition from the weakly connected three-legged G_3 to the strongly connected vertex Γ_3 eliminates radiative corrections from the external lines and is therefore sometimes called "amputation" of the external lines. For higher Green's functions such a transition is not confined, as is seen from (15), to the amputation of the external lines.

27.4. Reduction formulas. The Green's functions considered above may be connected by simple relations with the matrix elements of the scattering matrix.

It is quite clear that the matrix element of the transition $m \to n$,

$$\langle \overset{*}{\Phi}\ldots(\boldsymbol{p}_1, \ldots, \boldsymbol{p}_n)\,S\Phi(\boldsymbol{q}_1, \ldots, \boldsymbol{q}_m\rangle_0, \quad (19)$$

where

$$\Phi_{\alpha\ldots\omega}(\boldsymbol{q}_1, \ldots, \boldsymbol{q}_m) = \overset{+}{a}_\alpha(\boldsymbol{q}_1)\ldots\overset{+}{a}_\omega(\boldsymbol{q}_m)\,\Phi_0,$$

is described by the same Feynman diagrams as the connected $(n+m)$-legged Green's function G_{n+m}. The difference consists in the factors corresponding to the external lines. According to the Feynman rules from Section 20, to the external lines in the matrix element (19) there correspond the amplitudes of the initial and final particles $f_\sigma^{\pm}(\boldsymbol{k})$. For example,

$f^{\pm}(\boldsymbol{k}) = \pm(16\pi^3k_0)^{-1/2}$ for a spinless field,
$f_s^{\pm}(\boldsymbol{p}) = (2\pi)^{-3/2}v_s^{\pm}(\boldsymbol{p})$ for a spinor field,
$f_\nu^{\pm}(\boldsymbol{q}) = \pm(16\pi^3q_0)^{-1/2}e_\nu(\boldsymbol{q})$ for the electromagnetic field,

where the negative-frequency amplitudes f^- correspond to particles in the initial state, and the positive-frequency ones f^+ to particles in the final state.

At the same time, for Green's function taken in the momentum representation

$$\tilde{\Delta}_{(n)}(k_1, \ldots, k_n) = \int e^{i\sum k_jx_j}\Delta_{(n)}(x_1, \ldots, x_n)\,dx_1\ldots dx_n, \quad (20)$$

free propagators correspond to the external lines. Therefore one may demonstrate that

$$\langle \overset{*}{\Phi}(p_1, \ldots, p_n)\, S\Phi(q_1, \ldots, q_m)\rangle_0 = i\eta \prod_{1\leqslant k\leqslant n} [f_k^{+}(p_k)\,\Delta_k^{-1}(p_k)]\times$$

$$\times\tilde{\Delta}_{(n+m)}(p_1, \ldots, p_n; -q_1, \ldots, -q_m) \prod_{1\leqslant l\leqslant m} [f_l^{-}(q_l)\,\Delta_l^{-1}(q_l)]. \qquad (21)$$

The factor i on the right-hand side is due to the standard definition of Green's functions, while $\eta = \pm 1$ is a sign factor which takes into account the change of signs under permutation of the fermion operators.

The relation (21) is a *reduction formula* in the momentum representation.

28. THE RENORMALIZATION PROCEDURE

28.1. Renormalization of contributions to Green's functions. Our first goal will be to carry out subtraction of the divergences occurring in the dressed Green's functions and to obtain for them expressions not containing ultraviolet infinities (i.e., expressions for the so-called *renormalized Green's functions*). Such expressions will still bear the stamp of arbitrariness caused by the nonuniqueness of the procedure of subtracting infinites. However, the dependence of the renormalized Green's functions on the corresponding indefinite finite constants for renormalizable theories is quite simple and at the same time specific. This allows us to transfer the arbitrariness to such parameters as masses, coupling constants, and norms of field operators, and ultimately to obtain for the renormalized Green's functions expressions depending on the so-called renormalized masses and coupling constants and not containing any additional indefinite constants. The renormalized masses and coupling constants, in turn, are uniquely expressed in terms of quantities observed directly.

We shall initiate this program by analysing one-particle Green's functions.

As was established in Section 27.1, in the momentum representation the one-particle Green's function $G(p)$ is expressed in terms of the strongly connected two-legged vertex $\Gamma_2(p; g)$, which is equal to the mass operator, by the algebraic relation

$$G(p;\, g) = G_0(p) + G_0\Gamma_2(p;\, g)\, G_0 + G_0\Gamma_2 G_0\Gamma_2 G_0 + \ldots = (G_0^{-1} - \Gamma_2)^{-1}. \quad (1)$$

As a result of subtraction of divergences the function Γ_2, formally defined by the Feynman rules in terms of strongly connected diagrams, is replaced by $\tilde{\Gamma}_2(p)$, not containing divergences. The structure of $\tilde{\Gamma}_2(p)$ depends significantly on the type of the particular quantum field theory.

For renormalizable interaction models, when the maximum vertex index equals zero, the degree of divergence of the diagrams is independent of the perturbation order and for diagrams with two external boson lines is equal to 2, while for diagrams with two fermion lines it is equal to 1. Therefore in the first case, $\tilde{\Gamma}_2$ has the following structure:

$$\tilde{\Gamma}_2(p) = ap^2 + bm^2 + \Gamma_2^{\text{fin}}(p^2), \tag{2a}$$

while in the second case

$$\tilde{\Gamma}_2(p) = a\hat{p} + bm + \Gamma_2^{\text{fin}}(p), \tag{2b}$$

i.e., in both the cases, due to its relativistic invariance, $\tilde{\Gamma}_2$ involves only two finite arbitrary constants.

On the other hand, in nonrenormalizable models the degree of divergence depends on the number of vertices and increases indefinitely with the order of perturbation theory. In each finite order the function $\tilde{\Gamma}_2$ contains a finite polynomial in powers of p^2 or $\hat{p}$ with arbitrary coefficients, the degree of which increases with the order of the diagram. Therefore the vertex $\tilde{\Gamma}_2$, which is an infinite sum of contributions of various orders of perturbation theory, depends on an infinite number of arbitrary parameters (i.e., on an arbitrary function).

For the scalar case, with the aid of (1) and (2) we obtain

$$\tilde{G}(p) = [(1-b)\, m^2 - (1+a)\, p^2 - \Gamma_2^{\text{fin}}\ (p^2)]^{-1}. \tag{3}$$

We define the square of the renormalized mass of a scalar partticle, $\tilde{m}^2$, as the position of the pole of the propagator with respect to p^2, i.e.,

$$\tilde{G}^{-1}(\tilde{m}^2) = (1-b)\, m^2 - (1+a)\, \tilde{m}^2 - \Gamma_2^{\text{fin}}\ (\tilde{m}^2) = 0. \tag{4a}$$

The latter relation may be represented in the form

$$\tilde{m}^2 = m^2 + \Delta m^2,$$

where

$$\Delta m^2 = -\frac{a+b}{1+a} m^2 - \frac{\Gamma_2^{\text{fin}}\ (\tilde{m}^2)}{1+a} \tag{5a}$$

is the shift of the mass squared due to quantum (radiative) corrections. For the actual determination of $\tilde{m}^2$ and Δm^2 it is necessary to solve the transcendental equation (4a). In the vicinity of the pole $p^2 = \tilde{m}^2$ the renormalized propagator $\tilde{G}$ may be represented in the form

$$\tilde{G}(p) = \frac{Z_2}{\tilde{m}^2 - p^2}, \tag{6a}$$

where

$$Z_2^{-1} = 1 + a + \Gamma_2^{\prime\text{fin}}(\tilde{m}^2), \quad \Gamma'(m^2) = \left.\frac{\partial \Gamma(k^2)}{\partial k^2}\right|_{k^2 = m^2}.$$

The same reasoning for the spinor progagator leads us to find

$$G^{-1}(\hat{p} = m) = m(1 - b) - \tilde{m}(1 + a) - \Gamma_2^{\text{fin}}(\tilde{m}) = 0, \tag{4b}$$

$$\tilde{m} = m + \Delta m, \quad \Delta m = -\frac{a + b}{1 + a}m - \frac{\Gamma_2^{\text{fin}}(\tilde{m})}{1 + a}, \tag{5b}$$

$$\tilde{G}(\hat{p})\big|_{\hat{p} \sim m} \simeq \frac{Z_2}{\tilde{m} - \hat{p}}, \tag{6b}$$

$$Z_2^{-1} = 1 + a + \Gamma_2^{\prime\text{fin}}(\tilde{m}).$$

We see that in each case two arbitrary constants ultimately are involved in the renormalization of the mass Δm and the renormalization of the propagator Z_2 (i.e., the renormalization of the field function).

Here there may be established a simple correspondence between the renormalization procedure and the ambiguities arising from the insertion of the finite counterterms, which were considered in Section 25.3, into the Lagrangian. For the scalar field, counterterms of the following form are meant:

$$\Delta \mathscr{L} = \frac{Z_2 - 1}{2}(\partial_\mu \varphi)^2 - \frac{(Z_2 Z_m - 1)m^2}{2}\varphi^2, \qquad Z_m m^2 = \tilde{m}^2, \tag{7a}$$

whereas for spinor fields

$$\Delta \mathscr{L} = i(Z_2 - 1)\bar{\psi}\hat{\partial}\psi - (Z_2 Z_m - 1)m\bar{\psi}\psi, \qquad Z_m m = \tilde{m}, \tag{7b}$$

Below, for brevity, we shall consider only the second case. With the aid of arguments similar to the ones of Section 25.3 it is not difficult to establish that the effect of the counterterms (7b), which, for this purpose, are conveniently presented in the form

$$\Delta \mathscr{L} = (Z_2 - 1)\mathscr{L}_0 + Z_2(\tilde{m} - m)\bar{\psi}\psi,$$

$$\mathscr{L}_0 = i\bar{\psi}\hat{\partial}\psi - m\bar{\psi}\psi, \qquad \tilde{m} = Z_m m,$$

reduces to the multiplicative renormalization of the field function

$$\psi \to \psi' = \sqrt{Z_2}\psi,$$

and, as a consequence of this, to the corresponding renormalization of the one-particle propagator

$$G \to G' = Z_2^{-1}G$$

as well as to a shift of the mass

$$m \to \tilde{m} + (Z_m - 1)m,$$

which may also be represented by the multiplicative transformation

$$m \to \tilde{m} = Z_m m.$$

We now pass to the higher vertices. Concretely, we shall consider in parallel two objects: the four-vertex Γ_4 for scalar particles, and the three-legged boson–fermion vertex Γ_3 for models involving particles with integer and half-integer spin (something similar to spinor electrodynamics).

In models with zero maximum vertex indices (renormalizable models), the indices of diagrams do not depend on the perturbation order and for the two aforementioned vertices are equal to zero. Therefore the structures of the indefiniteness in both cases are the same:

$$\tilde{\Lambda}_4(p_1, \ldots, p_4) = d + \Lambda_4^{\text{fin}}(\ldots p \ldots), \tag{8a}$$

$$\tilde{\Lambda}_3(p_1, \ldots, p_3) = cO + \Lambda_3^{\text{fin}}(\ldots p \ldots). \tag{8b}$$

Here c, d are arbitrary finite constants, O is the Dirac matrix corresponding to the Lorentz structure of the three-vertex, and Λ_n are the sums of radiative corrections which together with the initial (skeleton) contribution of the lowest perturbation order form the complete vertices Γ_n. Thus, in the $h\varphi^4$ model

$$\Gamma_4^{\text{fin}} = h + \Lambda_4^{\text{fin}}, \tag{9a}$$

whereas in spinor electrodynamics

$$\Gamma_\mu^{\text{fin}} = e\gamma_\mu + (\Lambda_3^{\text{fin}})_\mu. \tag{9b}$$

The respective counterterms have the following forms: in the φ^4 model

$$\Delta\mathscr{L} = (Z_1 - 1)h\varphi^4,$$

and in spinor electrodynamics

$$\Delta\mathscr{L} = (Z_1 - 1)e\bar{\psi}\hat{A}\psi.$$

These counterterms lead to the multiplicative renormalization of the strongly connected vertex functions

$$\Gamma'_4 = Z_1\Gamma_4$$

and

$$\Gamma'_\mu = Z_1\Gamma_\mu,$$

which is equivalent to the finite renormalization of the coupling constants h and e. As was established in Section 25.3 (see (25.19)), renormalization of the field operator functions (i.e., of the one-particle propagators) also contributes to the renormalization of the coupling constants.

Thus in terms of Green's functions, the renormalization effect in the $h\varphi^4$ model may, with the aid of the counterterms

$$\Delta\mathscr{L} = (Z_1 - 1)h\varphi^4 + (Z_2 - 1)\mathscr{L}_0 - Z_2(Z_m - 1)\frac{m^2}{2}\varphi^2,$$

be written down as follows:

$$\Delta(h, m) \to Z_2^{-1}\Delta(\tilde{h}, \tilde{m}),$$

$$\Gamma_4(h, m) \to Z_1\Gamma_4(\tilde{h}, \tilde{m}), \qquad (10)$$
$$\tilde{h} = Z_1 Z_2^{-2} h, \qquad \tilde{m} = Z_m m.$$

Such transformations are called Dyson transformations.

We shall not discuss vertices with more than four external lines, as, according to (26.22), diagrams with so many external lines are overall convergent in renormalizable models, and no new abritrary constants correspond to them.

28.2. Theorem of renormalizability. The reasoning presented above was of a schematic nature. It did not, in particular, even touch upon the existence of divergences in the subdiagrams of a given diagram, and the structure of the corresponding arbitrariness was also not discussed.

To illustrate this fact, we point out, for example, that to evaluate in the scalar $h\varphi^4$ model the fifth-order contribution to the four-vertex Γ_4 of the diagram presented in Figure 28.1a, it is necessary to calculate the quadratically divergent self-energy subdiagram (Figure 28.1b) as well as the logarithmically divergent scattering subdiagram (Figure 28.1c). Therefore, the second term on the right-hand side of (8a) depends both on the arbitrary constants a, b introduced in (2) and on the constant d.

The restricted size of this book does not allow us to go into details of these important problems. We refer the interested reader to more technical literature (see, for example, Chapter V of the *Introduction* and Chapters II, III of the book by Zav'yalov (1979)); here we note only that the general assertion concerning the renormalizability of perturbation theory represents the essence of the *theorem of renormalization* (The Bogoliubov–Parasyuk theorem), while a convenient practical prescription for a unique subtraction of the divergences follows from the so-called R-operation of Bogoliubov.

From a practical point of view, the most important aspect of the renormalization procedure consists in the necessity of keeping to certain simple rules which provide for its self-consistency. According to these rules, within any given divergent diagram subtraction must always be performed in the same way, whether it is to be considered as an independent diagram or as a subdiagram of some complicated diagram. For example, both arbitrary constants appearing in the course of the evaluation of the finite part of the second-order "walnut" diagram in the scalar $h\varphi^4$ model (see Figure 25.3) must be determined in the same manner when the second-order contribution to the propagator is being calculated as in the course of any evaluations of diagrams, no matter how complicated, including even the diagram of Figure 28.1a. Under these conditions the arbitrary constants appearing during the removal of infinities enter into the final expressions for the Green's functions and the matrix elements in the form of a small number of combinations

$$\tilde{m}_i = f_i(m, g; a, b, \ldots), \quad \tilde{g}_k = \psi_k(m, g; a, b, \ldots),$$

which may be identified with the physical masses and the physical coupling constants.

Thus, it turns out to be possible in all orders of perturbation theory to realize in a consistent manner the transition from the initial parameters (masses and coupling constants) occurring in the Lagragian and in the Feynman rules to new renormalized parameters

$$\{g, m\} \rightarrow \{\tilde{g}, \tilde{m}\}. \tag{11}$$

The renormalized vertices and the observed matrix elements now depend only on the renormalized masses and coupling constants and do not separately depend on the initial masses, the coupling constants, and the arbitrary parameters $a, b, c, \ldots$.

28.3. A recipe for subtraction on the mass shell. We shall now present a method, most convenient for practical calculations, of determining the arbitrary parameters appearing in the course of the subtraction of divergences. Note, first, that according to (5a), (5b) the parameters a, b may be dealt with so that the shift in mass, Δm, turns out to be equal to zero. In the scalar case, it is necessary for this to assume

$$m^2 b = -m^2 a - \Gamma_2^{\text{fin}}(m^2). \tag{12}$$

By inserting this relation into (2) we obtain

$$\tilde{\Gamma}_2(p) = a(p^2 - m^2) + \Gamma_2^{\text{fin}}(p^2) - \Gamma_2^{\text{fin}}(m^2).$$

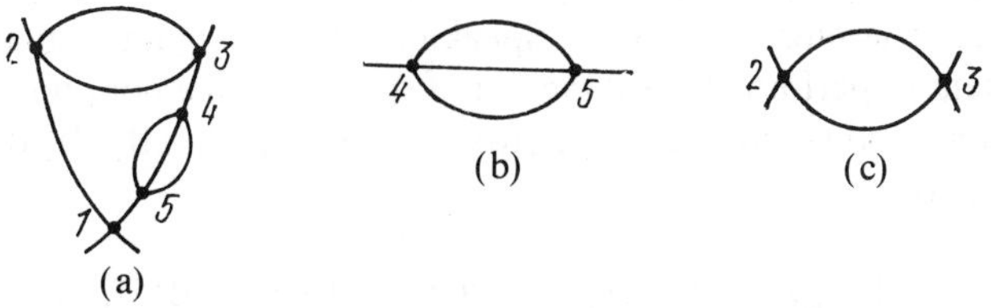

Fig. 28.1. Logarithmically divergent scalar diagram of the fifth order (a) and its divergent subdiagrams (b) and (c).

If we now deal with the constant a in such a way that the renormalized $\tilde{\Gamma}$ acquires at $p^2 = m^2$ a second order zero,

$$a = -\tilde{\Gamma}_2'^{\text{fin}}\,(m^2) = -\frac{d\Gamma_2^{\text{fin}}(p^2)}{dp^2}\bigg|_{p^2=m^2}, \tag{13}$$

then, in accordance with (7), the renormalization constant of the field function Z_2 will turn out to be equal to unity.

Note that the recipe (12), (13), which is equivalent to the definition

$$\Gamma_2^{\text{fin}}(p^2) = \Gamma_2^{\text{fin}}(p^2) - \Gamma_2^{\text{fin}}(m^2) - (p^2 - m^2)\Gamma_2'^{\text{fin}}(m^2), \tag{14}$$

reduces to the subtraction from the finite part of Γ^{fin} of the first two terms of its expansion in a Taylor series at the point $p^2 = m^2$, i.e., on the mass shell. This recipe for subtraction on the mass shell at the same time reduces to zero both the renormalization of the mass and the change in the norm of the field function

$$\tilde{m}^2 = m^2, \quad Z_2 = 1,$$

which is very convenient from a practical point of view.

In the spinor case the corresponding formula has the form

$$\Gamma_2^{\text{fin}}(\hat{p}) = \Gamma_2^{\text{fin}}(\hat{p}) - \Gamma_2^{\text{fin}}(m) - (\hat{p} - m)\Gamma_2'^{\text{fin}}(m). \tag{15}$$

Here we imply that the function Γ_2^{fin}, which depends on the matrix argument $\hat{p}$ and on the scalar argument p^2, may, if we assume the identity $p^2 = (\hat{p})^2$, be regarded as a function of a single argument $\hat{p}$. It is in this sense that the derivative in the last term on the right-hand side is understood: $\Gamma'^{\text{fin}} = (d/d\hat{p})\Gamma^{\text{fin}}$.

As will be established below in Section 29.4, the derivative of the mass operator of the electron, Σ', at the point $\hat{p} = m$ becomes infinite because of the so-called *infrared* divergence, which is due to the photons being massless. Therefore, generally speaking, it may happen that it is more convenient to perform the subtaction of the derivative not on the mass shell, but at some other point $\hat{p} = p_*$ (for example, at $\hat{p} = 0$). In this case (15) takes the form

$$\Gamma_2^{\text{ren}}(\hat{p}) = \Gamma_2^{\text{fin}}(\hat{p}) - \Gamma_2^{\text{fin}}(m) - (\hat{p} - m)\Gamma'^{\text{fin}}(p_*). \tag{16}$$

This version of the subtraction corresponds to a renormalization of the spinor field equal to

$$Z_2^{-1} = 1 + \Gamma_2'^{\text{fin}}(m) - \Gamma_2'^{\text{fin}}(p*). \tag{17}$$

28.4. Nonuniqueness of the vertex renormalization. We now proceed to consider the renormalization of vertex functions. As is shown in Appendix VII, the scalar four-legged vertex $\Gamma_4(p_1, \ldots, p_4)$ is a function of six independent scalar variables. We shall, however, write explicitly the dependence on seven variables:

$$\Gamma_4^{\text{fin}} = F\left(p_1^2,\ p_2^2,\ p_3^2,\ p_4^2;\ s = (p_1 + p_2)^2,\ t = (p_1 + p_3)^2,\right.$$
$$\left. u = (p_1 + p_4)^2\right), \tag{18}$$

bearing in mind the linear connection between them (see (AVII.7)). The transition to the mass shell fixes the first four variables and leaves the latter three connected by the relation

$$s + t + u = 4m^2. \tag{19}$$

The point at which the four-vetex is subtracted is not, thus, fixed uniquely when one goes to the mass shell. The remaining two-parameter arbitrariness is limited only by the condition of reality. As was shown in Section 26.2, the arbitrary constant d on the right-hand side of (8) is equivalent to the four-meson counterterm $d\varphi^4$ occurring in the Lagrangian and must, thus, be real. Therefore, the constant d may be determined through the value of the vertex

$$\Gamma_4^{\text{fin}}(s,\ u,\ t) = F\left(m^2,\ m^2,\ m^2,\ m^2,\ s,\ u,\ t\right) \tag{20}$$

for values of the arguments s, u, t within the range defined by the conditions

$$s \leqslant 4m^2, \quad t \leqslant 4m^2, \quad u \leqslant 4m^2. \tag{21}$$

The origin of these conditions is explained in Appendix VII.

A possible subtraction point is the point corresponding to the *reaction threshold* in one of the channels, for example, the point $s = 4m^2$, $u = t = 0$. The value of the four-vertex and of the scattering amplitude at this point are proportional to the so-called s-wave scattering length and may be directly measured experimentally.

If one assumes

$$\Lambda_4^{\text{reg}}(\ldots p \ldots) = \Lambda_4^{\text{fin}}(s, u, t) - \Lambda_4^{\text{fin}}(4m^2, 0, 0), \tag{22}$$

then the total vertex function in accordance with (9) will satisfy the relation

$$\Gamma_4^{\mathrm{reg}}\ (4m^2,\ 0,\ 0;\ g) = g, \tag{23}$$

which may be regarded as the definition of the *renormalized coupling constant.*

Often another, so-called symmetric, subtraction point $s = u = t = 4m^2/3$ is made use of. In that case

$$\Lambda_4^{\mathrm{reg}}(\ldots p\ldots) = \Lambda_4^{\mathrm{fin}}(s,\ u,\ t) - \Lambda_4^{\mathrm{fin}}(4m^2/3,\ 4m^2/3,\ 4m^2/3) \tag{24}$$

and

$$g = \Gamma_4^{\mathrm{reg}}\ (4m^2/3,\ \ 4m^2/3,\ \ 4m^2/3;\ g). \tag{25}$$

The coupling constants (23) and (25) may be expressed in terms of each other. Let us denote by Γ_4^{sym} the vertex function normalized at the symmetric point, i.e., that satisfies relation (25). The corresponding coupling constant will be denoted by g^{sym}, i.e.,

$$\Gamma_4^{\mathrm{sym}}\left(\frac{4m^2}{3},\ \frac{4m^2}{3},\ \frac{4m^2}{3};\ g^{\mathrm{sym}}\right) = g^{\mathrm{sym}}. \tag{26}$$

To the subtraction at the threshold we shall assign $\Gamma_4^{\mathrm{thresh}}$ and g^{thresh}:

$$\Gamma_4^{\mathrm{thresh}}(4m^2,\ 0,\ 0;\ g^{\mathrm{thresh}}) = g^{\mathrm{thresh}}. \tag{27}$$

Then

$$\Gamma_4^{\mathrm{thresh}}\left(\frac{4m^2}{3},\ \frac{4m^2}{3},\ \frac{4m^2}{3};\ g^{\mathrm{thresh}}\right) = g^{\mathrm{sym}} \tag{28}$$

and vice versa:

$$\Gamma_4^{\mathrm{sym}}(4m^2,\ 0,\ 0;\ g^{\mathrm{sym}}) = g^{\mathrm{thresh}}. \tag{29}$$

29. RENORMALIZATION IN SPINOR ELECTRODYNAMICS

29.1. The condition of gauge invariance. In quantum field theory describing electromagnetic interactions, an important role is played by the requirement of gauge invariance. As was shown in Section 4, a gauge transformation of the potential of the electromagnetic field,

$$A_\mu(x) \to A'_\mu(x) = A_\mu(x) + \partial_\mu f(x) \tag{1}$$

does not alter the values of the components of the electric and magnetic fields and thus should not lead to any observable effects.

The requirement of invariance of the matrix elements of the S-matrix under the transformation (1) is called the condition of gauge invariance of the scattering matrix. To formulate it explicitly, we note that under the transformation (1), where f is an arbitrary infinitesimal function, the term of nth order in the scattering matrix,

$$\int S_n(x, \ldots, x_n)\, dx_1 \ldots dx_n$$

will undergo an increase, the main part of which has the form

$$\sum_{1 \leqslant i \leqslant n} \int \frac{\partial S_n}{\partial A_\mu(x_i)} \frac{\partial f(x_i)}{\partial x_i^\mu}\, dx_1 \ldots dx_n.$$

By integrating this expression by parts we find that it becomes equal to zero if the following identity is satisfied:

$$\operatorname{div}\left(\frac{\partial S_n}{\partial A}\right) \equiv \frac{\partial}{\partial x_i^\mu} \frac{\partial S_n(x_1, \ldots, x_n)}{\partial A_\mu(x_i)} = 0. \tag{2}$$

It is not difficult to verify that the condition obtained provides also for invariance of the S-matrix under finite gauge transformations.

The condition (2) may be considered from the point of view of the formal definition of the S-matrix as a time-ordered exponential function of interaction Lagrangians. It may be shown (see, for example, the *Introduction*, Sections 28.2, 33.2, and 33.3) that the gauge invariance of the S-matrix represented, according to the Feynman rules, by products of propagators is due to the equation of continuity of the fermion current in the interaction representation.

We shall examine the condition of gauge invariance as applied to the scattering matrix not containing divergences, i.e., we shall assume that the coefficient functions S_n are expressions obtained *after* subtraction of divergences.

29.2. Gauge transformation of the photon propagator. Notice now that if the function $f(x)$ on the right-hand side of the gauge transformation (1) is considered to be an operator depending on the operator potentials A_μ, then it is possible to obtain for the time-ordered pairing of the electromagnetic field a more general expression than the diagonal one used in the Feynman rules, which we shall now write in the form

$$\overline{A_\mu(k)\, A_\nu}(q) \equiv \langle T A_\mu(k)\, A_\nu(q)\rangle_0 = \frac{-ig_{\mu\nu}}{k^2+i\varepsilon}\,\delta(k+q). \tag{3}$$

Let us consider gauge transforming of the potentials A_μ, satisfying (3), into the potentials

$$A'_\nu = A_\nu + k_\nu (kA(k))\, \mathcal{F}(k^2)/k^2,$$

where $\mathcal{F}$ is a certain c-function. In determining the time-ordered pairing of the new operators A' we obtain

$$\overline{A'_\mu(k)\, A'_\nu}(q) = \frac{1}{ik^2}\left\{\left(g_{\mu\nu} - \frac{k_\mu k_\nu}{k^2}\right) + \frac{k_\mu k_\nu}{k^2}\, d_l(k^2)\right\}\delta(k+q), \tag{4}$$

where

$$d_l(k^2) = (1+\mathcal{F}(k^2))^2.$$

The first term on the right-hand side of (4) is proportional to the transverse projection operator $P^{\text{tr}}_{\mu\nu}$, and the second to the longitudinal projection operator $P^{l}_{\mu\nu}$, i.e.,

$$P^{\text{tr}}_{\mu\nu}(k) = g_{\mu\nu} - \frac{k_\mu k_\nu}{k^2}, \quad P^{l}_{\mu\nu} = \frac{k_\mu k_\nu}{k^2}. \tag{5}$$

In accordance with the condition of gauge invariance, the matrix elements should be independent of $d_l(k^2)$.

Assuming $d_l = 0$, we obtain from (4) an expression that satisfies the condition of transversality,

$$k^\nu\, \overline{A^{\text{tr}}_\nu(k)\, A^{\text{tr}}_\mu}(q) = 0.$$

When $d_l = 1$ we come back to the usual diagonal pairing (3).

29.3. The photon propagator with radiative corrections. In Section 24 the simplest one-loop diagrams of spinor electrodynamics were evaluated. There it was demonstrated that to the diagram of vacuum polarization in Figure 24.1b there corresponds the quadratically divergent integral $\Pi_{\mu\nu}$ (see (24.7), (24.8)), which represents the lowest-order approximation to the polarization operator. The finite part of this integral, in accordance with (24.11), (24.12), may be written in the form

$$\Pi_{\mu\nu}(k) = a g_{\mu\nu} + (g_{\mu\nu}k^2 - k_\mu k_\nu)\left[b + I\left(\frac{k^2}{4m^2}\right)\right]\frac{1}{\pi}, \tag{6}$$

where a and b are arbitrary finite constants, and the integral

$$I(z) = 2\int_0^1 dx\, x(1-x)\ln[1 - 4x(1-x)z - i\varepsilon]$$

according to (24.13), may be represented as

$$I(z) = -\frac{z}{3}\int_1^\infty \frac{d\sigma(1+\frac{1}{2}\sigma)}{\sigma(\sigma - z + i\varepsilon)}\sqrt{1-\frac{1}{\sigma}} = -\frac{4}{9} + \frac{2}{3}[J(z) + J'(z)],$$

In the limit of large $|z|$

$$I(z) \to \frac{1}{3}\ln(-z).$$

Let us now make use of the condition (2) to define the indefinite constants occurring in $\Pi_{\mu\nu}$. Notice, to this end, that the polarization operator enters into $S_2(x, y)$ as follows:

$$-i:A^\mu(x)\,\Pi_{\mu\nu}(x-y)\,A^\nu(y):.$$

Therefore the condition of gauge invariance gives

$$\frac{\partial}{\partial x_\nu}\Pi_{\nu\mu}(x) = 0, \quad k^\nu\Pi_{\nu\mu}(k) = 0. \tag{7}$$

By inserting the right-hand side of (6) into the second of the two relations (7), we obtain $a = 0$. The vacuum-polarization operator, by virtue of the condition of gauge invariance, has the purely transverse form

$$\Pi^{\mathrm{reg}}_{\mu\nu}(k) = (k^2 g_{\mu\nu} - k_\mu k_\nu)\left[b + I\left(\frac{k^2}{4m^2}\right)\right]\frac{1}{\pi}. \tag{8}$$

Now consider the radiative correction to the photon propagator due to $\Pi_{\mu\nu}$, which we shall, in accordance with (4), write in an arbitrary gauge:

$$D_{\mu\nu}(k) = -\frac{1}{k^2}\left(g_{\mu\nu} - \frac{k_\mu k_\nu}{k^2}\right) - \frac{d_l}{k^2}\frac{k_\mu k_\nu}{k^2} = -\frac{P^{\mathrm{tr}}_{\mu\nu} + d_l P^{\mathrm{l}}_{\mu\nu}}{k^2}, \tag{9}$$

where P^{tr} and P^l are the projection operators introduced in (5). According to the conclusion of Section 29.2, the coefficient d_l may depend on k^2. However, here and henceforth we shall everywhere consider $d_l =$ const.

To calculate the correction, one must add the expression $i^2 D_{\mu\sigma}(k)\Pi^{\sigma\rho}(k)D_{\rho\nu}(k)$ to (9). Substituting (8) and (9) into it, we find

$$D_{\mu\nu}(k, \alpha) = -\frac{d(k^2, \alpha)}{k^2} P^{\mathrm{tr}}_{\mu\nu}(k) - \frac{d_l}{k^2} P^{\mathrm{l}}_{\mu\nu}(k), \tag{10}$$

where

$$d(k^2, \alpha) = 1 + \frac{\alpha}{\pi}\left[I\left(\frac{k^2}{4m^2}\right) + b\right].$$

Taking into account the repeated one-loop insertions to the photon propagator (Figure 29.1)

$$D - D\Pi D + D\Pi D\Pi D - D\Pi D\Pi D\Pi D + \ldots$$

we again arrive at (10), where now $d(k^2, \alpha)$ is represented by the geometric progression

$$d(k^2, \alpha) = 1 + \frac{\alpha}{\pi}(I + b) + \frac{\alpha^2}{\pi^2}(I + b)^2 + \ldots,$$

which after formal summation becomes

$$d(k^2, \alpha) = \frac{1}{1 - \pi(k^2, \alpha)}, \tag{11}$$

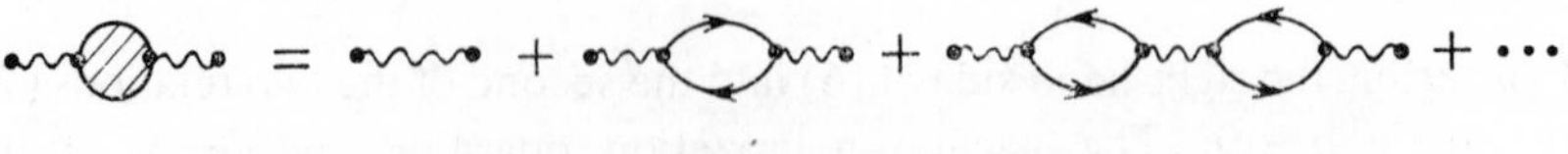

Fig. 29.1. The contribution of the simplest one-loop diagrams to the photon propagator.

where

$$\pi(k^2, \alpha) = \frac{\alpha}{\pi}\left[I\left(\frac{k^2}{4m^2}\right) + b\right].$$

Here it is appropriate to make some remarks. First, it may be seen that the condition of gauge invariance leads to the absence of renormalization of the photon ($a = 0$). However, the second subtraction parameter b, analogous to the parameter a in (28.2), remains arbitrary and contributes to the renormalization of the transverse part of the electromagnetic potential. Taking into account that, according to (24.13), $I(0) = 0$, we obtain

$$d(0, \alpha) = \frac{1}{1 - \frac{\alpha}{\pi} b} = Z_3.$$

By analogy with the arguments of Section 28, it is convenient to set the constant b equal to zero. In this case the photon propagator is not renormalized on the mass shell of the photon, and it does not contribute to the effective value of the coupling constant.

Secondly, notice that the formula (11) is of a very general nature. If we understand $\Pi_{\mu\nu}$ to be not the second-order diagram but the sum of all strongly connected polarization diagrams (Figure 29.2), then we shall again obtain (11), in which now

$$\pi(k^2, \alpha) = \frac{\alpha}{\pi} I + \alpha^2 I_2 + \alpha^3 I_3 + \ldots \tag{12}$$

Here $I_n(k^2)$ is the contribution of the n-loop strongly connected diagrams normalized, according to the remark just made, at zero: $I_n(0) = 0$. On the mass shell

$$d(0, \alpha) = 1.$$

Finally, it should be noted that the radiative correction studied in this section, representing the lowest-order approximation of the vacuum-polarization effect, becomes significant in the ultraviolet region when $|k^2/m^2| \gg 1$. In this limit we find

$$\pi(k^2, \alpha) \to \frac{\alpha}{3\pi}\left(\ln\frac{k^2}{m^2} + \text{const}\right)$$

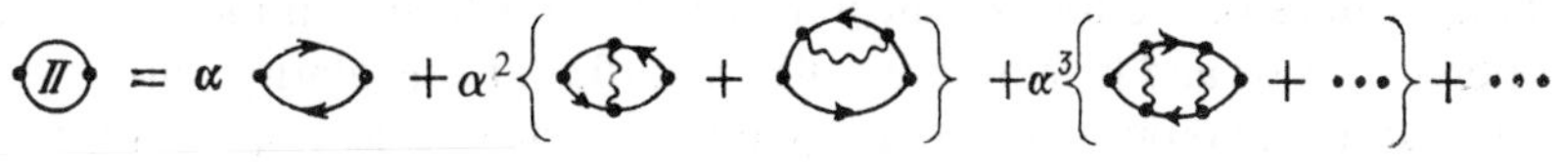

Fig. 29.2. Terms occurring in the polarization operator.

and thus

$$d(k^2, \alpha) \to \frac{1}{1-\frac{\alpha}{3\pi}\ln k^2}. \tag{13}$$

29.4. The dressed electron propagator. The electron self-energy operator Σ in the second order of perturbation theory is formally defined by the relations

$$\Sigma(x) \sim -ie^2\gamma^\nu S^c(x)\gamma^\mu D^c_{\nu\mu}(x), \tag{14a}$$

$$\Sigma(p) \sim \frac{ie^2}{(2\pi)^4}\int \frac{dk}{k^2+i\varepsilon}\left[P^{\mathrm{tr}}_{\mu\nu}(k)+d_l P^{\mathrm{l}}_{\mu\nu}(k)\right]\gamma^\nu \frac{m+\hat{p}-\hat{k}}{m^2-(p-k)^2}\gamma^\mu. \tag{14b}$$

We shall present without the computations the result of evaluating the integral (14b), which is linearly divergent in the ultraviolet region:

$$\Sigma^{\mathrm{reg}}(p) =$$
$$= \frac{\alpha}{4\pi}\left\{\frac{p^2-m^2}{m^2}\left[-3mA(p^2)+d_l\left(\frac{p^2+m^2}{p^2}\hat{p}-m\right)A(p^2)+d_l\hat{p}\frac{m^2}{p^2}\right]+\right.$$
$$\left.+c_1(\hat{p}-m)+c_2 m\right\}, \tag{15}$$

where

$$A(p^2) = \int_0^1 \frac{m^2\,dx}{xp^2-m^2} = \frac{m^2}{p^2}\ln\left(\frac{m^2-p^2}{m^2}\right).$$

If one performs subtraction on the mass shell, then assuming $c_2 = 0$, one obtains

$$\Sigma(m) = 0. \tag{16}$$

However, in the course of calculating the derivative

$$\frac{\partial\Sigma(p)}{\partial\hat{p}} = \frac{\partial\Sigma(\hat{p},p^2)}{\partial\hat{p}} + 2\hat{p}\,\frac{\partial\Sigma(\hat{p},p^2)}{\partial p^2}$$

at the point $\hat{p} = m, p^2 = m^2$, we shall encounter the logarithmic divergence of the function $A(p^2)$. This divergence is related to the behavior of the integrand at *small* values of the four-momentum k (i.e., to behavior in the vicinity of the point $k_\mu = 0$) and is therefore called an *infrared* singularity. Physically, it is due to the photon mass being equal to zero as well as to the possibility that a charged particle (electron or positron) can radiate a large number of very soft photons. We cannot discuss this extremely important problem here in detail, and so we refer the reader to more special literature (see, for example, *Introduction*, Section 35.4; Bjorken and Drell (1965), Section 17.10; Akhiezer and Berestetskii (1969), Sections 30, 51).

Here two different methods are possible. The first one consists in formally ascribing to the photon a small mass λ_0. By introducing the corresponding term $k^2 + i\varepsilon \to k^2 - \lambda_0^2 + i\varepsilon$ into the denominator of the integrand of (14b) we arrive at (15) in which $A(p^2)$ is replaced by

$$A(p^2, \lambda_0^2) = \int_0^1 \frac{dx\,(x-1)\,m^2}{(1-x)(m^2 - xp^2) + x\lambda_0^2} =$$

$$= \begin{cases} \dfrac{m^2}{p^2} \ln\left(\dfrac{m^2 - p^2}{m^2}\right), & |p^2 - m^2| \gg \lambda_0^2, \\ \dfrac{1}{2} \ln \dfrac{\lambda_0^2}{m^2}, & p^2 = m^2. \end{cases} \tag{17}$$

It is now possible to evaluate the constant c_1 from the condition

$$\left.\frac{d\Sigma(p)}{d\hat{p}}\right|_{\hat{p}=m} = \Sigma'(m) = 0, \tag{18}$$

which gives

$$c_1 = 2(3 - d_l)\, A(m^2, \lambda_0^2) - 2d_l. \tag{19}$$

In the final expressions for the observed probabilities of elastic processes the quantity $|\lambda_0|$ plays the role of a parameter describing the uncertainty in the momentum value of the charged particle, which is related to the possibility that soft photons undetected in the given experiment are emitted (see literature cited above).

The other method consists in carrying out the second subtraction of the electron mass operator off the mass shell. Assuming $c_2 = 0$, we write (15) in the form

$$\alpha\Sigma_2(\hat{p}) = (\hat{p} - m)\frac{\alpha}{4\pi}\, s(\hat{p}),$$

where

$$s(\hat{p}) = c_1 - 3\left(\frac{\hat{p}+m}{m}\right) A(p^2) + d_l\left[\frac{p^2 + \hat{p}m}{p^2} + \left(\frac{p^2}{m^2} + \frac{m\hat{p}}{p^2}\right) A(p^2)\right]. \tag{20}$$

We now see that the choice

$$c_1 = -3 - d_l, \qquad s(0) = 0 \tag{21}$$

leads to the electron propagator, represented with the aid of (27.4) in the form

$$G(p) = \frac{1}{i(m-\hat{p})}\left[1 - \frac{\alpha}{4\pi} s(\hat{p})\right]^{-1}, \tag{22}$$

having, in the vicinity of the mass shell, the form

$$\begin{gathered} G(p) \sim \frac{1}{i(m-\hat{p})} Z_2, \quad \hat{p} \sim m, \\ Z_2^{-1} \simeq 1 + \frac{\alpha}{4\pi}\left[2(3-d_l) A(m^2, \lambda_0^2) + 3 - d_l\right]. \end{gathered} \tag{23}$$

29.5. The vertex part and Ward's identity. The third-order contribution to the vertex part is represented by the expressions

$$\begin{aligned} &\Lambda^\nu(x, z; y) \sim ie^2 \gamma_\mu S^c(x-y)\gamma^\nu S^c(y-z)\gamma_\sigma D^{\mu\sigma}(x-z; d_l), \\ &\Lambda^\nu(p', p; k = p'-p; d_l) \sim \\ &\qquad \sim \frac{ie^2}{(2\pi)^4}\int dq D^{\mu\sigma}(p-q; d_l)\gamma_\mu S^c(q+k)\gamma^\nu S^c(q)\gamma_\sigma. \end{aligned} \tag{24}$$

The second of these corresponds to the arrangement of the momentum variables in Figure 29.3. The photon propagator $D^{\mu\sigma}$ is defined by (9). The integral (24) diverges logarithmically in the ultraviolet region.

We shall not present here the cumbersome finite part of the vertex function Λ^ν in an arbitrary gauge of the electromagnetic field. (The corresponding expression is given, for example, in Section 35.4 of the *Introduction*.) For our purpose it will be quite sufficient to use the expression for Λ^ν evaluated in Section 24.3 in the diagonal gauge for which $d_l = 1$.

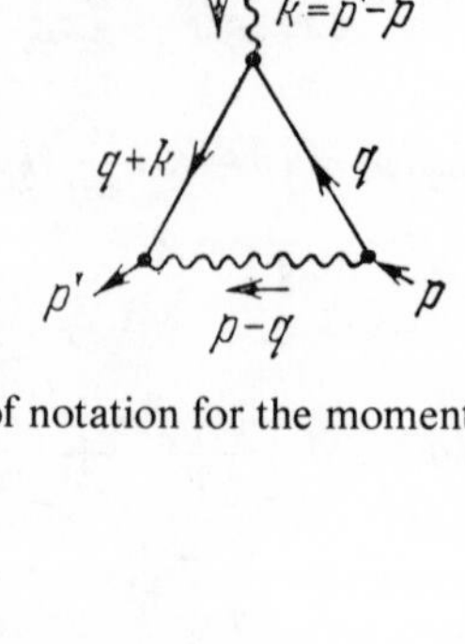

Fig. 29.3. Choice of notation for the momenta in the integral (24).

Assume

$$\Lambda_\nu^{\text{reg}}(p',\ p;\ k;\ d_l=1)=c_3\gamma_\nu+\Gamma_\nu^{\text{fin}}(p',\ p;\ k;\ \mu^2=m^2), \qquad (25)$$

where Γ_ν^{fin} is defined in (24.23), and c_3 is an arbitrary constant. This constant, however, is not independent and may be expressed, with the aid of the condition of gauge invariance (2), in terms of the constant c_1 from the expression (15) for the mass operator.

We apply this condition to the third-order terms of the scattering matrix $S_3(x,\ y,\ z)$, which may conveniently be divided into two groups. The first group includes terms containing the product of the three operators $A(x), A(y), A(z)$, while the second one contains expressions linear in A. Owing to different operator structures, the terms occurring in the first and second groups contribute to different matrix elements. Therefore the condition of gauge invariance may be investigated for them separately. Consider this condition for diagrams of the second group. The corresponding contribution to S_3 may be written as the sum of three terms

$$:A_\nu(x)\,J^\nu(x\,|\,y,\ z):+:A_\nu(y)\,J^\nu(y\,|\,z,\ x):+:A_\nu(z)\,J^\nu(z\,|\,x,\ y):,$$

differing by permutation of the integration variables x, y, z. The condition of gauge invariance takes the form

$$\frac{\partial}{\partial x^\nu}J^\nu(x\,|\,y,\ z)=0. \qquad (26)$$

The function J^ν contains four terms corresponding to the diagrams depicted in Figure 25.4, and four more terms differing from the first four by permutation of the arguments z and y.

We note first that the divergence of the term corresponding to the diagram in Fig. 25.4d,

$$\Pi^{\nu\rho}(x-z)\,D_{\rho\mu}(x-y)\,:\bar{\psi}(y)\,\gamma^\mu\psi(y):$$

vanishes after removal of divergences because of the condition of gauge invariance of the finite part of the polarization operator (7). Therefore, one should consider the sum of the first three diagrams in Figure 25.4. It may be represented in the form

$$:\bar{\psi}(z)\,\Lambda_\nu(z,\ y;\ x)\,\psi(y):+\bar{\psi}(z)\,\Sigma(z-y)\,S^c(y-x)\,\gamma_\nu\psi(x)+$$
$$+\bar{\psi}(x)\,\gamma_\nu S^c(x-z)\,\Sigma(z-y)\,\psi(y).$$

Applying the operation $i\partial/\partial x_\nu$ to this expression, after slight transformations we obtain, with the aid of the Dirac equations for $\psi(x)$, $\overline{\psi}(x)$ and the corresponding inhomogeneous equations for S^c, the following:

$$\overline{\psi}(z)\left\{i\frac{\partial\Lambda_\nu(z,\ y;\ x)}{\partial x_\nu}+\Sigma(z-y)\,\delta(y-x)-\delta(x-z)\,\Sigma(z-y)\right\}\psi(y).$$

Thus the condition of gauge invariance connects the contribution to the vertex part, Λ_ν, with the contribution to the self-energy of the electron, Σ:

$$i\,\frac{\partial\Lambda_\nu^{\rm reg}(z,\ y;\ x)}{\partial x_\nu}=\Sigma^{\rm reg}(z-y)\,\{\delta(x-z)-\delta(y-x)\}. \qquad (27)$$

Passing to the momentum representation by the formulas

$$\Sigma(y)=(2\pi)^{-4}\int e^{ipy}\,\Sigma(p)\,dp,$$
$$\Lambda_\nu(z,\ y;\ x)=(2\pi)^{-8}\int e^{ikx+ipy-ip'z}\Lambda_\nu(p',\ p;\ k)\,\delta(p+k-p')\,dp'\,dp\,dk,$$

we get

$$k^\nu\Lambda_\nu(p',\ p;\ k)=\Sigma(p')-\Sigma(p);\quad p'=p+k. \qquad (28)$$

We have proven this formula for lowest-order contributions to Λ_ν and Σ. It may, however, be demonstrated that a similar relation holds also for higher orders by adding up one by one each of the contributions to the vertex function

$$\Gamma^\nu(p',\ p;\ k\,|\,\alpha)=\gamma^\nu+\sum_{n\geqslant 1}\alpha^n\Lambda^\nu_{(n)}(p',\ p;\ k)$$

as well as the contributions to the mass operator $\Sigma_{(n)}$ having the same index n. The corresponding contributions will then be described by strongly connected vertex diagrams of the $(2n+1)$st order and strongly connected electron self-energy diagrams of the $2n$th order.

Summing (28) over all n and allowing for (27.4), we obtain

$$k_\nu\Gamma^\nu(p',\ p;\ k\,|\,\alpha)=G^{-1}(p;\ \alpha)-G^{-1}(p';\ \alpha). \qquad (29)$$

This formula is known as the *generalized Ward identity* (the Ward–Takahashi identity).

In the general case, (29) establishes the relation between the arbitrary finite constants in the counterterms of spinor electrodynamics:

$$\Delta\mathscr{L} = (Z_2 - 1)\mathscr{L}_0(\bar{\psi}, d\psi) + Z_2(Z_m - 1)m\bar{\psi}\psi + (Z_3 - 1)\mathscr{L}_0(H_{\mu\nu}) + (Z_1 - 1)e\bar{\psi}\hat{A}\,\psi \tag{30}$$

(here $\mathscr{L}_0(\bar{\psi}, \psi)$ is the free Lagrangian of the spinor field (5.5), and $\mathscr{L}_0(H_{\mu\nu})$ is the free transverse Lagrangian of the electromagnetic field (4.12)). This relation has the form

$$Z_2 = Z_1 \tag{31}$$

It is called the Ward identity.

The expression (30) contains no contribution proportional to $(\partial A)^2$, since such a term does not appear naturally by virtue of gauge invariance; but if, nevertheless, such a term is introduced, this leads merely to a change of the parameter d_l in the electromagnetic propagator.

Note, in this connection, that the dependence on d_l, as was established above explicitly in the one-loop approximation, enters into the electron Green's function $G(\hat{p};\, \alpha, d_l)$ and into the vertex $\Gamma_\nu(p, k;\, \alpha, d_l)$, while it does not enter into the photon polarization operator $\Pi_{\mu\nu}(k, \alpha)$.

Correspondingly, the finite Dyson transformations in spinor electrodynamics have the form

$$\begin{aligned} G(m, \alpha, \alpha d_l) &\to G' = Z_2^{-1}G(\tilde{m}, \alpha, \alpha d_l), \\ \Gamma_\nu(m, \alpha, \alpha d_l) &\to \Gamma'_\nu = Z_1\Gamma_\nu(\tilde{m}, \tilde{\alpha}, \alpha d_l), \\ \Pi(k, \alpha) &\to \Pi' = Z_3^{-1}\Pi(\tilde{m}^2, \alpha), \end{aligned} \tag{32}$$

$$\alpha \to \tilde{\alpha} = Z_1^2 Z_2^{-2} Z_3^{-1}\alpha = Z_3^{-1}\alpha, \qquad m \to \tilde{m} = Z_m m. \tag{33}$$

CHAPTER VIII

DESCRIPTION OF REAL INTERACTIONS

30. ELECTROMAGNETIC INTERACTIONS

The way we discussed of describing interactions of quantum fields with the aid of perturbation theory has a chance of being successful only if the interaction of the fields may be considered to be weak. Such a property is certainly an attribute of the electromagnetic interaction, to which there corresponds a dimensionless small parameter $\alpha = e^2/4\pi\hbar c \sim 10^{-2}$, the so-called fine-structure constant.

The field-theory model involving interacting spinor and electromagnetic fields has turned out to be the simplest one. Such a model is adequate for describing the interaction of electrons and positrons with photons, since within a wide range of energies it turns out to be possible to disregard weak and strong interactions, even those involved in virtual effects. We shall begin with an examination of this model, spinor electrodynamics.

30.1. Spinor electrodynamics. This term properly describes processes and effects involving electrons and positrons, and γ-quanta in initial, final, and intermediate states. These include such processes as scattering of photons by electrons (Compton scattering) and scattering of electrons by electrons (Möller scattering), as well as similar processes involving positrons, creation and annihilation of electron–positron pairs, interaction of electrons with an external magnetic field (anomalous magnetic moment), bremsstrahlung of an electron in an external electrostatic field (for example, the field of an atomic nucleus), and certain others, including the purely electromagnetic interactions of muons and hadrons with each other and with electrons.

We shall not here go into any details concerning these processes and effects, and shall restrict ourselves to describing common natural properties and to presenting the most impressive illustrations.

The phenomena under consideration may symbolically be divided into two categories. The first one includes processes which may largely be understood and evaluated by taking into account electromagnetic effects within the framework of nonrelativistic quantum mechanics, or more precisely, within the framework of the quantum theory of the interaction of matter with radiation. Here we have in mind such problems as the motion of an electron in the Coulomb field of a nucleus, scattering of light on free electrons, two-photon annihilation of an electron–positron pair, bremsstrahlung of an electron in the field of a nucleus, and certain others.

These problems have in common that the main contribution to the matrix elements comes from Feynman diagrams not containing closed loops or integrations over virtual four-momenta (so-called *tree* diagrams). The corresponding calculations were carried out at the end of the twenties just after the creation of quantum mechanics. It has turned out to be that theoretical results are in good agreement with experiments.

An example of this is presented by the Klein–Nishina–Tamm formula obtained in 1929–1930 for the differential scattering cross section of a photon with momentum $\boldsymbol{k}_1$ and energy $k_1 = |\boldsymbol{k}_1|$ by an electron at rest:

$$\frac{d\sigma}{d\Omega} = \frac{r_0^2}{2}\frac{k_2^2}{k_1^2}\left(\frac{k_1}{k_2} + \frac{k_2}{k_1} - \sin^2\theta\right). \tag{1}$$

Here $r_0 = \alpha/m$ is the so-called classical electron radius, θ is the scattering angle of the electron, and k_2 is the energy of the scattered photon, which is related to k_1 and θ by the Compton formula

$$k_2 = \frac{mk_1}{m + k_1(1 - \cos\theta)}.$$

The Klein–Nishina–Tamm formula corresponds to the square of the sum of the matrix elements corresponding to the two second-order diagrams presented in Figure 24.3.

Evaluation of the contribution of fourth-order diagrams (see Figure 24.4) initially encountered difficulties connected with ultraviolet divergences. To overcome these difficulties took about two decades. Only towards the second half of the forties was the so-called covariant perturbation theory constructed, in the framework of which the renormalization procedure was formulated.

The creation of the renormalization method opened up the way for calculating corrections to the formulas of the tree approximation. The radiative correction to (1), corresponding to fourth-order diagrams, was evaluated by Brown and Feynman in 1952.

The second category of phenomena may be considered to include those whose existence is due to radiative effects. To these belong the hyperfine splitting of electron levels in the hydrogen atom, the Lamb shift of levels, photon–photon scattering, and the anomalous magnetic moment of the electron.

30.2. The anomalous magnetic moment of the electron. It is well known that the Dirac equation for an electron interacting with an external electromagnetic field contains a term responsible for the magnetic interaction (AII.30). This term may also be derived from the matrix element corresponding to the Born interaction of the electron with the electromagnetic field A^{ext} of the magnetic type ($A_0^{\text{ext}} = 0$, $|\boldsymbol{k}| = 0$).

In the expression

$$-e\bar{u}^{+,s}(p')\gamma^{\nu}u^{-,s}(p)A_{\nu}(k) = e\bar{u}^{+,s}(p')\boldsymbol{\gamma}\boldsymbol{A}(\boldsymbol{k})u^{-,s}(p), \qquad \boldsymbol{p}' = \boldsymbol{p} + \boldsymbol{k}, \tag{2}$$

written in the interaction representation, we shall pass to the split representation of spinors and matrices (AII.22), (AII.23),

$$u(\boldsymbol{p}) = \begin{pmatrix} \varphi(\boldsymbol{p}) \\ \chi(\boldsymbol{p}) \end{pmatrix}, \quad \boldsymbol{\alpha} = \gamma^0\boldsymbol{\gamma} = \begin{pmatrix} 0 & \boldsymbol{\sigma} \\ \boldsymbol{\sigma} & 0 \end{pmatrix}$$

and make use of the free-field equations for the two-component spinors φ and χ. We eliminate the small positron components χ with the aid of the relations following from these equations:

$$\chi(\boldsymbol{p}') = \frac{(\boldsymbol{\sigma}\boldsymbol{p}')}{2m}\varphi(\boldsymbol{p}'), \quad \overset{*}{\chi}(\boldsymbol{p}') = \overset{*}{\varphi}(\boldsymbol{p}')\frac{(\boldsymbol{\sigma}\boldsymbol{p}')}{2m},$$

and represent (2) in the form

$$e\left[\overset{*}{\varphi}(\boldsymbol{p}')\boldsymbol{\sigma}\boldsymbol{A}\chi(\boldsymbol{p}) + \overset{*}{\chi}(\boldsymbol{p}')\boldsymbol{\sigma}\boldsymbol{A}\varphi(\boldsymbol{p})\right] = \frac{e}{2m}[\overset{*}{\varphi}(\boldsymbol{p}')[(\boldsymbol{\sigma}\boldsymbol{A}(\boldsymbol{k}))(\boldsymbol{\sigma}\boldsymbol{p}) + (\boldsymbol{\sigma}\boldsymbol{p}')(\boldsymbol{\sigma}\boldsymbol{A}(\boldsymbol{k}))]\varphi(\boldsymbol{p})].$$

Using thereafter the multiplication formulas for Pauli matrices (AII.28), we finally obtain

$$\frac{e}{2m}\overset{*}{\varphi}(\boldsymbol{p}')\{i\boldsymbol{\sigma}[\boldsymbol{k}\times\boldsymbol{A}(\boldsymbol{k})] + (\boldsymbol{p}+\boldsymbol{p}')\boldsymbol{A}(\boldsymbol{k})\}\varphi(\boldsymbol{p}).$$

The first term of the above expression has the form of the purely magnetic interaction

$$\frac{e}{2m}\overset{*}{\varphi}(\boldsymbol{p}')\boldsymbol{\sigma}\varphi(\boldsymbol{p})\cdot i[\boldsymbol{k}\times\boldsymbol{A}(\boldsymbol{k})] = \boldsymbol{M}_0\boldsymbol{H}(\boldsymbol{k}),$$

where

$$H(\boldsymbol{k}) = i\,[\boldsymbol{k} \times \boldsymbol{A}(\boldsymbol{k})], \quad \boldsymbol{M}_0 = \mu_0 \overset{*}{\varphi} \boldsymbol{\sigma} \varphi,$$

and

$$\mu_0 = \frac{e}{2m} \tag{3}$$

is the Bohr magneton.

The expression (2) corresponds to the diagram of the fundamental electromagnetic interaction presented in Figure 30.1a. Effects due to radiative corrections require that (2) be replaced by

$$-e\bar{u}^{+,\,s}(\boldsymbol{p}')\,\Gamma^{\nu}_{\text{ren}}(p',\,p;\,k)\,u^{-,\,s}(\boldsymbol{p})\,A^{\text{ext}}_{\nu}(k), \tag{2'}$$

where the matrix element $\bar{u}\Gamma u$ may be represented in a form similar to (24.24):

$$\begin{aligned}\bar{u}(\boldsymbol{p}')\,\Gamma^{\nu} u(\boldsymbol{p}) = f_1(k^2,\,\alpha)\,\bar{u}(\boldsymbol{p}')\,\gamma^{\nu} u(\boldsymbol{p}) + \\ + f_2(k^2,\,\alpha)\,\bar{u}(\boldsymbol{p}')\left[\frac{p^{\nu}+p'^{\nu}}{2m} - \gamma^{\nu}\right] u(\boldsymbol{p}).\end{aligned} \tag{4}$$

We now take into account that in accordance with (24.30) the second term on the right-hand side vanishes at $k = 0$, as a result of which the electric charge of the electron is determined by the first term. Thus, the form factor f_1 at the point $k = 0$ is usually normalized to unity:

$$f_1(0,\,\alpha) = 1. \tag{5}$$

(a) (b)

Fig. 30.1. Diagrams of the magnetic moment corresponding to the Bohr magneton and to the Schwinger correction.

Such a normalization is achieved by an appropriate choice of the indefinite finite parameters appearing in the course of subtracting ultraviolet divergences (for example, the choice of the parameter μ in (24.23)) and corresponds to the definition of the renormalized electric charge e as a quantity describing the interaction of an electron with an electromagnetic field of infinitesimal frequency.

It now follows from the condition (5) that the anomalous magnetic moment is determined entirely by the second term on the right-hand side of (4). With the aid of (24.28) the contribution of this term to (2′) may be represented as follows:

$$-\frac{e}{4m}f_2(k^2,\ \alpha)\,\bar{u}(\boldsymbol{p}')\,(\hat{A}_{\text{ext}}(k)\,\hat{k}-\hat{k}\hat{A}_{\text{ext}}(k))\,u(\boldsymbol{p})=$$

$$=-\frac{e}{2m}f_2(k^2,\ \alpha)\ \overset{*}{\varphi}(\boldsymbol{p}')\sigma\varphi(\boldsymbol{p})\boldsymbol{H}(\boldsymbol{k}).$$

Thus, the quantity $ef_2(0,\ \alpha)/2m$ represents a correction to the magnetic moment of the electron that may be written in the form

$$\mu=\mu_0(1-f_2(0,\ \alpha))=\mu_0\left[1+\frac{\alpha}{\pi}a_2+\left(\frac{\alpha}{\pi}\right)^2a_4+\left(\frac{\alpha}{\pi}\right)^3a_6+\ldots\right].$$

The first radiative correction to the Bohr magneton,

$$a_2={}^1/_2$$

corresponds to the diagram in Fig. 30.1b as well as to the computations of Section 24.3, and was calculated by J. Schwinger in 1948 (see Exercise M5). To calculate the next contribution required taking into account the five two-loop diagrams depicted in Figure 30.2. The corresponding result,

$$a_4=\frac{197}{144}+\frac{\pi^2}{12}-\pi^2\ln 2+\frac{3}{4}\zeta(3)\simeq 0.328479\ldots \tag{6}$$

was obtained by Peterman and Sommerfeld at the end of the fifties.

To determine the contribution a_6, it is necessary to consider 40 different three-loop diagrms. To perform calculations analytically by hand is practically impossible because of the length of the computations. The first calculation of a_6, which was carried out in 1972, contained a large uncertainty, $a_6=1.49\pm 0.25$, caused by the approximate nature of the numerical calculations of a significant fraction of the integrals. During the following years, by using computers for performing *analytical computations*, it has been possible to complete the evaluation of 30 of the 40 three-loop

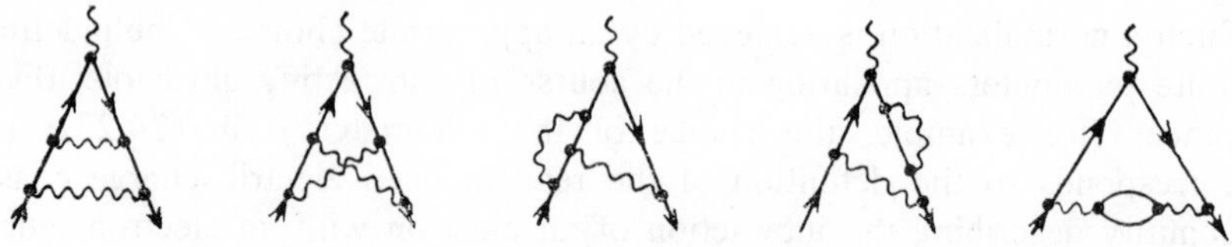

Fig. 30.2. Two-loop diagrams contributing to the anomalous magnetic moment of the electron.

diagrams, which has led to a significant improvement of the precision. The up-to-date result has the form*

$$a_6 = 1.183 \pm 0.011 = 1.183(11). \tag{7}$$

The precision achieved in theoretical computations of (7) makes important the corresponding precision in the determination of the fine-structure constant. The present-day value $\alpha^{-1} = 137.035987(29)$ turns out to be sufficiently accurate for these purposes, and together with (6) and (7) gives the following value for the anomalous magnetic moment:

$$(\mu/\mu_0)_{\text{theor}} = 1.00115965236(28). \tag{8}$$

Here the greater part of the uncertainty in the result is due to the uncertainty in the numerical value of the constant α.

The experimental value obtained in 1978,

$$(\mu/\mu_0)_{\text{exp}} = 1.00115965241(20), \tag{9}$$

is in agreement with the theoretical value (8). The degree of correspondence achieved between the experimental and theoretical values ($<1 \times 10^{-10}$ in relative units) represents a record in physics.

30.3. The limits of spinor electrodynamics. The example presented above is actually very impressive. For other effects in quantum electrodynamics, either the experimental precision is not so high or theoretical calculations point to the necessity of taking into account effects which are beyond the scope of pure electrodynamics.

Thus, for example, the experimental value of the hyperfine splitting of the level $1S_{1/2}$ in the hydrogen atom, which is due to the interaction of the

*Here the uncertainty in a numerical value is written in the form introduced in Section 1.

electron with the magnetic field of the proton, is now known with a record accuracy (to 13 decimal numbers):

$$\nu_{\mathrm{Hfs}}^{\mathrm{exp}} = 1420.4057517864(17)\ \mathrm{MHz}; \tag{10}$$

however, theoretical estimates give only 7 decimal places, even though the finite dimensions of the proton are taken into account.

The Lamb shift, i.e., the fine splitting of the $2S_{1/2}$ and $1P_{1/2}$ levels in the hydrogen atom, is known with an accuracy of the order of 10^{-6}:

$$\begin{aligned} s^{\mathrm{theor}} &= 1057.89(2)\ \mathrm{MHz}, \\ s^{\mathrm{exp}} &= 1057.89(2)\ \mathrm{MHz}, \end{aligned} \tag{11}$$

in which those effects beyond the scope of spinor electrodynamics are of the order of magnitude of 10^{-5} and are taken into account in the theoretical value.

On the whole, it is necessary to say that the whole set of experimental data on a large number of effects is in agreement with the results of theoretical calculations, performed within the framework of perturbation theory in spinor electrodynamics, in those cases when the strong interactions involved either are negligible or may be taken into account. There exists, however, a domain in which the description given by electrodynamics is incomplete. For this there are two reasons.

First, the electromagnetic interactions of hadrons influence the electrodynamics of leptons (electrons and positrons, and muons) through virtual states. Thus, for example, in processes going via one-photon annihilation

$$\begin{aligned} e^- + e^+ &\to e^- + e^+, \\ e^- + e^+ &\to \mu^- + \mu^+ \end{aligned} \tag{12}$$

and realized at accelerators with colliding beams, a significant contribution is provided by Feynman diagrams containing insertions to the virtual photon lines corresponding to the hadron polarization of the photon (Figures 30.3, 30.4). Owing to the resonant interaction of two pions (the so-called ρ-resonance), the cross section of the processes (12) at energies of an order of magnitude of 700–800 MeV in the center-of-mass system exceeds by more than one order of magnitude the value calculated in accordance with the electrodynamic perturbation theory.

Second, as we have seen in the course of calculating the lowest-order electrodynamic radiative corrections in Sections 24, 29, these corrections contain logarithmically increasing contributions proportional to the ultra-

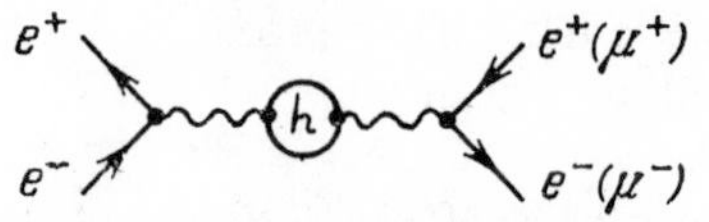

Fig. 30.3. The influence of the "vacuum hadron polarization" on processes of creation and annihilation of lepton pairs.

violet logarithm $L = \ln(p^2/m^2)$. Analysis of the higher radiative corrections reveals that actually the product αL, or αL^2, becomes the expansion parameter of perturbation theory. In a number of cases (in the high-energy region) terms of the form of αL^2, αL may introduce significant contributions to cross sections as well as to other observable quantities. They have to be evaluated and the corrections which they introduce must be taken into account in the analysis of processes measured with sufficiently high accuracy.

On the whole, however, the problem of taking into account higher radiative corrections is still largely only of a theoretical interest.

31. WEAK INTERACTIONS

31.1. Fermi interactions. Four-fermion interactions of the contact type are considerably weaker than electromagnetic interactions. As was pointed out in Section 10, the dimensionless parameter composed of the Fermi coupling constant G and the proton mass M_p turns out to be three orders of magnitude smaller than the fine-structure constant α.

The simplest examples of weak interactions were considered in Section 10.1. Therein it was mentioned that the Fermi-interaction Lagrangian has the form of the product current × current:

$$\mathscr{L} = \frac{G}{\sqrt{2}} j_\mu(x) \overset{+}{j}_\mu(x), \tag{1}$$

where j_μ is the so-called weak current, bilinear in the fermion fields. An expression of this type was proposed by Fermi in 1934 for describing the β-decay of the neutron. However, it took more than 30 years to establish all the

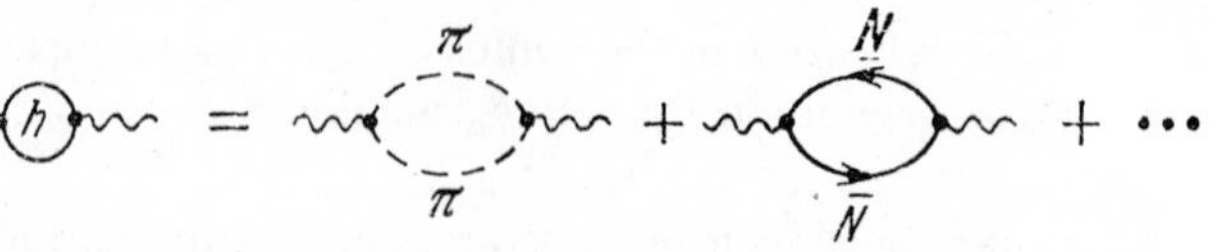

Fig. 30.4. The contribution of pion and nucleon loops to the "vacuum hadron polarization".

details of the structure of the weak current (more precisely, of its lepton part).

The weak current j_ν is a sum of quadratic operator structures of the form

$$\bar{u}_i(x)\, O_\nu^{(i)} v_i(x) \qquad (2)$$

(no summation is over i), where operators u_i and v_i, generally speaking, correspond to different fields. Some of the terms contain only lepton fields and form the lepton component of the weak current (in short, the *lepton current*) l_ν. The remaining terms involving hadron fields form the hadron current h_ν. Thus,

$$j_\nu = l_\nu + h_\nu, \qquad (3)$$

where each of the two components represents a sum of structures of the form of (2). Examples of such constructions were presented in (10.12).

The product of two lepton currents makes up the *lepton* part of the Fermi Lagrangian. They are responsible for the pure leptonic weak processes of the form

$$\begin{aligned} &\mu^+ \to e^+ + \nu_e + \bar{\nu}_\mu, \\ &\nu_\mu + e^- \to \nu_\mu + e^-, \\ &\bar{\nu}_e + e^- \to \bar{\nu}_e + e^- \end{aligned} \qquad (4)$$

etc.

Products of lepton and hadron components $h_\nu \overset{+}{l}{}^\nu$ correspond to so-called *semileptonic* processes which involve both leptons and hadrons. Examples of semileptonic interactions are the following:

$$\begin{aligned} &n \to p + e^- + \bar{\nu}_e, \\ &\nu_\mu + n \to p + \mu^-, \\ &\Lambda \to p + e^- + \bar{\nu}_e \end{aligned} \qquad (5)$$

and many others.

Finally, in nonleptonic weak interactions, corresponding to the product $h_\nu \overset{+}{h}{}^\nu$, only hadrons take part. The corresponding real processes are realized by combining strong and weak interactions. Important examples of such processes are the decays of strange hadrons:

$$\begin{aligned} &K^+ \to \pi^+ + \pi^+ + \pi^-, \\ &\Lambda^0 \to p + \pi^- \end{aligned} \qquad (6)$$

and so on.

31.2. The structure of the weak current. The structure of the matrices O occurring in the components of the weak current was established in the mid-fifties after the discovery of parity violation in weak interactions. It was revealed that the weak current represents the sum of a vector V and of an axial vector A. Mutual products of V and A occurring in the Lagrangian (1) lead to transitions in which parity is violated.

In the general case, to such a structure of the weak current there corresponds a matrix O of the following form:

$$O_\alpha^{(i)} = \gamma_\alpha (1 + g_i \gamma_5), \tag{7}$$

where g_i are certain constants. Taking into account the Hermiticity of γ_5, the constants g_i must be considered to be real.

Experiments have demonstrated that for the lepton terms all the constants g_i are equal to unity. To this important property of the lepton currents

$$l_\alpha = \bar{e}(x)\gamma_\alpha(1+\gamma_5)\nu_e(x) + \bar{\mu}(x)\gamma_\alpha(1+\gamma_5)\nu_\mu(x) + \ldots \tag{8}$$

there corresponds the property of neutrinos of being two-component (see Section 9.2).

Thus, the lepton part of the Fermi interaction contains only one parameter G. This parameter may be determined from the lifetime of the muon (see Exercise F4). It then turns out that

$$GM_p^2 = \frac{G}{\hbar c}\frac{M^2c^2}{\hbar^2} = 1.0262(1) \times 10^{-5} \tag{9}$$

As for the hadron component, it has a somewhat more complicated structure. Thus, for instance, in the nucleon component of the hadron current $\bar{p}O_\alpha n$ responsible for the β-decay of the neutron, the constant g_A, determined from the lifetime of the neutron and from (9), turned out to be equal to $g_A = 1.25$.

The fact, alluded to above, that the weak current is nondiagonal leads to the weak current changing the quantum numbers of fermions: it transforms a neutrino into an electron, a neutron into a proton, etc. The currents examined above change the electric charge Q. They are called *charged* currents. (In this terminology the usual electromagnetic electron–positron current is not charged, i.e., is neutral.) In the hadron component of the weak current there are constituents that change the strangeness of hadrons by one. The notation adopted for them is $h_\alpha(\Delta S = 1)$. Thus, the total hadron current may be represented by the sum

$$h_\alpha(\Delta S=0)+h_\alpha(\Delta S=1). \tag{10}$$

Here for the second term the rule $\Delta S=\Delta Q$ is fulfilled.

Some years ago the existence of *neutral* weak currents j^0_α that do not change the electric charge was proved. The product of neutral currents enters into the Lagrangian additively, i.e., (1) should be replaced by

$$\mathscr{L}(x)=\frac{G}{\sqrt{2}}\,[j^\alpha(x)\overset{+}{j}_\alpha(x)+j^0_\alpha(x)j^{0,\alpha}(x)]. \tag{11}$$

Neutral currents may also be represented as the sum $j^0=l^0+h^0$, where the lepton component has a relatively simple structure:

$$l^0_\alpha(x)=\frac{1}{2}\,\bar{\nu}_e(x)\,\gamma_\alpha(1+\gamma_5)\nu_e(x)+\frac{1}{2}\,\bar{\nu}_\mu(x)\,\gamma_\alpha(1+\gamma_5)\,\gamma_\mu(x)+ \\ +\bar{e}(x)\,\gamma_\alpha(g_V+g_A\gamma_5)\,e(x)+(e\leftrightarrow\mu), \tag{12}$$

where g_V, g_A are constants smaller than one.

31.3. Nonrenormalizability and ultraviolet behavior. On the whole, the Fermi theory of weak interactions is in a position reminiscent of that of quantum electrodynamics before the renormalization method appeared.

The interaction Lagrangian is known, and it contains a small number of coupling constants, while the matrix elements evaluated in the tree approximation are in good agreement with a large amount of experimental evidence. In those cases when it is possible to disregard strong-interaction effects one may hope that the accuracy of this agreement is limited only by radiative corrections of electromagnetic (the small parameter $\alpha/\pi\sim 1/400$) or weak (the small $GM^2\sim 10^{-5}$) origin. However, we would not be able to evaluate radiative corrections that are due to weak interactions.

The Lagrangian (1), (2) has an index $\Omega=+2$ and, in accordance with the classification introduced in Section 26.3, belongs to the nonrenormalizable type. This property is reflected in the dimension of the coupling constant G and in the exponential increase of the matrix elements in the ultraviolet region.

Thus, for example, the total elastic scattering cross section of two fermions caused by the weak interaction of the first order behaves like

$$\sigma(s)\sim G^2 s \qquad (s\gg m_i^2), \tag{13}$$

i.e., it increases to infinity with the total energy $W=\sqrt{s}$ in the center-of-mass system. It should also be expected that the radiative corrections to any processes due to weak interactions will behave in a similar fashion. In other words, the effective dimensionless combination containing the constant G

and representing the expansion parameter in perturbation theory depends on the energy and equals

$$GW^2 = G \cdot s. \tag{14}$$

Comparison of (14) and (9) reveals that at energies of an order of magnitude of

$$W^* \simeq 10^{2.5} M_p \simeq 300 \text{ GeV}, \tag{15}$$

the Fermi interaction ceases to be weak. The energy W^* is sometimes referred to as the *unitary limit*.

31.4. The intermediate-boson hypothesis. The structure of the Fermi interaction (current $\times$ current) gives rise to the idea that an expression of the form of (1) may not be fundamental, but is only an effective interaction which represents the weak currents j interacting with each other through a certain intermediate field.

To elucidate this idea, we return to spinor electrodynamics

$$\mathscr{L}_{\text{ed}}(x) = eJ_\nu(x)A^\nu(x)$$

and for a while mentally confine ourselves to considering initial and final states in which no photons are present. Then photons will remain only in the virtual states to which there correspond virtual lines in the Feynman diagrams. Imagine also that the virtual photons have mass M. Then (Figure 31.1) one may formally eliminate the electromagnetic field by replacing it with the nonlocal effective Lagrangian

$$e^2 J_\nu(x)\, D_M^{\nu\mu}(x-y)\, J_\mu(y) \rightarrow \mathscr{L}_{\text{eff}} = e^2 J_\nu(x) \int D_M^{\nu\mu}(x-y)\, J_\mu(y)\, dy.$$

Then in the momentum representation

$$D_M^{\nu\mu}(k) \sim \frac{g^{\nu\mu}}{M^2 - k^2}.$$

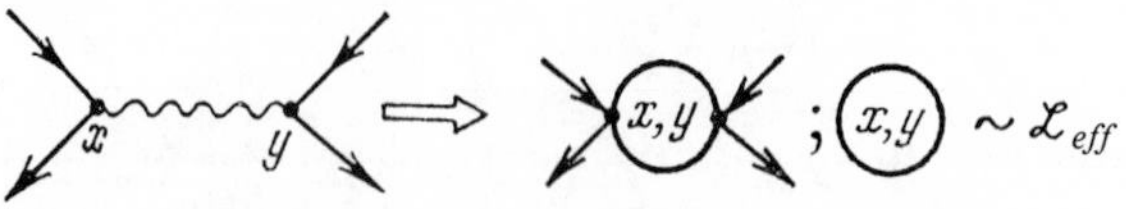

Fig. 31.1. The formal description of internal lines with the aid of a nonlocal effective Lagrangian.

If all the momenta and masses of interest of the particles are considerably smaller than the mass of the "heavy intermediate photon" M, then on the right-hand side of the latter expression one can neglect k^2 as compared to M^2 and obtain

$$D_M^{\mu\nu}(x-y) \to \frac{g^{\mu\nu}}{M^2}\delta(x-y), \qquad \mathscr{L}_{\text{eff}}(x) \to \frac{e^2}{M^2} J_\nu(x) J^\nu(x). \tag{16}$$

Therefore the image of an intermediate vector field of large mass M allows us to understand in a natural way the dimension of the Fermi constant and gives us hope that it will be possible to construct a new formulation of weak interactions such that they will turn out to be renormalizable. For this, of course, from the very beginning it is necessary to assume that the mass of the intermediate vector boson M is considerably greater than would be attainable at present energies (energies in the center-of-mass system are implied) of leptons, i.e., it must be of the order of magnitude of several tens, if not hundreds, of GeV.

At present great popularity has been acquired by the model of weak interactions based on the idea that the intermediate bosons are quanta of a non-Abelian gauge field—the Weinberg–Salam model. This model has quite a complicated structure, but its complications are fully compensated by the fact that, besides a renormalizable formulation of weak interactions, this model also intrinsically includes electromagnetic interactions, i.e., it is a unified theory of weak and electromagnetic interactions.

32. THE WEINBERG-SALAM MODEL

32.1. The main properties. Here we shall present a general outline of the unified model of weak and electromagnetic interactions put forward at the end of the sixties by Weinberg and Salam. This model is remarkable in several respects.

First, it is a Yang–Mills model of weak interactions of leptons which, in the region of "sufficiently low" energies ($\lesssim$50 GeV in the center-of-mass system), reduces to the Fermi interaction. The universality of the Fermi constant acquires a natural explanation in terms of the universality of the gauge interaction constant.

Second, this model unifies in a quite simple manner the weak and electromagnetic interactions of leptons and allows one to introduce massive intermediate bosons with the aid of spontaneous symmetry breaking, at the price of introducing only one Higgs boson. And, third, it predicted the existence of neutral currents which were subsequently observed experimentally.

Further, the condition of renormalizability of the "quark expansion" of the model, i.e., of the version containing the description of the weak interaction of hadrons in terms of the gauge interaction of the constituent quarks, led to the necessity of the existence of a fourth ("charmed") quark and of a new quantum number (charm). This prediction was also confirmed experimentally.

Finally, all the most significant numerical results of this model depend only on three parameters: two gauge coupling constants g and g_1, and the mass M_W of the charged intermediate bosons $W^{\pm}$. The electromagnetic interaction constant e and the Fermi constant G_F are expressed in terms of them as follows:

$$e = \frac{gg_1}{(g^2+g_1^2)^{1/2}}, \quad \frac{G_F}{\sqrt{2}} = \frac{g^2}{8M_W^2}. \tag{1}$$

If we utilize the known experimental values of e and G_F, there remains only one numerical degree of freedom. It is usually connected with the arc-sine of the ratio e/g:

$$\sin\theta_W = \frac{e}{g} = \frac{g_1}{(g^2+g_1^2)^{1/2}}, \tag{2}$$

where θ_W is called the *Weinberg angle.*

As a rule, the predictions of the model are expressed in terms of three quantities e, G_F, θ_W. Thus, with the aid of (1) and (2) one may obtain the following formula and estimate for the mass of the charged intermediate boson:

$$M_W = \left(\frac{e^2\sqrt{2}}{8G_F}\right)^{1/2} \frac{1}{\sin\theta_W} = \frac{37.7\,\text{GeV}}{\sin\theta_W} \geq 37.7\ \text{GeV}. \tag{3}$$

32.2. The structure of the boson sector. The model is based on the hypothesis that there exist two gauge fields. One of them (A_a, $a = 1, 2, 3$) has three components and corresponds to the adjoint representation of the group $SU(2)$, while the second one (B) has one component and its gauge group is the group $U(1)$.

Thus, the gauge group of the Weinberg–Salam model is the compact group $SU(2) \times U(1)$. It contains two numerical parameters, g and g_1. Four gauge vector particles are necessary for the description of three intermediate (massive) mesons and a photon. Of the four particles, two (the components A_1, A_2) are charged and two (A_3 and B) are neutral. As the fields A_3 and B have the same quantum numbers, mixing between them can take place. The physical neutral vector particles, the photon and the massive Z-meson, turn out to be superpositions of the fields A_3 and B. To make the vector mesons massive, one uses the mechanism of spontaneous symmetry breaking (see Section 11.3), for which one introduces an auxiliary two-component complex (four degrees of freedom) scalar field φ. Three of the four degrees of freedom are used for "providing", with the aid of the Higgs mechanism, a superfluous polarization state for each of the three vector components, while the fourth one leads to the physical massive Higgs boson.

Thus, the boson sector of the Weinberg–Salam model is based on the Lagrangian

$$\mathscr{L}_B = -\frac{1}{4} G^a_{\mu\nu} G^a_{\mu\nu} - \frac{1}{4} F_{\mu\nu} F_{\mu\nu} + |D_\mu \tilde{\varphi}|^2 - \frac{\lambda^2}{4} (|\tilde{\varphi}|^2 - \eta^2)^2. \tag{4}$$

Here F is the field strength tensor of the Abelian field

$$F_{\mu\nu} = \partial_\mu B_\nu - \partial_\nu B_\mu,$$

$G_{\mu\nu}$ is the field strength tensor of the non-Abelian field

$$G^a_{\mu\nu} = \partial_\mu A^a_\nu - \partial_\nu A^a_\mu + g\varepsilon^{abc} A^b_\nu A^c_\mu,$$

the field $\tilde{\varphi}$ represents the isotopic doublet (of the gauge $SU(2)$ group) of the complex fields

$$\tilde{\varphi} = \begin{pmatrix} \tilde{\varphi}_1 \\ \tilde{\varphi}_2 \end{pmatrix},$$

and $D_\mu \tilde{\varphi}$ are the covariant derivatives

$$D_\mu \tilde{\varphi}(x) = \left(\partial_\mu - \frac{ig}{2} \boldsymbol{\tau} \boldsymbol{A}_\mu - \frac{ig_1}{2} B_\mu \right) \tilde{\varphi}(x). \tag{5}$$

Spontaneous symmetry breaking is realized when the second component of the field $\tilde{\varphi}$ is shifted by a real constant η:

$$\tilde{\varphi} = \begin{pmatrix} \varphi_1 \\ \varphi_2 + \eta \end{pmatrix} = \varphi + \begin{pmatrix} 0 \\ \eta \end{pmatrix}, \quad \mathrm{Im}\, \eta = 0. \tag{6}$$

As a result of the shift, the term $|D_\mu \tilde{\varphi}|^2$ gives the following contribution to the "mass matrix" (i.e., to the terms bilinear in the components A^a, B):

$$\frac{g^2\eta^2}{4} [(A^1_\nu)^2 + (A^2_\nu)^2] + \frac{\eta^2}{4} [g A^3_\nu - g_1 B_\nu]^2. \tag{7}$$

The diagonalization of this contribution is performed by the linear transformation

$$(A_\nu^3,\ B_\nu) \to (Z_\nu,\ A_\nu),$$

where

$$Z_\nu = \frac{-gA_\nu^3 + g_1 B_\nu}{(g^2+g_1^2)^{1/2}}, \quad A_\nu = \frac{g_1 A_\nu^3 + g B_\nu}{(g^2+g_1^2)^{1/2}}. \tag{8}$$

Introducing the notation

$$M_W = \frac{g\eta}{\sqrt{2}}, \quad M_Z = \frac{\eta\,(g^2+g_1^2)^{1/2}}{\sqrt{2}} = \frac{M_W}{\cos\theta_W}, \tag{9}$$

we write the expression (7) as follows:

$$M_W^2 \overset{*}{W}{}^\nu W_\nu + (1/2)\, M_Z^2 Z_\nu Z^\nu. \tag{10}$$

Here, for reasons of future convenience, new mutually complex conjugate fields have been introduced for the intermediate charged mesons:

$$W_\nu^{\pm} = \frac{A_\nu^1 \mp iA_\nu^2}{\sqrt{2}}, \quad W_\nu \equiv W_\nu^-, \ \overset{*}{W}_\nu = W_\nu^+. \tag{11}$$

If the components of the doublet φ are represented in the form

$$\varphi_1(x) = \frac{\Phi_1(x) + i\Phi_2(x)}{\sqrt{2}}, \quad \varphi_2(x) = \frac{\sigma(x) + i\Phi_3(x)}{\sqrt{2}},$$

then the last term in (4) results in the expression

$$-\frac{\lambda^2}{16}(\Phi^2+\sigma^2)^2 - \frac{\lambda^2\eta}{2\sqrt{2}}\sigma(x)(\Phi^2+\sigma^2) - \frac{\lambda^2\eta^2}{2}\sigma^2(x),$$

$$\Phi^2 = \Phi_1^2 + \Phi_2^2 + \Phi_3^2,$$

from which it clearly follows that the field σ has the mass

$$m_\sigma = \lambda\eta, \tag{12}$$

while the components Φ_a represent Goldstone fields. They can be removed by a gauge transformation, as a result of which three components of the non-Abelian gauge field will acquire a superfluous polarization component (the Higgs effect). Thus, the latter term on the right-hand side of (4) reduces to

$$-\frac{g^2}{32}\frac{m_\sigma^2}{M_W^2}\sigma^4(x)-\frac{gm_\sigma^2}{4M_W}\sigma^3(x)-\frac{m_\sigma^2}{2}\sigma^2(x). \tag{13}$$

32.3. The fermion sector. As is well known, the weak interaction does not conserve parity. Right-hand-polarized neutrinos have not been observed experimentally. Therefore, the structure of the fermion sector should, in the first place, account for these properties. To this end, the left-hand-polarized component of the electron spinor $e(x)$ and the electron neutrino $\nu_e(x)$ are united into a "left-handed" doublet $L_e(x)$, while the right-hand-polarized part of $e(x)$ forms a "right-handed" singlet with respect to the $SU(2)$ group:

$$L_{(e)}(x)=\frac{1-\gamma_5}{2}\begin{pmatrix}\nu_e(x)\\ e(x)\end{pmatrix},\quad R_{(e)}(x)=\frac{1+\gamma_5}{2}e(x). \tag{14}$$

Analogous multiplets are introduced for the muon and for the muon neutrino:

$$L_{(\mu)}(x)=\frac{1-\gamma^5}{2}\begin{pmatrix}\nu_\mu(x)\\ \mu(x)\end{pmatrix},\quad R_{(\mu)}(x)=\frac{1+\gamma^5}{2}\mu(x).$$

Such a structure of the multiplets is convenient because it allows us to retain in the charged components of weak currents only the left-handed components of leptons united in the doublets L.

Assuming that the left-handed doublets take part in the gauge interaction of the $SU(2)\times U(1)$ group, while the right-handed ones are connected only with the Abelian subgroup, we arrive at the following terms in the Lagrangian:

$$\mathscr{L}_{\text{lep}}(x)=i\bar{L}(x)\gamma^\mu\left(\partial_\mu-\frac{ig}{2}\boldsymbol{\tau}A_\mu-\frac{ig_1}{2}B_\mu\right)L(x)+ \\ +i\bar{R}(x)\gamma^\mu(\partial_\mu-ig_1B_\mu)R(x). \tag{15}$$

Here we have dropped the lower indices (μ) and (ν), implying that it is necessary to perform independent summation over them.

Direct introduction of mass terms for the leptons

$$m(\bar{L}R+\bar{R}L)$$

breaks gauge invariance. However, if one assumes that the multiplets L and R interact, like the gauge fields A_μ and B_μ, with the scalar field $\widetilde{\varphi}$ introduced in (4), they then acquire mass as a result of spontaneous symmetry breaking.

Assuming that the corresponding interaction term is of the Yukawa form

$$-G\left[(\bar{L}(x)\tilde{\varphi}(x))R(x)+\bar{R}(x)\left(\overset{*}{\tilde{\varphi}}(x)L(x)\right)\right], \tag{16}$$

we find that the "shift" (6) of the field $\widetilde{\varphi}$ by a constant results in the appearance in the Lagrangian of lepton mass terms

$$-m_e\bar{e}(x)e(x)-m_\mu\bar{\mu}(x)\mu(x),$$

where

$$m_e=\eta G_e/2, \quad m_\mu=\eta G_\mu/2. \tag{17}$$

Thus, the complete expression for the lepton part of the Lagrangian, based on the terms (15), (16) and written in terms of the variables W, A, φ, will have the form

$$\mathcal{L}(x)=\mathcal{L}_{(e)}(x)+\mathcal{L}_{(\mu)}(x)+\mathcal{L}^0_{\text{lep}}(x), \tag{18}$$

where

$$\mathcal{L}_{(e)}(x)=$$
$$=-\frac{g}{2\sqrt{2}}\left[\bar{\nu}_e(x)\gamma^\mu(1-\gamma_5)e(x)W_\mu(x)+\bar{e}(x)\gamma^\mu(1-\gamma_5)\nu_e(x)\overset{*}{W}_\mu(x)\right]+$$
$$+\frac{gg_1}{(g^2+g_1^2)^{1/2}}\bar{e}(x)\gamma^\mu e(x)A_\mu(x)-\frac{G_e}{2}\bar{e}(x)e(x)\sigma(x)+\frac{(g^2+g_1^2)^{1/2}}{4}\times$$
$$\times\left[\bar{\nu}_e(x)\gamma^\mu(1-\gamma_5)\nu_e(x)-2\bar{e}(x)\gamma^\mu\left(\gamma^5+\frac{g^2-3g_1^2}{g^2+g_1^2}\right)e(x)\right]Z_\mu(x), \tag{19}$$

$\mathcal{L}_{(\mu)}$ follows from $\mathcal{L}_{(e)}$ by the substitution $e(x)\to\mu(x)$, $\mu_e\to\nu_\mu$, and $\mathcal{L}^0_{\text{lep}}$ is the free Lagrangian of the fields $e(x)$, $\mu(x)$, $\nu_e(x)$, $\nu_\mu(x)$:

$$\mathcal{L}^0_{\text{lep}}(x)=i\bar{\nu}_e(x)\hat{\partial}\nu_e(x)+i\bar{\nu}_\mu(x)\hat{\partial}\nu_\mu(x)+$$
$$+\bar{e}(x)(i\hat{\partial}-m_e)e(x)+\bar{\mu}(x)(i\hat{\partial}-m_\mu)\mu(x). \tag{20}$$

From (19) it follows that the coupling constants of the electromagnetic interaction e and of the Fermi weak interaction G_F are linked to g, g_1, and M by the formulas (1).

We recall also that the mass of the Higgs boson, m_σ, is expressed through the parameters of the Lagrangian by (12).

32.4. The Lagrangian and quantization. Collecting the results, we find that the total classical Lagrangian in the Weinberg–Salam model has the form

$$\mathcal{L}_{\text{WS}}=\mathcal{L}_B+\mathcal{L}^0_{\text{lep}}+\sum_{(i)}\mathcal{L}_{(i)}, \tag{21}$$

where $\mathcal{L}_B$ is the boson part determined by the relation (4), $\mathcal{L}_{1i)}$ are the lepton components of the form (19), and $\mathcal{L}^0_{\text{lep}}$ is the sum of the free lepton contributions (20). The first term of $\mathcal{L}_B$ written in the variables W_ν, A_ν, σ contains the following quadratic terms:

$$\mathcal{L}_B^0(x) = -\frac{1}{4}(\partial_\mu A_\nu - \partial_\nu A_\mu)^2 + \frac{1}{2}(\partial_\mu \sigma)^2 - \frac{m_\sigma^2}{2}\sigma^2 +$$
$$+ \frac{1}{2}(\partial_\mu \overset{*}{W}_\nu - \partial_\nu \overset{*}{W}_\mu)(\partial_\mu W_\nu - \partial_\nu W_\mu) + M_W^2 \overset{*}{W}_\nu W_\nu -$$
$$- \frac{1}{4}(\partial_\mu Z_\nu - \partial_\nu Z_\mu)^2 + \frac{M_Z^2}{2} Z_\nu Z_\nu. \qquad (22)$$

This expression, together with the free lepton Lagrangian $\mathcal{L}^0_{\text{letp}}$, described by (20), contains exactly the "free" (to be more precise, bilinear) constituent of the Weinberg–Salam Lagrangian which must be quantized in order to construct perturbation theory in the coupling constants g, g_1, $G_{(i)}$ (or their combinations). The quantization procedure for a singular Abelian Lagrangian of the electromagnetic field was considered in Section 8.5, and for a non-Abelian field its outline was presented in Section 19.4. Therein it was demonstrated that the Feynman rules for the Yang–Mills field involve additional elements which may be expressed through the Faddeev–Popov auxiliary fictitious ghost field. We shall not present the quantization procedure here, and we refer the reader interested in the Feynman rules for a massive Yang–Mills field to Appendix VIII.

We shall make one more remark concerning the ultraviolet divergences. The Weinberg–Salam model has quite a complicated structure and involves massive vector fields. The propagators of such fields (see (18.14)) contain in the numerator a second-degree polynomial in components of the momentum, and therefore, in accordance with the analysis of Section 26, lead, at first glance, to nonrenormalizable ultraviolet divergences. At the same time, however, the initial Lagrangian of the type (4), before the shift of the scalar fields by a constant is performed, contains massless vector particles and, as it may be demonstrated, is renormalizable. Since the initial Lagrangian and the final one, obtained as a result of spontaneous symmetry breaking, are physically equivalent (and differ only in choice of variables), the boson sector of the model turns out to be renormalizable.

A more detailed investigation shows (see, for instance, Chapters IV, V of the book by Faddeev and Slavnov (1978)) that a real difficulty from the viewpoint of renormalizability arises in the fermion sector. It is related to parity violation and manifests itself in the so-called anomalous triangular diagrams (Figure 32.1) in which all three internal lines are fermion lines, while the external ones are vector gauge lines, and in the vertices of which there are γ^5 matrices (compare with the first line on the right-hand side of (19)). The divergences of these diagrams cannot be renormalized so as to retain gauge invariance. They can, however, be compensated.

One compensation possibility comes from the quark extension of the model. If to the Weinberg–Salam lepton model described one adds weak

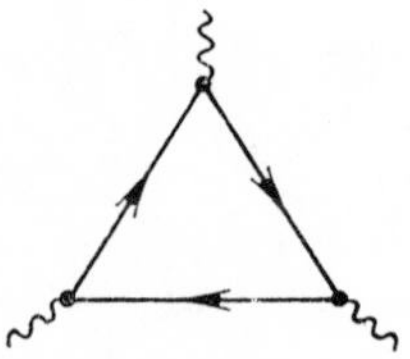

Fig. 32.1. Anomalous triangular diagram in non-Abelian gauge theory.

interactions of hadrons, introducing them by means of gauge interactions of fermion quark fields, then one may achieve renormalizability of the extended model, if the condition of balance between leptons and quarks is fulfilled. To four leptons (e, μ, μ_e, μ_μ) there should correspond four quarks. Just these considerations, supplemented by certain qualitative physical arguments, led to the hypothesis of the fourth quark with a new quantum number (charm), which was confirmed in a striking manner by the discovery in 1975 of the family of ψ-particles and the subsequent observation of hadrons with nonzero charm.

32.5. Physical content and experimental status. We shall now give a short description of the physical content of the Weinberg–Salam model and of its correspondence to experimental data.

A significant feature of the model is the existence of neutral currents (due to the neutral intermediate meson). These currents were discovered in 1973 in experiments on neutrino–nucleon interaction. The measured quantitative characteristics are in agreement with the predictions of the model for $\theta_W \simeq 30°$, $\sin^2 \theta_W \simeq 0.23 \pm 0.02$. Inserting this value into (3) and (9), we obtain approximate values for the masses of the intermediate bosons: $M_W \simeq 75$ GeV, $M_Z \simeq 87$ GeV. The interaction range of weak forces corresponding to these values equals $r_{\text{weak}} \simeq 3 \times 10^{-16}$ cm.

For the direct detection of effects linked to the existence of such heavy particles, rather high accelerator energies are necessary. Thus, for example, the neutral Z-boson can be created in colliding e^+e^- or $p\bar{p}$ beams, when the energy of each individual beam is not less than 40–50 GeV. Such accelerators may appear during the present decade.

The model involves the scalar Higgs boson. Its mass m_σ is determined by the formula (12), containing the free parameter λ. Therefore the mass m_σ may be "chosen" to be sufficiently large for the absence of the heavy scalar neutral boson not to cause any headache for a long time.

33. THE GENERAL PICTURE

33.1. Strong interactions. In the course of the discussion of interactions between fields in Section 10 we wrote down the pseudoscalar (10.17) and pseudovector (10.19) versions of the Yukawa Lagrangian. Therein it was pointed out that when the pion–pion interaction (10.22) is introduced, there is no serious reason to consider such expressions to be of a fundamental nature. They may be regarded purely from a phenomenological point of view, and one may try to determine the concrete interaction Lgrangian as well as the numerical values of the coupling constants from comparison with experimental data. Such attempts were repeatedly undertaken, mainly during the fifties. It was established that certain expressions of this type may serve as the basis of a renormalizable perturbation theory. Such, for example, are the Lagrangians (10.17) and (10.22). (At the same time, the Yukawa pseudovector interaction (10.19) is nonrenormalizable.) However, no satisfactory quantitative descriptions of strong-interaction processes have yet been obtained in this direction.

A quite clear (although not the only) reason for the failures is that the coupling constants g and h in the strong interactions, and consequently the expansion parameters of perturbation theory, are not small compared to unity. Thus, for example, the coupling constant g of the Yukawa pion–nucleon pseudoscalar interaction estimated through dispersion relations (see Section 33.3 below) turned out to equal

$$g^2/4\pi \simeq 14.7. \qquad (1)$$

During the seventies, in connection with the development of the composite quark model of hadrons, expressions like (10.17), (10.22) gradually "went out of fashion", and the efforts of theorists have of late been concentrated on the so-called quark–gluon model, in which the leading part is played by hypothetical "truly fundamental" constituents of hadrons, called *quarks*. Fermion quark fields are linked to each other by a gauge field, the quanta of which are vector particles and are called *gluons*.

Owing to the property of asymptotic freedom of gauge fields (see Section 33.3 below and Appendix IX), the quark–gluon interaction becomes weaker with increasing energy, and as a consequence, quarks and gluons in the ultraviolet limit behave as quasifree pointlike objects. Experimentally, in high-energy, so-called deep-inelastic processes, hadrons display a complicated structure and seem to be made up of pointlike constituents called partons. The observed picture may be explained naturally by the property of asymptotic freedom. The role of partons is then played by gluons and quarks, whose existence was first postulated at the beginning of the sixties in connection with the phenomenological classification of the hadrons and hadron resonances based on the concept of unitary symmetry.

The quantum-mechanical model describing interactions of fermion quark fields by means of the gauge gluon field is called *quantum chromodynamics*.

We cannot go into any details here of the physics of strong interactions and of quantum chromodynamics. We merely point out that, in accordance with modern methodology, the application within this book of the method of renormalized perturbation theory to certain physical situations turns out to be justified. To be precise, success in quantum chromodynamics is achieved by the simultaneous application of perturbative methods and of the renormalization group (see Appendix IX). Nevertheless, the main physical problems of the quark–gluon model connected with the absence of free-state quarks and gluons have still to be solved. They cannot be investigated by the weak-coupling formalism (i.e., within the ordinary perturbation theory).

Thus, the field where methods of renormalizable perturbation theory may be applied is restricted in a natural way to quantum field theory models with numerically small values of coupling constants. There also exist other limitations.

33.2. Restrictions of the perturbation-theory method. It turns out that even in the case of numerically small coupling constants, there exist ranges of values of the kinematic variables within which the subsequent terms of the perturbative expansion may not diminish. As was established above, one-loop corrections to propagators evaluated in the region of large momenta p (the ultraviolet asymptotics) increase logarithmically (see, for example, (24.6), (29.6)), and therefore their relative contribution to the propagator is proportional to the product g^2L of the square of the coupling constant g and the large "ultraviolet" logarithm

$$L = \ln (p^2/\lambda^2). \tag{2}$$

An analysis of the higher radiative corrections shows that in the ultraviolet region n-loop contributions increase as $(g^2L)^n$. An analogous situation arises in the region of small momenta (the infrared asymptotics) in the framework of certain quantum-field models involving massless fields. Thus, for instance, in spinor electrodynamics the radiative correction to the electron propagator, which is due to the emission and absorption of a massless photon, contains (see (29.17), (29.20)) the product αl of the expansion parameter $\alpha = e^2/4\pi$ and the "infrared" logarithm

$$l = \ln \frac{m^2 - p^2}{m^2}, \tag{3}$$

which increases infinitely when the electron momentum p approaches the mass shell ($p^2 \to m^2$). The higher radiative corrections contain the products $(\alpha l)^n$.

Thus in the ultraviolet and infrared regions, products of the type $g^2 L$ and $g^2 l$ turn out to be actual expansion parameters of perturbation theory. In these regions one goes effectively beyond the limits of weak coupling, even when the numerical values of the interaction coupling constants are small. The above holds not only for one-particle propagators, but also for higher vertex functions.

There arises the problem of determining and summing such large higher-order contributions. In some individual cases this problem can be partially solved by a certain extension of the perturbation theory technique. An example is provided by the formula (29.13) for the ultraviolet asymptotics of the photon propagator. However, in the general case one is obliged to make use of methods that are beyond the scope of ordinary perturbation theory or, at least, to combine such methods with results obtained by perturbation theory (see Section 33.3 below).

Thus, the field of applications of the renormalizable perturbation theory technique presented in Chapters IV–VII of the present book turns out to be limited in several respects.

First, this technique may be applied only to the so-called renormalizable interactions. In particular, it may not be utilized in the case of weak interactions in the Fermi form (as mentioned in Section 31.3) or for gravitational interactions (as the graviton spin equals 2 and the propagator increases greatly in the ultraviolet region).

Second, in renormalizable quantum field theory models the method of perturbation theory may be used only in case the coupling constants are small. This condition is satisfied by spinor electrodynamics and by its generalizations describing weak interactions like the Weinberg–Salam model. However, in most of the renormalizable models of strong interactions of the old-fashioned type, perturbation theory cannot be made use of, owing to the large value of the coupling constant (compare with (1)).

Finally, as has just been demonstrated, even in renormalizable models with numerically small coupling constants there exist kinematic restrictions. To overcome them, various devices and methods are utilized which allow one, at least partially, to go beyond the scope of the "standard" perturbation theory discussed in the present book.

33.3. Other methods of quantum field theory. Since, as was noted, the possibilities are limited for application of perturbation theory to real processes, especially to problems of hadron physics, great importance is attached to any successful theoretical means of analysing problems of the theory of particles not connected directly with the smallness of the coupling constant and exponential expansions. There exist three such approaches that have proven useful. They are the method of dispersion relations, the method of functional (or path) integration, and the renormalization group method.

The method of *dispersion relations* is based on the axiomatic formulation of quantum field theory, the outline of which was presented in Chapter IV. The axioms formulated in Section 15 were used in Section 16 for constructing the expansion of the scattering matrix in powers of the interaction strength, i.e., in powers of the coupling constant (or constants). The results obtained turn out to be equivalent to the solution of the Schrödinger equation by the iterative method presented in Section 14.

A somewhat extended version of the axioms of Section 15 (the Bogoliubov axiomatics) was used by one of the authors in the mid 1950s to substantiate the method of dispersion relations (see *Introduction*, Chapter X). The decisive elements of the rigorous mathematical argument are the formulation of the condition of causality in its differential form (15.16), and the uniting of the methods of the theory of functions of several complex variables together with the theory of distributions.

The dispersion relations formally represent a consequence of the Cauchy theorem, from the theory of functions of a complex variable, applied to the scattering amplitude $f(E)$ analytically continued to the complex plane of variable energy $E \to E + iV$. Usually they are written down with the aid of integral relations of the form

$$\mathrm{Re}\, f(E) = \frac{1}{\pi} \int_{-\infty}^{\infty} \frac{\mathrm{Im}\, f(E')}{E' - E} dE' + \ldots$$

between the real and imaginary parts of the two-particle forward scattering amplitude

$$f(E) = f(E, \theta = 0) = \mathrm{Re}\, f(E) + i\, \mathrm{Im}\, f(E),$$

each of which can be expressed by algebraic relations (see the formulas of Section 21.4) in terms of two directly measured quantities, the total effective cross section $\sigma(E)$ and the differential forward scattering cross section $\partial\sigma(E, \theta = 0)/\partial\Omega$. Analytical continuation of the scattering amplitude $f(E)$ to the complex energy plane turns out to be possible owing to the condition of causality.

The dispersion relations follow rigorously from the main axiomatic principles of local quantum field theory formulated in the form of energy-integral relations for observed quantities. These relations do not form a system of equations sufficient for the theoretical definition of the functions occurring in them. They may, however, be verified directly by experiments, and they also represent the basis for obtaining rigorous qualitative as well as approximate quantitative results.

During the subsequent decades the method of dispersion relations was developed extensively and right up to the 1970s was the only means of quantitative analysis of strong and electromagnetic interaction processes involving hadrons.

The *functional integral* method is based on Feynman's formulation of quantum mechanics in the form of path integral (see Section 138 of the textbook by Blokhintsev (1976)). Transition amplitudes and Green's functions are expressed in terms of certain formal expressions (functional integrals) which may be imagined as limits of multiple integrals when the number of integrations tends to infinity. It turns out to be possible to introduce certain rules for dealing with functional integrals and, in some cases, devices for their explicit evalaution. So far the functional integral method has proved to be a good method for obtaining general results in quantum field theory. With its aid non-Abelian gauge fields have been quantized (see the book by Faddeev and Slavnov (1978)), a very elegant derivation was achieved of the formulas for gauge transformations of the complete Green's functions in quantum electrodynamics (see Chapter VIII of the *Introduction*), and certain other important results were obtained as well. Recently a number of interesting properties, linked to the structure of the perturbation theory series as a whole, were established by the approximate evaluation of the functional integral by using the functional saddle point method, which is the analog of the saddle-point method known in the theory of functions of a complex variable. In this way it turned out to be possible to demonstrate that the quantum field perturbation theory series are asymptotic expansions (and consequently have no domain of convergence).

The *renormalization group* method is based on the group character of the finite renormalizations of the Green's functions and coupling constants corresponding to various ways of removing divergences (for example, to different subraction points of the R-operation; see Section 28). The mathematical formulation of group properties (see Appendix IX in this book and Chapter IX of the *Introduction*) leads to functional relationships linking Green's functions for different renormalization procedures to each other. The corresponding group differential equations (Lie equations) turn out to be very convenient for investigating the ultraviolet and infrared asymptotics of Green's functions, i.e., for the analysis of those cases when, as was pointed out in Section 33.2, one effectively goes beyond the scope of weak coupling.

It must be noted that, contrary to the main bulk of results obtained with the aid of dispersion relations and the functional integral representation, here the main physical results are obtained by means of a kind of symbiosis of perturbation theory calculations and rigorous renormalization group relations. They include the property of *asymptotic freedom* of a number of quantum field models involving Yang–Mills fields.

The essence of the phenomenon of asymptotic freedom is expressed by the formula

$$\bar{g}^2(k^2, g) = \frac{g^2}{1 + cg^2 \ln(k^2/\lambda^2)}, \qquad c > 0, \quad |k^2| \to \infty \tag{4}$$

for the so-called effective (or invariant) interaction coupling constant $\bar{g}$ depending on the square of the four-momentum and describing the effective change of the interaction intensity in the region of large momenta (i.e., at small distances). The formula (4) represents the sum of the "highest" ultraviolet logarithmic contributions $(g^2L)^n$ of all orders of perturbation theory, and differs from the analogous formula of spinor electrodynamics for the invariant electric charge,

$$\bar{\alpha}(k^2, \alpha) = \frac{\alpha}{1 - \dfrac{\alpha}{3\pi} \ln \dfrac{k^2}{\lambda^2}}, \qquad |k^2| \to \infty \tag{5}$$

in the algebraic sign of the logarithmic term in the denominator. (Equation (5) is a direct consequence of the formula (29.13) for the photon propagator and corresponds to the sum of the ultraviolet asymptotics of the contributions of the interations of a simple electron–positron loop; see Figure 29.1.)

This difference leads to important physical consequences. Equation (5) corresponds to the "normal picture" of screening (similar to screening of a pointlike charge submerged in a polarizable medium), when the effective charge $\bar{\alpha}$ increases with k^2, i.e., when the distance decreases (Figure 33.1). Contrary to this, equation (4) leads to a decrease of the effective interaction

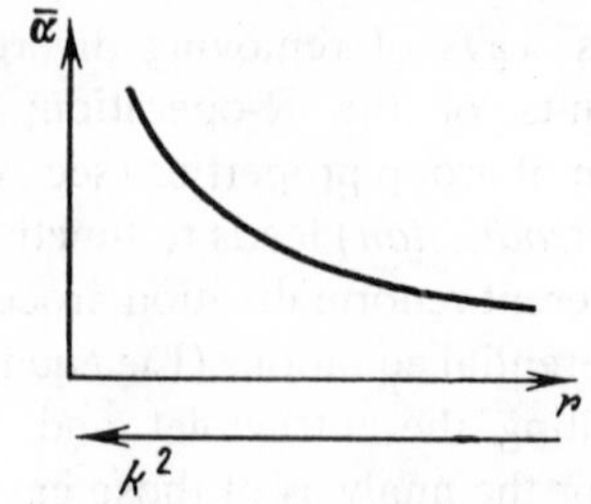

Fig. 33.1. The effect of vacuum polarization in spinor electrodynamics leading to the "normal screening" picture.

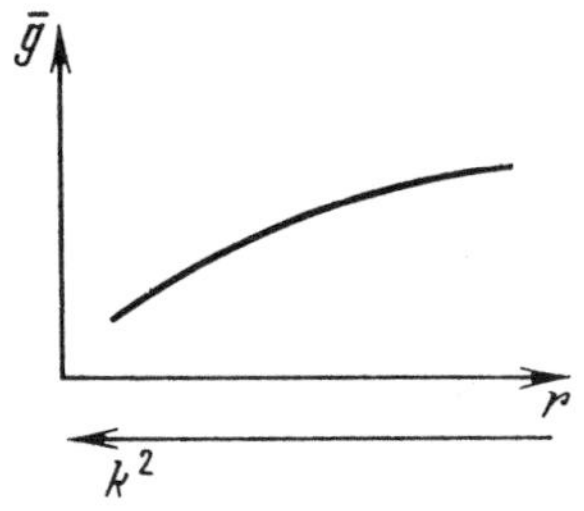

Fig. 33.2. The decrease of the effective interaction intensity with increasing momentum in the non-Abelian gauge theory.

intensity, as the momentum increases, which in the asymptotic ultraviolet limit tends to zero, with the fields taking part in the interaction tending towards free ones (Figure 33.2). This phenomenon of "self-switching-off" of the interaction at small distances is what is meant by asymptotic freedom. For a time it used to be a decisive argument for establishing th gauge mechanism of interaction of fields as the only possible one for strong interactions.

33.4. Conclusion. The method of renormalization in perturbation theory presented in this book is the primary tool of quantum field theory. It allows one to obtain reliable quantitative results describing a wide variety of physical phenomena in the microcosm. An important element of the method is the renormalization procedure for removing ultraviolet divergences. Sometimes the opinion is expressed that the existence of these divergences represents a serious defect of quantum field theory, a defect which has to be "concealed" by a mathematically unsatisfactory renormalization procedure.

In this connection two points must be stressed. The existence of ultraviolet divergences is intrinsically linked to the quantum field theory being a local theory. Thus, divergences and renormalization will inevitably accompany any modification of the modern formulation of quantum field theory provided that its local nature is to be conserved.

The renormalization procedure at present represents, in turn, a mathematically rigorous operation which is performed according to a prescription based on the theory of distributions. The physical meaning of renormalization is that the influence of "small" distances on the physics of "large" distances may be effectively allowed for with the aid of a restricted number of finite parameters.

APPENDICES

APPENDIX I

ISOTOPIC SPIN FORMALISM

The isotopic spin formalism was introduced by Heisenberg in order to describe the electric charge independence of nuclear forces. In the framework of this formalism the neutron and proton are regarded as two different (charge) states of the same particle (the nucleon).

The nucleon wave function is then assigned the properties of a two-component spinor under rotational transformations in a certain fictitious three dimensional space, namely, the isotopic-spin space (the isospin space). Analogous properties are assigned to the wave functions of the other charge multiplets. For example, the pion triplet π^+, π^0, π^- is described by three functions that are the components of a vector in the isospin space, and so on.

Invariance under rotations in isospin space corresponds physically to the equivalence of particles within each multiplet. It is clear that this equivalence is violated by the electromagnetic interaction. However, if we ignore the latter, the property of isospin invariance is found to be valid for strong (nuclear) interactions.

Below we shall give a brief review of the basic formulas of the isosping formalism for nucleons and pions. A more detailed discussion can be found in Chapter 3 of the book by Perkins (1975).

1. The nucleon doublet. Consider, to begin with, the nucleon isodoublet. A spinor in three-dimensional (isotopic) space has two components corresponding to the proton (ψ_p) and neutron (ψ_n) wave functions:

$$\Psi = \begin{pmatrix} \psi_p \\ \psi_n \end{pmatrix}, \quad \overset{*}{\Psi} = (\overset{*}{\psi}_p, \overset{*}{\psi}_n). \tag{1}$$

Operators of the rotations in isospin space are constructed from matrices of rank two. They are identical with the Pauli matrices (App.II.8) but are labeled by different symbols:

$$\tau_1 = \begin{pmatrix} 0 & 1 \\ 1 & 0 \end{pmatrix}, \quad \tau_2 = \begin{pmatrix} 0 & -i \\ i & 0 \end{pmatrix}, \quad \tau_3 = \begin{pmatrix} 1 & 0 \\ 0 & -1 \end{pmatrix}. \tag{2}$$

The matrices τ anticommute with one another:

$$\tau_a \tau_b + \tau_b \tau_a = 0 \qquad (a \neq b), \qquad \tau_a^2 = 1 \tag{3}$$

and obey the relations

$$\tau_1 \tau_2 = i\tau_3 \qquad \text{(and cyclic permutations of indices 1, 2, 3).} \tag{4}$$

The operators for isospin rotations of nucleons are related to the matrices τ by

$$I_a = \tau_a/2. \tag{5}$$

By virtue of (3), (4) they satisfy the commutation relations for the rotation operators:

$$[I_1, I_2] = iI_3 \qquad \text{(and cyclic permutations of indices 1, 2, 3)}$$

and the relation

$$I^2 = I_1^2 + I_2^2 + I_3^2 = {}^3/_4,$$

so that the eigenvalue of the total isotopic angular momentum I given by the quantum-mechanical relation

$$I^2 = I\,(I+1),$$

turns out to be $I = \frac{1}{2}$.

The linear combintions of the matrices τ_1 and τ_2,

$$\tau_+ = \tau_1 + i\tau_2 = \begin{pmatrix} 0 & 1 \\ 0 & 0 \end{pmatrix}, \quad \tau_- = \tau_1 - i\tau_2 = \begin{pmatrix} 0 & 0 \\ 1 & 0 \end{pmatrix} \tag{7}$$

transform a neutron state into a proton state, and vice versa:

$$\tau_+ \begin{pmatrix} 0 \\ \psi \end{pmatrix} = \begin{pmatrix} \psi \\ 0 \end{pmatrix}, \quad \tau_- \begin{pmatrix} \psi \\ 0 \end{pmatrix} = \begin{pmatrix} 0 \\ \psi \end{pmatrix},$$

so that τ_+ and τ_- are occasionally referred to as "lowering" and "raising" operators of the electric charge. The transformation formulas for the spinor Ψ under rotation in isospin space around the z_a-axis through an angle φ have the form analogous to (AII.18):

$$\Psi' = \Lambda_a(\varphi)\,\Psi, \quad \overline{\Psi}' = \Lambda_a^{-1}(\varphi)\,\overline{\Psi},$$

where

$$\Lambda_a(\varphi) = e^{-iI_a\varphi} = e^{-i\tau_a\varphi/2} = \cos\frac{\varphi}{2} - i\tau_a \sin\frac{\varphi}{2}.$$

the bilinear forms $\bar{\Psi}\tau_a\Psi$ ($a = 1, 2, 3$) form the spatial density of the isospin vector

$$I = \int \bar{\Psi}(x)\frac{\boldsymbol{\tau}}{2}\Psi(x)\,dx,$$

the third component of which is related to the electric charge and the hypercharge

$$Q = I_3 + \frac{B+S}{2} = I_3 + \frac{Y}{2}. \tag{8}$$

2. The pion triplet. The triplet of pseudoscalar pions π^+, π^0, π^- forms an isotopic triplet that can be described by three pseudoscalar field functions forming an isotopic vector. Two representations of these functions are commonly used. In the first, the field functions are real:

$$\pi = \{\pi_1, \pi_2, \pi_3\}, \qquad \overset{*}{\pi}_a(x) = \pi_a(x). \tag{9}$$

In the second, the two real functions π_1, π_2 are used to form the complex combinations

$$\varphi_1 = \frac{\pi_1 - i\pi_2}{\sqrt{2}}, \quad \varphi_2 = \pi_3, \quad \varphi_3 = \frac{\pi_1 + i\pi_2}{\sqrt{2}}. \tag{10}$$

The transformation of (9) to the representation

$$\varphi = \{\varphi_1, \varphi_2, \varphi_3\} \tag{11}$$

can be written in the matrix form

$$\varphi = O\pi, \qquad O = \begin{pmatrix} \frac{1}{\sqrt{2}} & \frac{-i}{\sqrt{2}} & 0 \\ 0 & 0 & 1 \\ 1/\sqrt{2} & i/\sqrt{2} & 0 \end{pmatrix}, \tag{12}$$

where O is a unitary matrix:

$$\overset{+}{O}O = 1.$$

The transformation formulas for the isotopic vector (9) under rotations of three-dimensional isotopic space have the form

$$\pi \to \pi' = \Lambda(\alpha)\pi, \qquad \Lambda(\alpha) = e^{-i\omega_n\alpha_n},$$

where α_n ($n = 1, 2, 3$) are the angles of rotation and ω_n are the isospin matrices in the representation (9):

$$\omega_1 = \begin{pmatrix} 0 & 0 & 0 \\ 0 & 0 & -i \\ 0 & i & 0 \end{pmatrix}, \quad \omega_2 = \begin{pmatrix} 0 & 0 & i \\ 0 & 0 & 0 \\ -i & 0 & 0 \end{pmatrix}, \quad \omega_3 = \begin{pmatrix} 0 & -i & 0 \\ i & 0 & 0 \\ 0 & 0 & 0 \end{pmatrix}. \tag{13}$$

The square of the matrix vector ω is given by

$$\omega(\omega+1)=\omega_1^2+\omega_2^2+\omega_3^2=2.$$

Therefore, ω is the angular-momentum operator with eigenvalues 1, 0, -1.

The operator of the transformation of rotation through an angle α about the z_3-axis can be written in the form

$$\Lambda(\alpha)=e^{-i\omega_3\alpha}=1-\omega_3^2+\omega_3^2\cos\alpha-i\omega_3\sin\alpha=\begin{pmatrix}\cos\alpha & -\sin\alpha & 0\\ \sin\alpha & \cos\alpha & 0\\ 0 & 0 & 1\end{pmatrix}. \tag{14}$$

The components of the matrices (13) can be expressed in terms of a completely antisymmetric unit tensor of rank three:

$$(\omega_a)_{bc}=-i\varepsilon_{abc}.$$

The matrices ω_a satisfy the commutation relations

$$[\omega_1, \omega_2]=i\omega_3 \quad \text{(and cyclic permutations of indices 1, 2, 3).} \tag{15}$$

The transition to the representation given by (11) is achieved by the unitary transformation

$$T_n=O\omega_n\overset{+}{O}.$$

The matrices T_n have the form

$$T_1=\frac{1}{\sqrt{2}}\begin{pmatrix}0 & -1 & 0\\ -1 & 0 & 1\\ 0 & 1 & 0\end{pmatrix},\quad T_2=\frac{1}{\sqrt{2}}\begin{pmatrix}0 & i & 0\\ -i & 0 & -i\\ 0 & i & 0\end{pmatrix},\quad T_3=\begin{pmatrix}1 & 0 & 0\\ 0 & 0 & 0\\ 0 & 0 & -1\end{pmatrix} \tag{16}$$

and satisfy the commutation relations (15). An important point is that the matrix T_3 is diagonal. This property reflects the fact that, in the representation (11), the components φ_ν correspond to definite values of the electric charge. On the basis of (8) we should make the following identifications:

$$\varphi_1\sim\pi^+,\quad \varphi_2\sim\pi^0,\quad \varphi_3\sim\pi^-. \tag{17}$$

APPENDIX II

DIRAC MATRICES AND THE DIRAC EQUATION

This appendix contains technical material concerning Dirac matrices, the Dirac equation, and transformation properties of spinor functions. It serves as an introduction to Section 5, and may be useful for evaluating matrix elements. Exercises on this material are to be found in the "September" assignment below.

1. Dirac matrices. The properties of the four hypercomplex numbers γ_ν ($\nu = 0, 1, 2, 3$) are defined by the basic anticommutation relation

$$\gamma_\mu\gamma_\nu + \gamma_\nu\gamma_\mu = 2g_{\mu\nu}. \tag{1}$$

From this relation it first follows that the following 16 quantities form an algebra over the field of complex numbers:

$$\left.\begin{array}{rl} \text{the unit matrix} & I = g^{\nu\nu}\gamma^\nu\gamma^\nu \quad \text{(no summation)}, \\ \text{the four quantities} & \gamma_\nu, \\ \text{the six quantities} & \sigma_{\nu\mu} = (\gamma_\nu\gamma_\mu - \gamma_\mu\gamma_\nu)/2i \quad (\nu < \mu), \\ \text{the single quantity} & \gamma_5 = i\gamma_0\gamma_1\gamma_2\gamma_3 \\ \text{and the four quantities} & D_\nu = \gamma_\nu\gamma_5 \end{array}\right\} \tag{2}$$

In other words, the set of elements defined by all posible linear combinations of the quantities (2) with complex coefficients, as well as by all possible mutual products for such quantities and their combinations, represents a set closed with respect to the operations of addition of elements to each other and multiplication of elements by each other or by a complex number.

Note that the expression for $\sigma_{\nu\mu}$ is bilinear in γ, the expression for D_ν is cubic, and γ_5 is the only fourth-order expression. All other expressions of the fourth order, as well as those of the fifth and higher orders, reduce, with the aid of the basic relation (1), to one of the quantities (2).

Making use of a number of algebraic theorems, one may demonstrate that the rank r of an irreducible matrix representation of a system of hypercomplex numbers is connected with the number n of linearly independent elements by the relationship

$$r^2 = n,$$

which is absolutely natural if the hypercomplex numbers are represented by square matrices.

It is not difficult to show that the sixteen matrices (2) are linearly independent. To this end, it is necessary first to examine the traces of all these matrices and verify that all of them, with the exception of the trace of the unit matrix Sp $I = r$, are equal to zero. The proof is based on the possibility of cyclic permutation of matrix factors under the trace sign. As an example, consider the trace of the product $\gamma_\nu \gamma_\mu$ when $\nu \neq \mu$. Making use of the cyclic property and of the basic relation (1) consecutively, we obtain

$$\mathrm{Sp}\ \gamma_\nu \gamma_\mu = \mathrm{Sp}\ \gamma_\mu \gamma_\nu = (1/2)\ \mathrm{Sp}\ (\gamma_\nu \gamma_\mu + \gamma_\mu \gamma_\nu) = 0.$$

At the second stage it is necessary to consider the linear form

$$F = aI + b^\nu \gamma_\nu + c^{\nu\mu} \sigma_{\nu\mu} + d^\nu D_\nu + e\gamma_5$$

and, by evaluating the traces of F and of the products $F\gamma_\nu$, $F\sigma_{\nu\mu}$ FD_ν, $F\gamma_5$, to verify successively that the assumption $F = 0$ yields the following sequence of relations:

$$a = 0, \quad b^\nu = 0, \quad c^{\nu\mu} = 0, \quad d^\nu = 0, \quad e = 0.$$

Thus, in our case $n = 16$ and from the formula $n = r^2$ it now follows that the hypercomplex numbers γ^ν defined by the relation (1) may be represented by square matrices of the fourth rank, the Dirac matrices.

From the basic relation (1) it also follows that all the four matrices γ_ν may be chosen to be unitary if the conditions of Hermitian conjugation are imposed as follows:

$$\overset{+}{\gamma}{}^\nu = \gamma_\nu, \quad (\overset{+}{\gamma}{}^\nu)_{\alpha\beta} \equiv \overset{*}{\gamma}{}^\nu_{\beta\alpha}. \tag{3}$$

Noting also that the matrix γ_5 anticommutes with γ_ν ($\nu = 0, 1, 2, 3$), and that its square is equal to $+1$, we can write

$$\gamma_m \gamma_n + \gamma_n \gamma_m = 2g_{mn}, \quad m, n = 0, 1, 2, 3, 5, \tag{4}$$

where by definition

$$g_{55} = 1.$$

Also

$$\overset{+}{\gamma}_5 = \gamma_5 = \gamma^5.$$

The basic relations (1), as well as the subsequent properties, are invariant under the unitary transformation

$$\gamma_\nu \to O\gamma_\nu O^{-1},$$

where O is an arbitrary nonsingular matrix (i.e., one having an inverse) which may, taking into account (3), be regarded as unitary.

Thus, the Dirac matrices in general are determined up to a unitary transformation. It is customary to use that representation of the Dirac matrices in which γ^0 is diagonal:

$$\gamma^0 = \begin{pmatrix} 1 & \cdot & \cdot & \cdot \\ \cdot & 1 & \cdot & \cdot \\ \cdot & \cdot & -1 & \cdot \\ \cdot & \cdot & \cdot & -1 \end{pmatrix}, \quad \gamma^1 = \begin{pmatrix} \cdot & \cdot & \cdot & 1 \\ \cdot & \cdot & 1 & \cdot \\ \cdot & -1 & \cdot & \cdot \\ -1 & \cdot & \cdot & \cdot \end{pmatrix}, \quad \gamma^2 = \begin{pmatrix} \cdot & \cdot & \cdot & -i \\ \cdot & \cdot & i & \cdot \\ \cdot & i & \cdot & \cdot \\ -i & \cdot & \cdot & \cdot \end{pmatrix},$$
$$\gamma^3 = \begin{pmatrix} \cdot & \cdot & 1 & \cdot \\ \cdot & \cdot & \cdot & -1 \\ -1 & \cdot & \cdot & \cdot \\ \cdot & 1 & \cdot & \cdot \end{pmatrix}, \quad \gamma^5 = \begin{pmatrix} \cdot & \cdot & -1 & \cdot \\ \cdot & \cdot & \cdot & -1 \\ -1 & \cdot & \cdot & \cdot \\ \cdot & -1 & \cdot & \cdot \end{pmatrix} \tag{5}$$

(the dots stand for zeros). This turns out to be especially convenient for examining the nonrelativistic limit (see below). We shall refer to the representation (5) as to the *standard* representation of the Dirac matrices.

This representation of the matrices γ is connected with the frequently used matrices $\boldsymbol{\alpha}$, $\boldsymbol{\beta}$, $\boldsymbol{\sigma}$, and $\boldsymbol{\rho}$ by the relations

$$\begin{gathered} \alpha_n = \gamma_0\gamma^n \quad (n=1,\ 2,\ 3); \qquad \beta = \gamma^0; \\ \sigma_n = \gamma^5\gamma^n\gamma^0 \quad (n=1,\ 2,\ 3), \qquad \sigma_1 = \sigma^{23}, \quad \sigma_2 = \sigma^{31}, \quad \sigma_3 = \sigma^{12}; \\ \rho_1 = -\gamma^5, \quad \rho_2 = i\gamma^0\gamma^5, \quad \rho_3 = \gamma^0. \end{gathered} \tag{6}$$

The standard representation (5) may be written in the so-called "split-up" form

$$\gamma^0 = \begin{pmatrix} I & 0 \\ 0 & -I \end{pmatrix}, \quad \gamma^n = -\begin{pmatrix} 0 & \sigma_n \\ -\sigma_n & 0 \end{pmatrix}, \quad \gamma^5 = \begin{pmatrix} 0 & -I \\ -I & 0 \end{pmatrix} \tag{7}$$

through the Pauli matrices of second rank

$$\sigma_1 = \begin{pmatrix} 0 & 1 \\ 1 & 0 \end{pmatrix}, \quad \sigma_2 = \begin{pmatrix} 0 & -i \\ i & 0 \end{pmatrix}, \quad \sigma_3 = \begin{pmatrix} 1 & 0 \\ 0 & -1 \end{pmatrix}, \tag{8}$$

satisfying the relations

$$\sigma_i\sigma_j + \sigma_j\sigma_i = 2\delta_{ij}, \quad \sigma_k\sigma_j = \delta_{kj} + i\varepsilon_{kjl}\sigma_l, \tag{9}$$

and through the unit and zero matrices of rank two,

$$I = \begin{pmatrix} 1 & 0 \\ 0 & 1 \end{pmatrix}, \quad 0 = \begin{pmatrix} 0 & 0 \\ 0 & 0 \end{pmatrix}.$$

Let us return to the traces of the Dirac matrices and of their products. The trace of a unit matrix equals to the rank of the representation, i.e.,

$$\mathrm{Sp}\, I = \mathrm{Sp}\, \gamma_\nu\gamma^\nu = 4 \qquad (\text{no summation over } \nu).$$

Now, taking into account the fact proved above that the traces of the matrices $\sigma_{\mu\nu}$ and D_ν vanish, we have

$$\begin{gathered}
\mathrm{Sp}\,\gamma_\nu = 0, \quad \mathrm{Sp}\,\gamma_\nu\gamma_\mu = 4g_{\nu\mu}, \quad \mathrm{Sp}\,\gamma_\nu\gamma_\mu\gamma_\rho = 0, \\
(1/4)\,\mathrm{Sp}\,\gamma_\nu\gamma_\mu\gamma_\rho\gamma_\sigma = g_{\nu\mu}g_{\rho\sigma} + g_{\nu\sigma}g_{\mu\rho} - g_{\nu\rho}g_{\mu\sigma}, \\
\nu,\ \mu,\ \rho,\ \sigma = 0,\ 1,\ 2,\ 3.
\end{gathered} \tag{10}$$

As a general rule, we find that the traces of the products of an odd number of matrices γ are always equal to zero, while the traces of an even number of γ may, up to a factor 4, be expressed in terms of the antisymmetrized products of the factors $g_{\mu\nu}$, with the sign of the term $4g_{\mu_1\mu_2}g_{\mu_3\mu_4}\cdots g_{\mu_{2n-1}\mu_{2n}}$ in the expression for $\mathrm{Sp}(\gamma_{\nu_1}, \gamma_{\nu_2}\cdots\gamma_{\nu_{2n}})$ being determined by whether the permutation of indices in $P^{\mu_1,\mu_2,\ldots,\mu_{2n}}_{\nu_1,\nu_2,\ldots,\nu_{2n}}$ is odd or even.

We shall now present some more formulas for traces of the contractions of the matrices γ with the four-vectors $\hat{a} = a^\nu\gamma_\nu$. These formulas follow from (10):

$$\mathrm{Sp}\,\hat{a}\hat{b} = 4\,(ab), \tag{11a}$$

$$\mathrm{Sp}\,\hat{a}\hat{b}\hat{c}\hat{d} = 4\,\{(ab)(cd) + (ad)(bc) - (ac)(bd)\}, \tag{11b}$$

$$\begin{aligned}
&(1/4)\,\mathrm{Sp}\,[(\hat{k}_1 + m_1)(\hat{k}_2 + m_2)(\hat{k}_3 + m_3)(\hat{k}_4 + m_4)] = \\
&\qquad = (K_1K_2)(K_3K_4) + (K_1K_4)(K_2K_3) - (K_1K_3)(K_2K_4),
\end{aligned} \tag{11c}$$

where

$$K_iK_j = k_ik_j + \eta_{ij}m_im_j, \quad \eta_{ij} = (-1)^{i-j+1}.$$

A relationship between traces involving masses and the massless traces, analogous to the relationship between (11c) and (11b), holds when the number of factors is larger.

2. The Dirac equation. This equation is a matrix linear partial differential equation of the first order:

$$[(i\hat{\partial}_x - m)\,\psi(x)]_\alpha = \left(i\gamma_\nu\frac{\partial}{\partial x_\nu} - mI\right)_{\alpha\beta}\psi_\beta(x) = 0. \tag{12}$$

The field function ψ_α is four-component ($\alpha = 1, 2, 3, 4$). Applying the operator $(i\hat{\partial} + m)$ from the left to (12), we find, on taking into account (1), that

$$(i\hat{\partial} + m)(i\hat{\partial} - m)\,\psi = [-\partial^2 - m^2]\,\psi = (\Box - m^2)\,\psi = 0.$$

Therefore each component ψ_α satisfies the Klein–Gordon equation.

The adjoint equation is obtained from the Dirac equation in two steps. By taking the Hermitian conjugate of (12) and using (3), we obtain

$$i\overset{*}{\gamma}_{\nu}\frac{\partial\overset{*}{\psi}}{\partial x_{\nu}}+m\overset{*}{\psi}=i\frac{\partial\overset{*}{\psi}}{\partial x_{\nu}}\overset{+}{\gamma}_{\nu}+m\overset{*}{\psi}=i\frac{\partial\overset{*}{\psi}}{\partial x_{\nu}}\gamma^{\nu}+m\overset{*}{\psi}=0.$$

This expression does not have the "correct" Lorentz-symmetric form. Multiplying this equation on the right by γ^0, and taking into account that

$$\gamma_0\gamma^{\nu}\gamma_0=\gamma_{\nu}=\overset{+}{\gamma}{}^{\nu}, \tag{13}$$

we obtain

$$i\,\frac{\partial\bar{\psi}\,(x)}{\partial x^{\nu}}\,\gamma_{\nu}+m\bar{\psi}\,(x)=0\ , \tag{14}$$

the adjoint Dirac equation. The spinor

$$\bar{\psi}=\overset{*}{\psi}\gamma^0 \tag{15}$$

occurring in it is referred to as the *adjoint* spinor with respect to ψ.

The spinors ψ and $\bar{\psi}$ are often represented in the form of a four-component column and row, respectively. Despite its simple appearance, such a representation may, however, lead to a certain amount of confusion, for example, in the introduction of the operation of charge conjugation (see Section 9.3). In practice, it is sufficient, if ambiguity arises, to refer to the second form in which equation (12) may be written, the component form.

3. Transformation properties. We refer the reader interested in the derivation to the vast available literature and restrict ourselves here to presenting the final formulas for transformations of the spinor field functions.

On the basis of the general arguments of Section 1.3 we have, under translation transformations,

$$\psi\,(x)=\psi'\,(x'),\quad x'=x+a.$$

In the case of spatial rotations in the plane $x_m x_n$ through an angle φ,

$$x_m'=x_m\cos\varphi-x_n\sin\varphi,\quad x_n'=x_n\cos\varphi+x_m\sin\varphi$$

the corresponding transformation has the matrix form

$$\psi\,(x)\to\psi'\,(x')=\Lambda_{(m,\,n)}\,(\varphi)\,\psi\,(x), \tag{16}$$

where the rotation matrix Λ has the form

$$\Lambda_{(m,\,n)}\,(\varphi)=\exp\left(-\,i\sigma_{mn}\,\frac{\varphi}{2}\right),$$

and σ_{mn} is the "spin tensor matrix" introduced in (2). Thus,

$$\psi'(x') = \cos\frac{\varphi}{2}\,\psi(x) - i\sin\frac{\varphi}{2}\,\sigma_{mn}\psi(x). \tag{17}$$

For a Lorentz rotation in the x^0x^n $(n = 1, 2, 3)$,

$$x'_0 = x_0 \operatorname{ch}\varphi - x_n \operatorname{sh}\varphi, \qquad x'_n = x_n \operatorname{ch}\varphi - x_0 \operatorname{sh}\varphi \qquad (\operatorname{tg}\varphi = v = v/c)$$

we have, correspondingly,

$$\Lambda_{(0n)}(\varphi) = \exp\left(i\sigma_{0n}\frac{\varphi}{2}\right) = \operatorname{ch}\frac{\varphi}{2} + i\sigma_{0n}\operatorname{sh}\frac{\varphi}{2}.$$

The formulas obtained may be combined:

$$\Lambda_{(\mu\nu)}(\varphi) = \exp\left(-i\sigma^{\mu\nu}\frac{\varphi}{2}\right). \tag{18}$$

It may be demonstrated that the above formulas are consistent with the Dirac equation. To this end, one must first verify that the matrices $\Lambda_{(\mu\nu)}$ corresponding to the transformation

$$x' = Lx, \quad x'_\nu = L_{\nu\mu}x^\mu,$$

possess the property

$$\Lambda^{-1}_{(\nu\mu)}\gamma_\nu\Lambda_{(\nu\mu)} = L_{\nu\mu}\gamma_\mu, \quad \Lambda\gamma_\mu\Lambda^{-1} = L^{-1}_{\mu\nu}\gamma_\nu. \tag{19}$$

Multiplying then the Dirac equation (12) on the left by Λ, we obtain

$$\Lambda(i\gamma_\nu\partial^\nu - m)\psi = [i(\Lambda\gamma_\nu\Lambda^{-1})\partial^\nu - m]\Lambda\psi =$$
$$= (iL^{-1}_{\nu\mu}\gamma_\mu\partial^\nu - m)\psi'(x') = (i\hat{\partial}' - m)\psi'(x') = 0.$$

Here we have taken into account that, in accordance with (19),

$$\partial^\nu = \frac{\partial}{\partial x_\nu} = \partial'^\mu L_{\mu\nu}, \quad \partial'^\nu = \partial^\mu L^{-1}_{\mu\nu}.$$

It is well known that an equation is called *covariant* if after a transformation, i.e., in terms of the transformed functions and variables, it has the same form as before the transformation. Thus, we have verified that by virtue of the matrix transformation (16) the Dirac equation is covariant under transformations from the Lorentz group.

A similar conclusion for the adjoint Dirac equation can be drawn with the aid of the transformation formula for the adjoint spinor,

$$\bar{\psi}(x') = \bar{\psi}(x)\Lambda^{-1}(\varphi), \tag{20}$$

which is obtained from (16) by taking the Hermitian conjugate and multiplying on the right by γ_0, on taking into account that

$$\gamma_0\overset{+}{\Lambda}\gamma_0=\gamma_0\exp\left(i\overset{+}{\sigma}{}^{\mu\nu}\frac{\varphi}{2}\right)\gamma_0=\exp\left(i\sigma^{\mu\nu}\frac{\varphi}{2}\right)=\Lambda^{-1}.$$

The nonuniqueness of spinor functions follows directly from the formulas of spatial rotation (17). Setting $\varphi = 2\pi$ in these formulas, we find that the transformation matrix $\Lambda(2\pi) = -1$ corresponds to a complete spatial rotation of the coordinate system, i.e., the field function changes sign under such a transformation. However, as the transformation of rotation through 2π brings the coordinate system into its original position, i.e., coincides with the identity transformation, it follows that the spinor wave functions are always determined to within their sign.

We also exhibit the form of the matrix λ for the transformations of reflection of the coordinate axes. Noting that the transformation formulas for reflections of an even number of different space axes, which reduce to rotations, follow from (17), we restrict ourselves to the transformation of reflection of all three space axes (the P-transformation):

$$\begin{aligned} &x'^n=-x^n\ (n=1,\ 2,\ 3), \quad x'^0=x^0,\\ &\psi'(x')=\eta(P)\,\Lambda_{123}\psi(x), \quad \Lambda_{123}=\gamma^0. \end{aligned} \tag{21}$$

As the spinor representation is double-valued, the phase factor obeys the condition $\eta^2(P) = \pm 1$.

The formulas for "Lorentz transformations" of the matrices γ, taken together with the transformation formulas for ψ and $\bar{\psi}$, permit us to obtain the transformation laws for bilinear forms of the type of $\bar{\psi}O\psi$. Based on (16) and (20), we obtain

$$\bar{\psi}'(x')O\psi'(x')=\bar{\psi}(x)\,\Lambda^{-1}O\Lambda\psi(x).$$

With the aid of (19) we now find

(a) $O = 1$, $\bar{\psi}'(x')\psi'(x') = \bar{\psi}(x)\psi(x)$,
(b) $O = \gamma_\nu$, $\bar{\psi}'(x')\gamma_\nu\psi'(x') = L_{\nu\mu}\bar{\psi}(x)\gamma_\mu\psi(x)$,
(c) $O = \sigma_{\nu\mu}$, $\bar{\psi}'\sigma_{\nu\mu}\psi' = L_{\nu\nu'}L_{\mu\mu'}\bar{\psi}\sigma_{\nu'\mu'}\psi$, etc.

Thus, the quadratic forms under consideration transform in accordance with tensor representations of the Lorentz group. We shall separately consider these cases where the matrix γ_5 occurs as a factor in O. It may easily be verified that the marix (18) corresponding to rotations commutes with γ_5:

$$\Lambda^{-1}(\varphi)\,\gamma_5\Lambda\,(\varphi)=\gamma_5.$$

At the same time the matrix (21) of the reflection of an odd number of space axes anticommutes with γ_5:

$$\Lambda_{123}^{-1}\gamma_5\Lambda_{123}=\gamma_0\gamma_5\gamma_0=-\gamma_5.$$

As a result, we obtain the following classification of bilinear forms:

$\bar{\psi}\psi$ — is a scalar, $\bar{\psi}\gamma_5\psi$ — is a pseudoscalar,
$\bar{\psi}\gamma_\nu\psi$ — is a four-vector, $\bar{\psi}\gamma_\nu\gamma_5\psi$ — is a pseudovector,
$\bar{\psi}\sigma_{\mu\nu}\psi$ — is a tensor of rank two.

4. Nonrelativistic limit. The nonrelativistic limit for the Dirac equation corresponds to the case when $|p| \ll m$. To pass to the nonrelativistic limit, it is convenient to represent the four-component spinor of the four-dimensional space-time through two two-component spinors χ and φ in the three-dimensional Euclidean space:

$$u(p)=\begin{pmatrix}\varphi(p)\\ \chi(p)\end{pmatrix},\quad \varphi=\begin{pmatrix}\varphi_1\\ \varphi_2\end{pmatrix},\quad \chi=\begin{pmatrix}\chi_1\\ \chi_2\end{pmatrix} \tag{22}$$

and, fixing the standard representation (5), to pass over to the matrices $\boldsymbol{\alpha}$, β:

$$\boldsymbol{\alpha}=\begin{pmatrix}0 & \boldsymbol{\sigma}\\ \boldsymbol{\sigma} & 0\end{pmatrix},\quad \beta=\begin{pmatrix}I & 0\\ 0 & -I\end{pmatrix}. \tag{23}$$

The matrices σ of rank two, the Pauli matrices, which occur in $\boldsymbol{\alpha}$ were defined in (8).

Starting with the Dirac equation (12) written in the momentum representation

$$(p^0\gamma_0-\boldsymbol{p}\boldsymbol{\gamma}-mI)\,u(\boldsymbol{p})=0 \tag{24}$$

(here $u(\boldsymbol{p})$ stands for the negative-frequency part of the spinor $u^-(\boldsymbol{p})$), we obtain, by multiplying it on the left by $\gamma_0=\beta$ and making use of (22) and (23), two equations for the two-component spinors φ and χ:

$$(p^0-m)\,\varphi(\boldsymbol{p})-(\boldsymbol{p}\boldsymbol{\sigma})\,\chi(\boldsymbol{p})=0, \tag{25}$$
$$(\boldsymbol{p}\boldsymbol{\sigma})\,\varphi(\boldsymbol{p})-(p^0+m)\,\chi(\boldsymbol{p})=0. \tag{26}$$

In the nonrelativistic approximation, as is seen, the spinor χ is much smaller than the spinor φ. Indeed, from (26) we obtain

$$\chi(\boldsymbol{p})=\frac{(\boldsymbol{\sigma}\boldsymbol{p})}{m+p^0}\,\varphi(\boldsymbol{p})\simeq\frac{(\boldsymbol{\sigma}\boldsymbol{p})}{2m}\,\varphi(\boldsymbol{p}). \tag{27}$$

Therefore, ordinarily the spinor φ describes the electron components (compare with formula (5.12)), while χ describes the positron components.

On substituting (27) into (25), (26), we obtain the equation for φ:

$$p^0\varphi(p)=\left[m+\frac{(\sigma p)(\sigma p)}{2m}\right]\varphi(p).$$

If one takes advantage of the matrix relation

$$(\sigma a)(\sigma b)=(ab)-i(\sigma[a\times b]), \tag{28}$$

then one can see that the operator in the right-hand side turns out to be diagonal in the spinor indices. Passing to the coordinate representation and isolating the factor $\exp(imt)$,

$$\varphi(x,t)e^{-imt}=\int e^{-ip^0t+ipx}\varphi(p)\,dp,$$

we obtain for $\varphi(p, t)$

$$i\frac{\partial\varphi(x,t)}{\partial t}=-\frac{1}{2m}\nabla^2\varphi(x,t)$$

i.e., the Schrödinger equation.

If an external electromagnetic field is present, then, in accordance with the general rule for covariantizing derivatives (see Section 10.3), instead of (24) one should write

$$[(p^0-eA_0)\gamma_0-(p-eA)\gamma-mI]\,u(p)=0. \tag{29}$$

Repeating the transition to the two-component spinor of the electron and neglecting eA_0 as compared to $2m$, we obtain

$$i\frac{\partial\varphi}{\partial t}=\left[-\frac{1}{2m}(\nabla-ieA)^2+eA_0-\frac{e}{2m}(\sigma H)\right]\varphi. \tag{30}$$

Here $H=\operatorname{curl}A$ is the magnetic-field vector.

This equation is called the *Pauli equation*. Its form is the same as that of a Schrödinger equation involving the Hamiltonian

$$H=\frac{1}{2m}(p-eA)^2+eA_0-(\mu H), \tag{31}$$

where the last terms describes the energy of a magnetic dipole

$$\mu=\mu_0\sigma,\quad \mu_0=\frac{e}{2m}=\frac{e\hbar}{2mc} \tag{32}$$

in an external magnetic field H. The quantity μ_0 represents the electron magnetic moment and is called the *Bohr magneton*.

APPENDIX III*

CONTINUOUS GROUPS

In this appendix information is presented on the theory of continuous groups. For details of formulations and proofs of the presented facts we refer the reader to available texts on group theory.

We shall first give some definitions concerning abstract groups, i.e., groups whose elements and group operations are unconstrained.

1. General definitions. A set G (the elements of which we shall, as a rule, denote by g_1, g_2, and so on), in which an operation is defined for assigning to each pair of elements of the group g_1 and g_2 a certain element g of the same set, is called a group. This group operation is usually referred to as multiplication and is written in the form

$$g = g_1 g_2.$$

The quantity g is called the product of the elements g_1 and g_2 of the group. For concrete groups, the elements of which can be either numbers, or mappings, or matrices, or operators, etc., group operations can respectively be represented by the addition or multiplication of numbers, or by the multiplication of matrices and operators, or by compositions of mappings, and so on. The group operation must exhibit the following properties:

1. Associativity holds for any three elements g_1, g_2, g_3 of the group, i.e.,†

$$\forall g_1,\ g_2,\ g_3 \in G \qquad (g_1 g_2) g_3 = g_1 (g_2 g_3).$$

2. In the group G there exists a single element, e, called the identity element, or simply the unit element, of the group G, which has the property that for any element g of the group G

$$ge = eg = g.$$

*The authors are grateful to N.A. Sveshnikov, who wrote this Appendix.

† In this Appendix we shall make use of some abbreviations, common in mathematical literature: stands for 'for any"; $\in$ stands for "belongs to"; $\exists$ stands for "there exists".

3. For any element g of the group G there exists one and only one element of that group, which is called the inverse of g and is denoted by g^{-1}, such that

$$g^{-1}g = gg^{-1} = e.$$

Note that the listed properties of the group operation (which we shall henceforth always call multiplication) do not imply the property of commutativity, that is, generally speaking, $g_1g_2 \neq g_2g_1$ for arbitrary elements of the group G.

If the product for a given group is commutative, i.e.,

$$\forall g_1, g_2 \in G \quad g_1g_2 = g_2g_1,$$

then the group is said to be *Abelian*, or commutative. The simplest example of an Abelian group is the set of real numbers, denoted by the symbol **R**, in which ordinary addition is chosen to be the group multiplication. It is clear that here the unit of the group is the number 0, while the element that is inverse to x is $-x$.

Any subset of G closed with respect to the group operations is called a *subgroup* K of the group G, i.e.,

$$\forall g_1, g_2 \in K \quad g_1g_2 \in K, \quad \forall g \in K \quad g^{-1} \in K.$$

Thus, in the above example one can point to the subgroup $\mathbb{Z}$ of the group of real numbers, which consists of all integers. Any group G contains two so-called trivial subgroups. One of them coincides with the group G itself, while the other consists of a single element, the unit e of the group.

In group theory an important role is played by the notions of simple and semisimple groups. To define them, we first introduce the concept of an *invariant subgroup.* A subgroup K of the group G is called an invariant subgroup, (or, which is the same, a *normal divisor*) of the group G, if for any element g of the group G the set consisting of the elements gkg^{-1}, where k runs through the whole subgroup K, coincides with K, i.e.,

$$\forall g \in G \quad gKg^{-1} = K.$$

A group G is said to be *simple* if it has no invariant subgroups. A group G is called *semisimple* if it has no Abelian invariant subgroups. Note that such important groups for elementary particle theory as the Lorentz group and the unitary symmetry group fall into these categories.

Historically, the first groups that appeared in theoretical physics were the symmetry groups of crystals. An example of such a group is the group of rotations of a plane through angles which are multiples of 60°. In this case the group multiplication consists of composition of rotations. Another example of a finite group, i.e., a group with a finite number of elements, is the set of transformations of field functions formed by the operations C, P, and T introduced in Section 9, the identity transformation I, and all possible combinations of them. It is not difficult to verify that, by virtue of the property

$$C^2 = P^2 = T^2 = I$$

and the commutativity of the transformations C, P, T, the total number of elements of this particular group is equal to 8.

2. Lie groups. In quantum theory the most interesting groups are infinite groups, i.e., groups containing an infinite number of elements. Infinite groups may be divided into two large classes: discrete and continuous groups. The structure of infinite discrete groups is very close to the structure of finite groups: they have the property that all their elements differ "significantly" from each other. The Abelian group of integers, which has already been mentioned, may serve as an example of such groups. In contrast, a continuous group involves an additional structure (topology) that permits one to introduce the notions of nearness of elements of the group, of continuity of a limit, and so on. The existence of such a structure makes possible the continuous parameterization of the elements of the group by points in a certain topological space. The group operations must then be continuous transformations in the space of paameters. One says, for example, that the elements g_1 and g_2 of the group $\mathbf{R}$ are close to each other if the real numbers x_1 and x_2 corresponding to them are. If for a given group one can locally choose a finite-dimensional Euclidean space as the space of parameters, then this group is called a *Lie group*. To be more precise, a continuous group G is called an n-parameter Lie group (or a Lie group of dimension n) if a certain vicinity of any of its elements may be mutually uniquely and mutually continuously mapped into a certain region of the n-dimensional real Euclidean space $\mathbb{R}^n$. In practice, all continuous groups appearing in studies devoted to particle physics are Lie groups. This leads to the theory of Lie groups being of paramount importance for quantum theory.

The simplest physically interesting example of a Lie group is the Abelian group of rotations of a plane through arbitrary angles $\varphi \leq 2\pi$, the group $U(1)$. Elements of the $U(1)$ group may be continuously and uniquely parametrized with the aid of points of the interval $[0, 2\pi]$ of the real axis with ends identified according to the rule:

$$\text{rotation through an angle } \varphi \quad \leftrightarrow \quad \text{the number } \varphi \in [0, 2\pi]. \tag{1}$$

Thus, in accordance with the above definition the group $U(1)$ is a Lie group of dimension 1. Another parameterization of the elements of the group $U(1)$ arises when the rotation through an angle is put in correspondence with a point on the unit circle in the complex plane $\mathbb{C}$:

$$\text{“rotation through an angle } \varphi\text{”} \leftrightarrow \text{the number } z = e^{i\varphi} \in \mathbb{C}. \tag{2}$$

In particle physics the group $U(1)$ appears as the Abelian group of gauge transformations connected with the conservation of additive charges.

An important class of Lie groups is represented by the groups of nondegenerate linear transformations of finite-dimensional vector spaces to which one may uniquely relate the groups of square matrices. The elements of such groups are matrices with nonzero determinants, while the multiplication of matrices serves as the group operation. It is customary to reflect the properties of the group elements in its name. Thus, if there are no constraints on the form of the nondegenerate matrices (elements of the group), then the letter L (for linear) occurs in the name of the group; unitarity of the group elements is denoted by the letter U (unitary), and orthogonality (for real matrices) by the letter O. If matrices (elements of a group) have a unit determinant, then this fact is signaled by the letter S (special) in the name of the group. The parentheses coming after the name, as a rule, contain the dimensionality (the number of rows) of the matrices forming the group. For example, the group $SU(N)$ consists of all unitary matrices of dimension N, the determinants of which are equal to 1. By tradition the names of a number of groups of matrices, like $SU(N)$, $SO(N)$, $SL(N)$, etc., are transferred to the respective abstract groups, i.e., Lie groups with the same group and topological properties. Thus, the $SL(2, \mathbb{C})$ group is a Lie group such that there can be established a one-to-one continuous correspondence between the set of all its elements and all two-row complex matrices with unit determinant, such that to the product of elements of the group there will correspond the product of corresponding matrices, while to the inverse element there will correspond the inverse matrix.

An important role in theoretical physics is played by the $SU(2)$ spin group, which represents the simplest example of a non-Abelian Lie group. It is easy to check that for the two row unitary matrices A and B the relation $AB = BA$, generally speaking, does not hold. A complex 2×2 matrix, clearly, is parametrized by eight real numbers. The unitarity condition imposes on these numbers four independent conditions, while the determinant being equal to one imposes one more condition. Thus, an arbitrary element of the group $SU(2)$ may be parametrized by three real numbers, i.e., the dimension of $SU(2)$ is equal to 3.

In a similar manner it may be shown that the dimension of $SU(3)$ equals

8, and that the dimension of $SU(N)$ is equal to $N^2 - 1$. An arbitrary element g of the group $SU(2)$ may be written in the form

$$g = \exp\left(i \sum_{k=1}^{3} \tau_k \alpha_k\right) = \cos\alpha + \frac{i}{\alpha} \sin\alpha \sum_{k=1}^{3} \tau_k \alpha_k, \tag{3}$$

where τ_k are the Pauli matrices, and α_k are real numbers such that $\alpha \equiv (\alpha_1^2 + \alpha_2^2 + \alpha_3^2)^{1/2} \leq \pi$. Note that, as in the case of the group $U(1)$ (see (1), (2)), the range of the parameters in (3) is closed and bounded in the corresponding Euclidean space. As it turns out, this fact is not a consequence of the actual parametrization chosen, but reflects a common property of $U(1)$ and $SU(2)$, which is called compactness. Generally, a Lie group G is said to be *compact* it the set of its elements forms a compact manifold, i.e., if from any infinite sequence of its elements there can be isolated a convergent subsequence. A consequence of compactness is the possibility of a continuous parameterization of a Lie group by the points of a certain closed and bounded (i.e., compact) region of the corresponding Euclidean space. The groups $SU(N)$ and $SO(N)$ for any n represent important examples of compact Lie groups. The theory of compact semisimple Lie groups is the best developed part of the general theory of Lie groups.

Examples of noncompact Lie groups are the group T of space-time translations and the Lorentz group L.

An element a_T of the group T is defined as a parallel displacement of points of the space-time M by a vector $a \in M$:

$$(a_T x)_\mu = x_\mu + a_\mu.$$

It is clear that the product of the elements a_T and b_T of the group T is given by the formula

$$a_T b_T = (a + b)_T,$$

where the vector $a + b$ in the space M corresponds to the element $(a + b)_T$ of the group T. The inverse of a_T is the element $(-a)_T$. Clearly, the group T is Abelian. The concept of nearness (and, consequently, of convergence) of the elements of the group T coincides with the concept of nearness of the vectors corresponding to them in the space-time. Thus the noncompactness of the group T is a direct consequence of M being unbounded.

An arbitrary element Λ of the Lorentz group may be represented by a composition of a space rotation R and a pure Lorentz transformation (often referred to as a *boost*) B, which transforms a system from a rest state into a state of uniform motion along a straight line with a velocity v:

$$\Lambda = B(v) R.$$

The space rotations form a compact subgroup of the Lorentz group, the group $SO(3)$. The noncompactness of the group is connected with the obvious fact that the value of the hyperbolic angle $\chi = \text{arcth}(v/c)$, which parametrizes the boost continuously, can vary within infinite limits:

$$-\infty < \chi < +\infty.$$

The Poincaré group P, the essential group of all of relativistic physics, is a combination of the groups T and L in the sense that any of its elements p may be represented in the form

$$p = a_T \Lambda, \tag{4}$$

where the product of the elements $p_1, p_2 \in P$ is given by the formula

$$p_1 p_2 = (a_{1T} \Lambda_1)(a_{2T} \Lambda_2) = (a_1 + \Lambda_1 a_2)_T (\Lambda_1 \Lambda_2).$$

The meaning of (4) is that any linear transformation of the space-time which conserves the intervals between events may be represented as the result of performing in succession a Lorentz transformation Λ of the reference frame and a displacement of the origin of the coordinate system by the vector a. It is clear that the same result can be achieved by performing first a displacement by the vector Λ_a^{-1} and then a Lorentz rotation through Λ. Therefore, the transformation p can also be written in the form

$$p \equiv a_T \Lambda = \Lambda (\Lambda^{-1} a)_T.$$

It is easy to check that the group of translations, T, is an Abelian normal divisor of P. Because of this the Poincaré group is neither simple nor even semisimple.

3. Representations of Lie groups. In physical investigations we are usually interested in concrete realizations of a given abstract group. If we restrict ourselves to considering the realizations of groups by linear operators, we then arrive at the concept of a *linear representation* of a group. To be more precise, the continuous representation of a group G, commuting with the group multiplication, by the group of linear operators in a certain vector space $\mathscr{L}$ is referred to as a linear representation of the group G. Thus, if $g \in G$ is an arbitrary element of the group and $A(g)$ is the corresponding linear operator in $\mathscr{L}$, then

$$\forall g_1, g_2 \in G \quad A(g_1 g_2) = A(g_1) A(g_2)$$

and

$$\forall g \in G \quad A(g^{-1}) = A^{-1}(g).$$

The formulae demonstrate that the group multiplication and the inversion of an element of the group are replaced for the operators of the representation by the multiplication of operators in the space $\mathscr{L}$ and by taking the inverse operator, respectively. The dimension of the space $\mathscr{L}$ is called the dimension of the representation.

For a representation of dimension N, one may replace the operators by matrices corresponding to them (in a certain basis of the space $\mathscr{L}$) and study the representation of the initial group as the group of $N \times N$ matrices. For any group there exists a trivial one-dimensional representation which establishes the correspondence between each element and the number 1.

A representation A of the group G is called *faithful* if to different elements of the group g_1 and g_2 there correspond different operators $A(g_1)$ and $A(g_2)$, i.e., if there is established a one-to-one correspondence between the set of elements of the group and the set of operators forming its representation. A faithful representation reproduces completely the structure of the Lie group.

A linear representation A of the group G in the space $\mathscr{L}$ is called *reducible* if in $\mathscr{L}$ there exists a proper (i.e., differing form $\{0\}$ and $\mathscr{L}$) subspace $\mathscr{L}_1$ which is transformed into itself by all the operators of the representation:

$$A(g)\,\mathscr{L}_1 \in \mathscr{L}_1 \quad \forall g \in G.$$

Otherwise the representation A is called *irreducible*. In particle physics it is of utmost interest to study unitary linear representations, i.e., representations whose elements are unitary operators and conserve, therefore, the normalization of the state vectors. It may be shown that any reducible unitary representation U of a group G breaks up into a sum of irreducible representations, i.e., all the matrices $U(g)$ are, in an appropriate basis, of a block-diagonal form

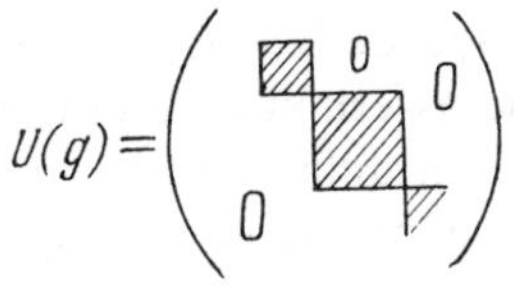

Thus, the investigation of an arbitrary unitary representation reduces to the study of irreducible representations.

We shall present several examples. For any Abelian group all irreducible representations are one-dimensional. In particular, for the group $\mathbb{R}$ the irreducible unitary representations have the form

$$A(x) = e^{ipx} \quad \forall x \in \mathbb{R},$$

where p is a real number which fixes the choice of the representation. Above, we mentioned that the group and topological properties of a number of groups

($SL(N)$, $SU(N)$, $SO(N)$, and so on) are identical with the properties of the corresponding groups of matrices. Obviously, in this case the group of matrices, the name of which coincides with the name of the given group, realizes its faithful irreducible representation. For $SU(2)$ and $SU(3)$, as well as for certain other groups often utilized in quantum field theory, this representation is customarily referred to as the *fundamental* representation.

An important role in the theory of Lie groups is played by the notion of an *adjoint* representation of a group. It turns out that any n-parameter Lie group has a representation of dimension n, the form of which is entirely determined by the local structure of the given group. We shall explain how to construct this representation after introducing the concept of a Lie algebra (see Section 4 of this appendix). We point out here that for compact simple groups (such as, in particular, the groups $SU(N)$ and $SO(N)$ for any value of N), the adjoint representation is irreducible. Thus, for example, for the group $SU(2)$ the adjoint representation is of dimension 3.

In concluding this section we note that in field theory the group $SU(2)$ is often associated with the group of the spin (or isospin) while the group $SU(3)$ is associated with the group of unitary symmetry. In this case vectors of the space $\mathscr{L}$, in which the irreducible representation of the group $SU(2)$ is realized, may be identified with the spin multiplets, and in the case of the group $SU(3)$, with the unitary multiplets. There exist spin multiplets of every dimension. At the same time unitary multiplets can have only the selected dimensions 1, 3, 8, 10, . . . , i.e., only for certain dimensions can irreducible representations of the group $SU(3)$ be constructed in the space $\mathscr{L}$.

4. Generators and the Lie algebra. From the very definition of the Lie group it follows that this group locally (i.e., within an infinitesimal vicinity of any element) has the structure of a Euclidean space. This permits us to introduce within a group such concepts as the difference between infinitesimally close elements, the derivative, etc. A remarkable result of the theory of Lie groups is the assertion that an element of a Lie group is not only a continuous but also a differentiable and even analytic function of the parameters. Consider the immediate vicinity of the identity of the group. Without loss of generality we may consider that the identity element corresponds to the value zero of the parameters. Then an arbitrary element g from the vicinity under discussion may, up to terms of the second order, be written in the form

$$g(a_1, a_2, \dots a_n) = e + i \sum_{k=1}^{n} X_k a_k, \tag{5}$$

where the quantities

$$X_k = \frac{1}{i} \frac{\partial g(\dots a \dots)}{\partial a_k} \bigg|_{a_1 = a_2 = \dots = a_n = 0} \tag{6}$$

are called infinitesimal generators of the Lie group, or simply *generators*. The introduction of the imaginary unit into the formula (6) gives the advantage that in physically interesting cases generators thus defined turn out to be Hermitian operators of physically important quantities like momentum, energy, spin, charge, and so on.

It may be demonstrated that the laws of group multiplication require the following relations for generators to be satisfied:

$$[X_l, X_j] = i \sum_{k=1}^{n} C_{lj}^{k} X_k, \tag{7}$$

$$[X_l, [X_j, X_k]] + [X_j, [X_k, X_l]] + [X_k, [X_l, X_j]] = 0, \tag{8}$$

the latter of which is called the Jacobi identity. Here $[X, Y] = XY - YX$ is the commutator of two generators, and C_{lj}^{k} are numbers referred to as *structure constants* of the Lie group. The properties of generators pointed out above are usually formulated as the assertion that generators of the Lie group form the basis of the Lie algebra. A set A is called a *Lie algebra* if it is a real or complex linear space, i.e., $\forall\ X, Y \in A$ and $\forall$ numbers α, β

$$\alpha X + \beta Y \in A,$$

and is closed with respect to the operation of commutation of elements, i.e., $\forall\ X, Y \in A$

$$[X, Y] \in A.$$

Note that the latter property distinguishes a Lie algebra from an ordinary algebra which is closed with respect to the operation of multiplication of elements that possesses the associative property. It may be shown that a set of structure constants, i.e., of numbers with the properties

$$C_{lj}^{k} = -C_{jl}^{k},$$

$$\sum_{l=1}^{n} \left(C_{ik}^{l} C_{lj}^{m} + C_{ji}^{l} C_{lk}^{m} + C_{kj}^{l} C_{li}^{m}\right) = 0, \tag{9}$$

(the first of which follows from (7), while the second may be derived from the Jacobi identity), determines completely the structure of the Lie algebra, i.e., by virtue of (5), the local structure of the Lie group.

In the case of an Abelian group the Lie algebra has the simplest possible form. In this case all structure constants are equal to zero. In particular, the generators of the group T constitute the momentum four-vector P_μ, and

$$[P_\mu, P_\nu] = 0.$$

The Lie algebra of the group $SU(2)$ is given by the commutation relations between three of its enerators I_k (they may be the operators of the components of the angular momentum, the spin, isospin, and so on):

$$[I_l, I_j] = i \sum_{k=1}^{3} \varepsilon_{ljk} I_k,$$

where ε is a completely antisymmetric tensor. Similar commutation relations are satisfied for the generators of the group $SO(3)$, i.e., the groups $SU(2)$ and $SO(3)$ coincide locally.

As for the Lie group, one may introduce for Lie algebras the concept of a linear representation, i.e., of the mapping of a Lie algebra into a algebra of operators in a certain vector space. The representation of the Lie algebra is connected with the corresponding representation of the Lie group by the formulas (5), (6), in which the elements of the abstract group and algebra should be replaced by their concrete realizations in the form of operators of the corresponding space. Note that if the representation of the group is unitary, then by virtue of (6) the elements of the Lie algebra are Hermitian operators. In particular, for the group $SU(2)$ the generators of the fundamental representation are represented by the 2×2 matrices

$$(I_k)^{\text{fund}} = \frac{1}{2}\tau_k,$$

and the generators of the adjoint representation are the 3×3 matrices

$$(I_k)^{\text{adj}}_{lj} = i\varepsilon_{ljk}.$$

Generally, the name *adjoint representation of the Lie algebra* of dimension n is given to the $n \times n$ matrix representation whose generators have matrix elements given by the structure constants

$$(X_k^{\text{ajd}})^l_m = iC^l_{mk}$$

It is easy to verify, using the property (9) of the structure functions, that the generators thus defined satisfy (7) and therefore define representations of the Lie algebra. The representation of groups corresponding to this representation in accordance with (5) is called the *adjoint representation of the Lie group.* Note that the relative simplicity of the theory of compact semisimple Lie groups is related to the fact that any element of such a group (and not only one infinitesimally close to the identity element) may be represented in the so-called normal form

$$g(a_1, \ldots, a_n) = \exp\left(i \sum_{k=1}^{n} X_k a_k\right).$$

In conclusion we write out the commutation relations defining the Lie algebra of the Poincaré group P. The total number of generators of the group equals 10; four of them, operators of the momentum components P_μ, are connected with the subgroup T, while the rest, components of the angular momentum $M_{\mu\nu}$, are connected with the Lorentz group. It can be demonstrated that

$$\begin{aligned} &[P_\mu, P_\nu] = 0, \\ &[P_\mu, M_{\nu\lambda}] = i(g_{\mu\lambda}P_\nu - g_{\mu\nu}P_\lambda), \\ &[M_{\mu\nu}, M_{\lambda\rho}] = i(g_{\mu\lambda}M_{\nu\rho} - g_{\nu\lambda}M_{\mu\rho} + g_{\nu\rho}M_{\mu\lambda} - g_{\mu\rho}M_{\nu\lambda}). \end{aligned} \tag{10}$$

APPENDIX IV

TRANSFORMATIONS OF OPERATORS

1. Linear continuous transformations. Let there be given a certain linear continuous transformation of the operators a_i, which depends on the parameter φ:

$$a_i \to a_i' = L_{ij}(\varphi)\, a_j, \quad a' = U^{-1}(\varphi)\, aU(\varphi)$$

and which has the group property

$$L(\varphi_1)\, L(\varphi_2) = L(\varphi_1 + \varphi_2).$$

Then the explicit form of the unitary operator U, which by virtue of the abovementioned property is by necessity of the exponential form

$$U(\varphi) = \exp(i\varphi V), \quad \overset{+}{V} = V,$$

may be found by means of the following general procedure.

Consider the infinitesimal transformation ($\varphi \to \varepsilon \ll 1$)

$$a' \simeq a + \varepsilon A, \quad U \simeq 1 + i\varepsilon V.$$

Equating

$$a + \varepsilon A \simeq (1 - i\varepsilon V)\, a\, (1 + i\varepsilon V),$$

we obtain

$$A = i\,[a,\ V], \tag{1}$$

the basic equation for defining the explicit form of the operator V. We shall give a number of important examples:

(a) *The transformation of shift by a c-number*:

$$a' = a + \varphi, \quad \overset{*}{a}{}' = \overset{*}{a} + \overset{*}{\varphi}$$

or

$$a' = a + \varphi + i\psi, \quad \overset{*}{a}{}' = \overset{*}{a} + \varphi - i\psi,$$
$$U = \exp(i\varphi u + i\psi v).$$

The basic equation gives

$$1=i\,(au-ua),\qquad 1=i\left[\overset{*}{a},\,u\right],$$
$$1=[a,\,v],\qquad -1=\left[\overset{*}{a},\,v\right].$$

It can be seen that u and v must be linear in the Bose operators a, $\overset{*}{a}$:

$$\left[a,\,\overset{*}{a}\right]=a\overset{*}{a}-\overset{*}{a}a=1,\quad [a,\,a]=\left[\overset{*}{a},\,\overset{*}{a}\right]=0.$$

We obtain

$$u=i\left(a-\overset{*}{a}\right),\quad v=\left(a+\overset{*}{a}\right),$$

as a result of which

$$U=\exp\left(\overset{*}{a}\varphi-a\overset{*}{\varphi}\right). \tag{2}$$

Note that a corresponding solution for Fermi operators does not exist. In this case the operation of shift requires the replacement of the c-number objects φ, $\overset{*}{\varphi}$ by the generators of the Grassman algebra.

(b) *The "rotation" transformation* $\{a=(a_1,\,a_2)\}$:

$$a_1'=a_1\cos\varphi-a_2\sin\varphi,\quad \text{(likewise for } \overset{*}{a_1'}\text{)},$$
$$a_2'=a_2\cos\varphi+a_1\sin\varphi,\quad \text{(likewise for } \overset{*}{a_2'}\text{)}.$$

The rotation transformation in the isotopic space, for example, reduces to such transformations.

The basic equations give

$$a_1=i\,[a_2,\,V],\quad -a_2=i\,[a_1,\,V], \tag{3}$$

as well as the Hermitean conjugate relations

$$\overset{*}{a}_1=i\left[\overset{*}{a}_2,\,V\right],\quad -\overset{*}{a}_2=i\left[\overset{*}{a}_1,\,V\right]. \tag{3'}$$

Contrary to the preceding, the operator V should be bilinear in a_k and $\overset{*}{a}_k$:

$$V=c\overset{*}{a}_1a_1+d\overset{*}{a}_2a_2+e\left(\overset{*}{a}_1a_2+\overset{*}{a}_2a_1\right)+if\left(\overset{*}{a}_1a_2-\overset{*}{a}_2a_1\right).$$

Inserting this most general Hermitian expression into (3) and (3′), we find $c=d=e=0,\ \ f=1$, i.e.,

$$U=\exp\varphi\left(\overset{*}{a}_2a_1-\overset{*}{a}_1a_2\right). \tag{4}$$

The expression obtained holds both for Bose and Fermi operators.

(c) *The Bogoliubov transformation for Bose operators*:

$$\begin{aligned} b_1\to\beta_1&=b_1\operatorname{ch}\varphi+\overset{*}{b}_2\operatorname{sh}\varphi, & \overset{*}{\beta}_1&=\overset{*}{b}_1\operatorname{ch}\varphi+b_2\operatorname{sh}\varphi,\\ b_2\to\beta_2&=\overset{*}{b}_1\operatorname{sh}\varphi+b_2\operatorname{ch}\varphi, & \overset{*}{\beta}_2&=b_1\operatorname{sh}\varphi+\overset{*}{b}_2\operatorname{ch}\varphi,\end{aligned} \tag{5}$$

The basic equation gives

$$\overset{*}{b}_2 = i\,[b_1, V], \quad \overset{*}{b}_1 = i\,[b_2, V]$$

as well as the corresponding Hermitian-conjugate equations. The solution has the form

$$V = i\left(b_1 b_2 - \overset{*}{b}_2 \overset{*}{b}_1\right) = i\left(\beta_1 \beta_2 - \overset{*}{\beta}_2 \overset{*}{\beta}_1\right),$$

i.e.,

$$U(\varphi) = \exp\left[\varphi\left(\overset{*}{b}_2 \overset{*}{b}_1 - b_1 b_2\right)\right]. \tag{6}$$

(d) *The Bogoliubov transformation for Fermi operators*:

$$\begin{aligned} \alpha_1 &= a_1 \cos\varphi + \overset{*}{a}_2 \sin\varphi, & \overset{*}{\alpha}_1 &= \overset{*}{a}_1 \cos\varphi + a_2 \sin\varphi, \\ \alpha_2 &= -\overset{*}{a}_1 \sin\varphi + a_2 \cos\varphi, & \overset{*}{\alpha}_2 &= -a_1 \sin\varphi + \overset{*}{a}_2 \cos\varphi. \end{aligned} \tag{7}$$

The basic equation gives

$$\overset{*}{a}_1 = -i\,[a_2, V], \quad \overset{*}{a}_2 = i\,[a_1, V]$$

and the Hermitian conjugate relations. The solution has the form

$$\begin{aligned} V &= i\left(a_1 a_2 - \overset{*}{a}_2 \overset{*}{a}_1\right) = i\left(\alpha_1 \alpha_2 - \overset{*}{\alpha}_2 \overset{*}{\alpha}_1\right), \\ U(\varphi) &= \exp\left[\varphi\left(\overset{*}{a}_2 \overset{*}{a}_1 - a_1 a_2\right)\right]. \end{aligned} \tag{8}$$

(e) *Lorentz "rotations" of the Dirac matrices*: We construct the operator $U(\varphi)$ of the transformation of the Dirac matrices

$$\gamma_\mu \to \tilde{\gamma}_\mu = L_{\mu\nu}(\varphi)\,\gamma_\nu,$$

where L is the matrix of the Lorentz four-rotations of the coordinate vector $x' = L(\varphi)x$. For a space (Euclidean) rotation in the $x_1 x_2$ plane

$$\tilde{\gamma}_1 = \gamma_1 \cos\varphi - \gamma_2 \sin\varphi, \quad \tilde{\gamma}_2 = \gamma_1 \sin\varphi + \gamma_2 \cos\varphi, \quad \tilde{\gamma}_0 = \gamma_0, \ \tilde{\gamma}_3 = \gamma_3;$$

for a Lorentz rotation in the $x_0 x_1$ plane

$$\tilde{\gamma}_0 = \gamma_0 \operatorname{ch}\varphi + \gamma_1 \operatorname{sh}\varphi, \quad \gamma_1 = \gamma_0 \operatorname{sh}\varphi + \gamma_1 \operatorname{ch}\varphi, \quad \tilde{\gamma}_2 = \gamma_2, \ \tilde{\gamma}_3 = \gamma_3.$$

The basic equation gives

$$\begin{aligned} -\gamma_2 &= i\,[\gamma_1, V_{12}], & \gamma_1 &= i\,[\gamma_2, V_{12}], & [\gamma_0, V_{12}] &= [\gamma_3, V_{12}] = 0; \\ \gamma_1 &= i\,[\gamma_0, V_{01}], & \gamma_0 &= i\,[\gamma_1, V_{01}], & [\gamma_2, V_{01}] &= [\gamma_3, V_{01}] = 0. \end{aligned}$$

The solutions have the form

$$V_{12} = -(i/2)\,\gamma_1\gamma_2, \quad V_{01} = -(i/2)\,\gamma_0\gamma_1.$$

The obtained results may be combined to yield a single formula

$$U_{\mu\nu}=\exp\left(\frac{i}{2}\,\gamma_\mu\gamma_\nu\right). \tag{9}$$

2. The "untangling" of exponentials*. Let

$$U=\exp(A+B),$$

where A and B do not commute with each other. The dependences on the operators A and B are required to be factorized. In the particular case when A contains only creation operators and B only annihilation operators, such a necessity arises in the course of reducing the operator U to its normal form.

In the general case the operator U may be represented in the form

$$U=e^A Q e^B,$$

where the factor Q, generally speaking, contains the dependences on the operators A and B and their repeated commutators.

To determine the explicit form of Q, we take advantage of the method of infinitesimal increments. To this end we introduce an auxiliary parameter τ:

$$U\to U(\tau)=e^{\tau(A+B)}=e^{\tau A}Q(\tau,\ A,\ B,\ [A,\ B],\ \ldots)\,e^{\tau B},$$

and, upon differentiating with respect to τ, we write for Q the differential operator equation

$$Q'(\tau)=e^{-\tau A}Be^{\tau A}Q(\tau)-Q(\tau)\,B,\quad Q(\tau=0)=1.$$

The simplest is the one when the commutator of the operators A and B,

$$[A,\ B]=C,$$

is a c-number (see (2)). Then

$$e^{-\tau A}Be^{\tau A}=B-\tau C,\quad Q(\tau)=\exp(-\tau^2C/2),$$

and we arrive at the so-called Baker-Haussdorf formula

$$(C=[A,\ B]\sim c\text{-number})\ e^{A+B}=e^A e^{-C/2} e^B. \tag{10}$$

In order to untangle the formulae (6), (8) appearing as a result of the canonical Bogoliubov transformation, the particular case which turns out to

---

*The exposition of this section follows that of Appendix B in the Book by Kirzhnitz (1963), in which there are also several other useful formulae.

be important is when the commutator C is an operator, and the operators A, B, C form a closed commutative algebra of the special form:

$$[A, B]=C, \quad [A, C]=-2\xi A, \quad [B, C]=2\xi B. \tag{11}$$

This case may also be investigated by the differential method. We leave it to the reader to verify that the desired result has the form

$$e^{\tau(A+B)}=e^{\alpha(\tau)A}e^{\gamma(\tau)C}e^{\alpha(\tau)B}, \tag{12}$$

where

$$\alpha(\tau)=\frac{1}{\sqrt{\xi}}\operatorname{th}(\tau\sqrt{\xi}), \quad \gamma(\tau, \xi)=\frac{1}{\xi}\ln\operatorname{ch}(\tau\sqrt{\xi}). \tag{13}$$

In the case of Bose operators

$$A=\overset{+}{b}_1\overset{+}{b}_2, \quad B=-b_1b_2, \quad C=1-\overset{+}{b}_1b_1-\overset{+}{b}_2b_2, \quad \xi=1,$$

while for Fermi operators

$$A=\overset{+}{a}_2\overset{+}{a}_1, B=-a_1a_2, \quad C=1+a_1\overset{+}{a}_1+a_2\overset{+}{a}_2, \quad \xi=-1.$$

3. Commutators with the particle density operator $n(k)$. Let us further obtain some useful formulas for the commutators with the particle density operator

$$n(k)=\overset{*}{a}(k)\,a(k)=a^+(k)\,a^-(k).$$

We introduce the linear superposition of the operators $\overset{+}{a}$ and a^-:

$$A=A_++A_-, \quad A_\pm=\int\lambda_\pm(q)\,a^\pm(q)\,dq. \tag{14}$$

Then

$$[n(k), A_\pm]=\pm\lambda_\pm(k)\,a^\pm(k)$$

and

$$[n(k), A]=K(k)=\lambda_+(k)\,a^+(k)-\lambda_-(k)\,a^-(k),$$

while the commutator of the operators A and K is a c-number:

$$[A, K(k)]=-2\lambda^2(k), \quad \lambda^2(k)\equiv\lambda_+(k)\,\lambda_-(k).$$

Therefore

$$[n(k), A^\nu]=\nu K(k)\,A^{\nu-1}-\nu(\nu-1)\,\lambda^2A^{\nu-2}$$

and, consequently,

$$[n(k), e^{\pm A}]=\{\pm K(k)+\lambda^2(k)\}\,e^{\pm A}. \tag{15}$$

APPENDIX V

LIST OF SINGULAR FUNCTIONS

1. Auxiliary singular functions. The four-dimensional Dirac δ-function:

$$\delta(x) = \frac{1}{(2\pi)^4} \int e^{ikx}\, dk, \tag{1}$$

$$\delta(x) = \delta(x^0)\,\delta(\boldsymbol{x}) = \delta(x^0)\,\delta(x^1)\,\delta(x^2)\,\delta(x^3). \tag{2}$$

The step functions $\theta(\alpha)$ and $\varepsilon(\alpha)$:

$$\theta(\alpha) = \frac{1}{2\pi i} \int\limits_{-\infty}^{\infty} \frac{e^{i\alpha\tau}}{\tau - i\varepsilon}\, d\tau = \begin{cases} 1, & \alpha > 0, \\ 0, & \alpha < 0 \end{cases} \quad (\varepsilon \to +0), \tag{3}$$

$$\varepsilon(\alpha) = \frac{1}{\pi i}\, \mathscr{P} \int\limits_{-\infty}^{\infty} e^{i\alpha\tau} \frac{d\tau}{\tau} = \begin{cases} 1, & \alpha > 0, \\ -1, & \alpha < 0 \end{cases} \tag{4}$$

($\mathscr{P}$ denotes the principal value).

The frequency parts of the δ-function:

$$\delta_{\pm}(\alpha) = \frac{1}{2\pi} \int\limits_{0}^{\infty} e^{\pm i\alpha\tau}\, d\tau = \frac{1}{2}\left[\delta(\alpha) \pm \frac{i}{\pi}\, \mathscr{P} \frac{1}{\alpha}\right]. \tag{5}$$

Some useful relations:

$$\frac{1}{\alpha + i\varepsilon} = \frac{2\pi}{i}\, \delta_+(\alpha) = \frac{\pi}{i}\, \delta(\alpha) + \mathscr{P} \frac{1}{\alpha} \qquad (\varepsilon \to +0), \tag{6}$$

$$\frac{1}{\alpha - i\varepsilon} = 2\pi i \delta_-(\alpha) = \pi i \delta(\alpha) + \mathscr{P} \frac{1}{\alpha}, \tag{7}$$

$$\mathscr{P} \frac{1}{\alpha} = \frac{1}{2i} \int\limits_{-\infty}^{\infty} \varepsilon(\tau)\, e^{i\alpha\tau}\, d\tau. \tag{8}$$

2. Scalar field functions. The Pauli-Jordan commutation function D:

$$[\varphi(x),\ \varphi(y)] = \frac{1}{i} D(x - y); \tag{8.4}$$

$$D(x) = \frac{i}{(2\pi)^3} \int e^{-ikx} \varepsilon(k^0)\, \delta(k^2 - m^2)\, dk =$$

$$= \frac{1}{(2\pi)^3} \int \frac{d\boldsymbol{k}}{\sqrt{\boldsymbol{k}^2 + m^2}}\, e^{i\boldsymbol{k}\boldsymbol{x}} \sin\left(x^0 \sqrt{\boldsymbol{k}^2 + m^2}\right); \tag{9}$$

$$\left.\frac{\partial D(x^0, \boldsymbol{x})}{\partial x^0}\right|_{x^0=0} = \delta(\boldsymbol{x}); \tag{10}$$

$$D(x) = \frac{1}{2\pi}\varepsilon(x^0)\delta(\lambda) - \frac{m}{4\pi\sqrt{\lambda}}\varepsilon(x^0)\theta(\lambda)J_1(m\sqrt{\lambda}). \tag{18.16}$$

$$\lambda = x^2 = x_0^2 - \boldsymbol{x}^2.$$

In the neighborhood of the light cone D has the form

$$D(x) \sim \frac{1}{2\pi}\varepsilon(x^0)\delta(\lambda) - \frac{m^2}{8\pi}\varepsilon(x^0)\theta(\lambda). \tag{18.21}$$

The *frequency parts* of the Pauli–Jordan function D^+ and D^-:

$$[\varphi^-(x), \varphi^+(y)] = \langle\varphi(x)\varphi(y)\rangle_0 = \frac{1}{i}D^-(x-y); \tag{11}$$

$$[\varphi^+(x), \varphi^-(y)] = \frac{1}{i}D^+(x-y) = iD^-(y-x); \tag{12}$$

$$D^{\pm}(x) = \frac{\pm 1}{(2\pi)^3 i}\int e^{ikx}\theta(\pm k^0)\delta(k^2-m^2)\,dk =$$

$$= \frac{\mp i}{(2\pi)^3}\int \frac{d\boldsymbol{k}}{2\sqrt{\boldsymbol{k}^2+m^2}} e^{\pm ix^0\sqrt{\boldsymbol{k}^2+m^2} - \boldsymbol{x}\boldsymbol{k}} = \tag{13}$$

$$= \frac{1}{4\pi}\varepsilon(x^0)\delta(\lambda) - \frac{m\theta(\lambda)}{8\pi\sqrt{\lambda}}[\varepsilon(x^0)J_1(m\sqrt{\lambda}) \pm iN_1(m\sqrt{\lambda})] \mp$$

$$\mp \frac{mi\theta(-\lambda)}{4\pi^2|\lambda|^{1/2}}K_1(m\sqrt{-\lambda}). \tag{18.18}$$

In the neighborhood of the light cone D^+ and D^- have the form

$$D^{\pm}(x) = \frac{\varepsilon(x^0)\delta(\lambda)}{4\pi} \pm \frac{i}{4\pi^2\lambda} \mp \frac{im^2}{8\pi^2}\ln\frac{m|\lambda|^{1/2}}{2} - \frac{m^2}{16\pi^2}\varepsilon(x^0)\theta(\lambda). \tag{14}$$

They are even and real:

$$D^{\pm}(-x) = -D^{\mp}(x) = -(D^{\pm}(x))^*, \quad D(x) = -D(-x) = \overset{*}{D}(x). \tag{15}$$

The causal Green function D^c:

$$\langle T\varphi(x)\varphi(y)\rangle_0 = \frac{1}{i}D^c(x-y); \tag{18.12}$$

$$(\square - m^2)D^c(x) = -\delta(x),$$

$$D^c(x) = \frac{1}{(2\pi)^4}\int e^{-ikx}D^c(k)\,dk, \quad D^c(k) = \frac{1}{m^2-k^2-i\varepsilon}; \tag{18.9}$$

$$D^c(x) = \frac{mi}{4\pi^2}\frac{K_1(m\sqrt{-\lambda+i\varepsilon})}{\sqrt{-\lambda+i\varepsilon}}. \tag{18.19}$$

The behavior of D^c in the neighborhood of the light cone:

$$D^c(x) \simeq \frac{1}{4\pi}\delta(\lambda) - \frac{i}{4\pi^2\lambda} + \frac{im^2}{8\pi^2}\ln|\lambda|^{1/2} - \frac{m^2}{16\pi}\theta(\lambda). \qquad (18.22)$$

The causal function for zero mass:

$$D_0^c(x) = \frac{1}{4\pi}\left(\delta(\lambda) - \frac{i}{\pi\lambda}\right) = \frac{1}{4\pi^2 i(\lambda - i\varepsilon)}. \qquad (16)$$

The retarded and advanced Green functions D^{ret} and D^{adv}:

$$D^{\text{ret}}(x) = 0 \quad \text{при} \quad x^0 < 0, \qquad D^{\text{adv}}(x) = 0 \quad \text{when} \quad x^0 > 0,$$

$$D^{\text{ret}}(x) = \frac{1}{(2\pi)^4}\int \frac{e^{-ikx}}{m^2 - k^2 - i\varepsilon k^0}\,dk = \theta(x^0)\,D(x); \qquad (18.5),\ (18.6)$$

$$D^{\text{adv}}(x) = \frac{1}{(2\pi)^4}\int \frac{e^{-ikx}}{m^2 - k^2 + i\varepsilon k^0}\,dk = -\theta(-x^0)\,D(x); \qquad (18.7)$$

$$D^{\text{ret}}(x) = \frac{1}{2\pi}\theta(x^0)\left\{\delta(\lambda) - \theta(\lambda)\frac{m}{2\sqrt{\lambda}}J_1(m\sqrt{\lambda})\right\}; \qquad (18.20)$$

$$D^{\text{adv}}(x) = \frac{1}{2\pi}\theta(-x^0)\left\{\delta(\lambda) - \theta(\lambda)\frac{m}{2\sqrt{\lambda}}J_1(m\sqrt{\lambda})\right\}. \qquad (17)$$

The relations between the functions D, D^+, D^-, D^c, D^{ret}, D^{adv}:

$$D(x) = D^+(x) + D^-(x),$$

$$D^c(x) = \theta(x^0)\,D^-(x) - \theta(-x^0)\,D^+(x), \qquad (18.8)$$

$$D^{\text{ret}}(x) = \theta(x^0)\,D(x) = D^c(x) + D^+(x), \qquad (18)$$

$$D^{\text{adv}}(x) = -\theta(-x^0)\,D(x) = D^c(x) - D^-(x), \qquad (19)$$

$$D(x) = D^{\text{ret}}(x) - D^{\text{adv}}(x). \qquad (20)$$

3. Functions of the electromagnetic, vector, and spinor fields.

The electromagnetic field:

$$[A_\mu(x), A_\nu(y)] = +ig_{\mu\nu}D_0(x-y), \qquad (8.22)$$

$$\langle A_\mu(x)\,A_\nu(y)\rangle_0 = ig_{\mu\nu}D_0^-(x-y), \qquad (21)$$

$$\langle TA_\mu(x)\,A_\nu(y)\rangle_0 = ig_{\mu\nu}D_0^c(x-y). \qquad (22)$$

The functions D_0, D_0^+, D_0^c, etc., are obtained from the functions D, $D^\pm$, D^c of the scalar field for $m = 0$; for example

$$D_0^c(x) = D^c(x)\big|_{m=0} = -\frac{1}{(2\pi)^4}\int \frac{e^{-ikx}}{k^2 + i\varepsilon}\,dk. \qquad (23)$$

The vector field:

$$[U_\mu(x),\ U_\nu(y)] = iD_{\mu\nu}(x-y), \tag{24}$$

$$\langle U_\mu(x)\, U_\nu(y)\rangle_0 = iD^-_{\mu\nu}(x-y), \tag{25}$$

$$\langle TU_\mu(x)\, U_\nu(y)\rangle_0 = iD^c_{\mu\nu}(x-y). \tag{26}$$

The functions $D_{\mu\nu}$, $D^{\pm}_{\mu\nu}$, $D^c_{\mu\nu}$ etc., may be obtained from the corresponding functions of the scalar field by applying the differential operator $g_{\mu\nu} + \partial_\mu\partial_\nu/m^2$; for example

$$D^c_{\mu\nu}(x) = \frac{1}{(2\pi)^4}\int \frac{g_{\mu\nu} - \dfrac{k_\mu k_\nu}{m^2}}{m^2-k^2-i\varepsilon}\, e^{-ikx}\, dk. \tag{27}$$

The spinor field:

$$[\psi_\alpha(x),\ \bar\psi_\beta(y)]_+ = -iS_{\alpha\beta}(x-y), \tag{9.4}$$

$$\langle \psi_\alpha(x)\, \bar\psi_\beta(y)\rangle_0 = -iS^-_{\alpha\beta}(x-y), \tag{28}$$

$$\langle T\psi_\alpha(x)\, \bar\psi_\beta(y)\rangle_0 = -iS^c_{\alpha\beta}(x-y). \tag{29}$$

The functions S, $S^{\pm}$, S^c, etc., may be obtained from the corresponding functions of the scalar field by applying the operator

$$(i\hat\partial + m)_{\alpha\beta} = i\gamma^\nu_{\alpha\beta}\partial_\nu + mI_{\alpha\beta};$$

for example

$$S(x) = \frac{i}{(2\pi)^3}\int e^{-ikx}\varepsilon(k^0)(\hat k + m)\,\delta(k^2-m^2)\, dk, \tag{30}$$

$$S(x)_{x^0=0} = i\gamma^0\delta(\boldsymbol{x}), \tag{31}$$

$$S^c(x) = \frac{1}{(2\pi)^4}\int \frac{m+\hat p}{m^2-p^2-i\varepsilon}\, e^{-ipx}\, dp. \tag{32}$$

APPENDIX VI

FORMULAE FOR INTEGRATION OVER MOMENTUM SPACE

1. The α-representation. The formula for transition to the α-representation:

$$\frac{1}{m^2-p^2-i\varepsilon}=i\int_0^\infty e^{i\alpha(p^2-m^2+i\varepsilon)}\,d\alpha \tag{22.2}$$

and its generalization:

$$\frac{1}{(D-i\varepsilon)^k}=\frac{i^k}{\Gamma(k)}\int_0^\infty e^{i\alpha(-D+i\varepsilon)}\,\alpha^{k-1}\,d\alpha, \qquad k>0. \tag{22.8}$$

The basic Gaussian quadratures (a is a positive number, b is a four-vector):

$$\frac{i}{\pi^2}\int dk e^{i(ak^2+2bk)}=\frac{1}{a^2}\,e^{-ib^2/a}\equiv I(a,\ b^2), \tag{22.3}$$

$$\frac{i}{\pi^2}\int dk e^{i(ak^2+2bk)}\,[k^\nu]=\left[-\frac{b^\nu}{a}\right]I(a,\ b^2), \tag{22.4a}$$

$$\frac{i}{\pi^2}\int dk e^{i(ak^2+2bk)}\,[k^\nu k^\mu]=\left[\frac{2b^\nu b^\mu+iag^{\mu\nu}}{2a^2}\right]I(a,\ b^2), \tag{22.4b}$$

$$\frac{i}{\pi^2}\int dk e^{i(ak^2+2bk)}\,[k^2]=\left[\frac{b^2+2ia}{a^2}\right]I(a,\ b^2), \tag{22.4c}$$

$$\frac{i}{\pi^2}\int dk e^{i(ak^2+2bk)}\,[k^\nu k^2]=\left[-\frac{b^\nu(b^2+3ia)}{a^3}\right]I(a,\ b^2). \tag{1}$$

The isolation of infinite integration:

$$\int_0^\infty d\alpha_1\ldots\int_0^\infty d\alpha_n=\int_0^1\{dx\}_n\int_0^\infty da\,a^{n-1}, \tag{2}$$

where

$$a=\alpha_1+\alpha_2+\ldots+\alpha_n,\qquad x_\nu=\alpha_\nu/a,$$

$$\{dx\}_n=dx_1\,dx_2\ldots dx_n\delta(1-x_1-x_2-\ldots-x_n). \tag{3}$$

Typical integrals over a:

$$\frac{1}{i}\int_0^\infty da e^{ia(A+i\varepsilon)}=\frac{1}{A+i\varepsilon}, \tag{4}$$

$$\int_0^\infty \frac{da}{a}\left(e^{iaA}-e^{iaB}\right)e^{-\varepsilon a}=\ln\frac{B+i\varepsilon}{A+i\varepsilon}, \tag{5}$$

$$\frac{1}{i}\int_0^\infty \frac{da}{a^2}\left(e^{iaA}-e^{iaB}\right)\left(e^{iaC}-e^{iaD}\right)e^{-\varepsilon a}=$$
$$=(A+D)\ln(A+D+i\varepsilon)+(B+C)\ln(B+C+i\varepsilon)-$$
$$-(A+C)\ln(A+C+i\varepsilon)-(B+D)\ln(B+D+i\varepsilon). \tag{6}$$

In the case when B and D contain a large parameter Λ^2 ($B=b\Lambda^2$, $D=d\Lambda^2$), we obtain from (6) in the limit $\Lambda^2\to\infty$

$$\to\Lambda^2\left[b\ln b+d\ln d-(b+d)\ln(b+d)\right]+(A+C)\ln\Lambda^2+$$
$$+(A+C)\left[1-\ln(A+C)\right]+A\ln d+C\ln b. \tag{7}$$

2. Feynman parametrization. The initial formula:

$$\frac{1}{a_1a_2\ldots a_n}=(n-1)!\int_0^1\frac{\{dx\}_n}{\left(\sum_{1\leqslant\nu\leqslant n}a_\nu x_\nu\right)^n}, \tag{22.10}$$

where the symbol $\{dx\}_n$ is defined in (3).

Momentum quadratures of the type

$$J_L(D)=\frac{i}{\pi^2}\int\frac{dq}{(g^2-D+i\varepsilon)^L}=\frac{(-1)^{L+1}(L-3)!}{(D-i\varepsilon)^{L-2}(L-1)!},\qquad L\geqslant 3, \tag{22.16}$$

are conveniently generalized for practical use (by shifting the integration variable):

$$F_L(D,k)=J_L(k^2-D)=\frac{i}{\pi^2}\int\frac{dp}{\left[p^2-2kp+D\right]^L}=-\frac{(L-3)!}{(D-k^2+i\varepsilon)^{L-2}(L-1)!}, \tag{8}$$

$$F_L^{\mu}(D,k)\doteq\frac{i}{\pi^2}\int\frac{p^\mu dp}{\left[p^2-2kp+D\right]^L}=-\frac{k^\mu(L-3)!}{(D-k^2)^{L-2}(L-1)!}=k^\mu F_L(D,k), \tag{9}$$

$$F_L^{\mu\nu}(D,k)=\frac{i}{\pi^2}\int\frac{p^\mu p^\nu dp}{\left[p^2-2kp+D\right]^L}=\frac{g^{\mu\nu}(D-k^2)+2(L-3)k^\mu k^\nu}{2(L-3)}F_L(D,k), \tag{10}$$

$$F_L^{(2)}(D,k)=\frac{i}{\pi^2}\int\frac{p^2dp}{\left[p^2-2kp+D\right]^L}=\frac{2D+(L-5)k^2}{(L-3)}F_L(D,k) \tag{11}$$

etc.

3. Dimensional regularization:

$$\int dp=\int d^4p\to\int d^np=\mu^{2\varepsilon}\int_{\Omega(n)}d\Omega\int_0^\infty p^{n-1}\,dp,\qquad n=4-2\varepsilon. \tag{23.10}$$

The modification of Gaussian quadratures

$$\frac{i}{\pi^2}\int d^n k e^{i(ak^2+2bk)} = \left(\frac{ia\mu^2}{\pi}\right)^{\varepsilon}\frac{1}{a^2}e^{-ib^2/a} = I_\varepsilon(a,\ b^2), \tag{23.11}$$

$$\frac{i}{\pi^2}\int d^n k e^{i(ak^2+2bk)}[k^\nu] = \left[-\frac{b^\nu}{a}\right] I_\varepsilon(a,\ b^2), \tag{12}$$

$$\frac{i}{\pi^2}\int d^n k e^{i(ak^2+2bk)}[k^\nu k^\mu] = \left[\frac{2b^\nu b^\mu + iag^{\nu\mu}}{2a^2}\right] I_\varepsilon(a,\ b^2) \tag{13}$$

(these three formulae differ from the formulae (22.3) and (22.4a), (22.4b) by the substitution $I(a,\ b^2) \to I_\varepsilon(a,\ b^2)$);

$$\frac{i}{\pi^2}\int d^n k e^{i(ak^2+2bk)}[k^2] = \left[\frac{b^2 + ia(2-\varepsilon)}{a^2}\right] I_\varepsilon(a,\ b^2), \tag{14}$$

$$\frac{i}{\pi^2}\int d^n k e^{i(ak^2+2bk)}[k^2 k^\nu] = \left[-\frac{b_\nu[b^2+ia(3-\varepsilon)]}{a^3}\right] I_\varepsilon(a,\ b^2) \tag{15}$$

etc.

Note that the formula (14) may be derived either from (23.11) by differentiation with respect to the parameter a or from (13) by summation over the Lorentz indices. In the latter case it is necessary to take into account that

$$g_{\mu\nu}g^{\nu\mu} = \mathrm{Sp}\, g_{\mu\nu} = n = 4 - 2\varepsilon. \tag{16}$$

With this property, the operations with Dirac matrices should be performed by using the usual relation

$$\gamma_\mu\gamma_\nu + \gamma_\nu\gamma_\mu = 2g_{\mu\nu} \qquad (\mu,\ \nu = 0,\ 1,\ 2,\ 3).$$

The modification of formulae for Feynman integration:

$$\frac{i}{\pi^2}\int \frac{d^n p}{[p^2 - 2pk + D + i\varepsilon]^L} = -\left(\frac{i^2\mu^2}{\pi}\right)^{\varepsilon}\frac{\Gamma(L+\varepsilon-2)}{(D-k^2)^{L+\varepsilon-2}\Gamma(L)} \equiv F_L^\varepsilon(D,\ k), \tag{17}$$

$$\frac{i}{\pi^2}\int \frac{p^\nu d^n p}{[p^2 - 2pk + D]^L} = k^\nu F_L^\varepsilon(D,\ k), \tag{18}$$

$$\frac{i}{\pi^2}\int \frac{p^\nu p^\mu d^n p}{[p^2 - 2pk + D]^L} = \left\{k^\nu k^\mu + \frac{g^{\nu\mu}(D-k^2)}{2(L+\varepsilon-3)}\right\} F_L^\varepsilon(D,\ k), \tag{19}$$

$$\frac{i}{\pi^2}\int \frac{p^2 d^n p}{[p^2 - 2pk + D]^L} = \left\{k^2 + (D-k^2)\frac{2-\varepsilon}{L+\varepsilon-3}\right\} F_L^\varepsilon(D,\ k^2), \tag{20}$$

$$\frac{i}{\pi^2}\int \frac{p_\nu p^2 d^n p}{[p^2 - 2pk + D]^L} = k_\nu\left\{k^2 + \frac{3-\varepsilon}{L+\varepsilon-3}(D-k^2)\right\} F_L^\varepsilon(D,\ k^2). \tag{21}$$

From the above formulae, allowing for the fact that as $\varepsilon \to 0$, $\Gamma(\varepsilon) \to 1/\varepsilon - C$ (where $C = 0.5772\ldots$ is the Euler constant), in particular it follows that

$$\frac{i}{\pi^2}\int \frac{d^n p}{p^2-2pk+D} = \frac{D-k^2}{\varepsilon} + \left\{\ln\frac{\mu^2}{(k^2-D)\,\pi} + 1 - C\right\}(D-k^2), \tag{22}$$

$$\frac{i}{\pi^2}\int \frac{d^n p}{[p^2-2pk+D]^2} = -\frac{1}{\varepsilon} + \ln\frac{(k^2-D)\,\pi}{\mu^2} + C, \tag{23}$$

$$\frac{i}{\pi^2}\int \frac{p_\nu d^n p}{[p^2-2pk+D]^2} = k_\nu\left\{-\frac{1}{\varepsilon} + \ln\frac{(k^2-D)\,\pi}{\mu^2} + C\right\}, \tag{24}$$

$$\frac{i}{\pi^2}\int \frac{p_\nu p_\mu d^n p}{[p^2-2pk+D]^2} = \left\{-\frac{1}{\varepsilon} + \ln\frac{(k^2-D)\,\pi}{\mu^2} + C\right\}\left\{k_\mu k_\nu + g_{\mu\nu}\frac{k^2-D}{2}\right\}, \tag{25}$$

$$\frac{i}{\pi^2}\int \frac{p_\nu p_\mu d^n p}{[p^2-2pk+D]^3} = -\frac{g_{\mu\nu}}{4\varepsilon} + \frac{k_\nu k_\mu}{2(k^2-D)} + \frac{g_{\mu\nu}}{4}\left[\ln\frac{(k^2-D)\,\pi}{\mu^2} + C\right] \tag{26}$$

4. Regularization by means of a cutoff:

$$\int dp = i\int (d^4p)_E \to i\int_{\Omega(4)} d\Omega \int_0^{\Lambda} p^3\,dp \equiv \mathrm{reg}_\Lambda \int dp.$$

We list here the divergent integrals which are used most. The details of the corresponding calculations can be found, for example, in Section 36 of the book by Akhiezer and Berestetskii (1969).

$$\mathrm{reg}_\Lambda \frac{i}{\pi^2}\int \frac{dp}{p^2-2pk+D} = -\Lambda^2 + (D-k^2)\ln\frac{\Lambda^2}{k^2-D}, \tag{27}$$

$$\mathrm{reg}_\Lambda \frac{i}{\pi^2}\int \frac{dp}{(p^2-2pk+D)^2} = \ln\frac{k^2-D}{\Lambda^2} + 1, \tag{23.17}$$

$$\mathrm{reg}_\Lambda \frac{i}{\pi^2}\int \frac{p_\nu\,dp}{(p^2-2pk+D)^2} = k_\nu\left(\ln\frac{k^2-D}{\Lambda^2} + \frac{3}{2}\right), \tag{28}$$

$$\begin{aligned}\mathrm{reg}_\Lambda \frac{i}{\pi^2}\int \frac{p_\nu p_\mu\,dp}{(p^2-2pk+D)^2} = {} & \frac{g_{\nu\mu}}{4}\Lambda^2 - \left\{\frac{g_{\mu\nu}}{2}(k^2-D) + k_\mu k_\nu\right\}\ln\frac{\Lambda^2}{\mu^2} - \\ & - \frac{g_{\mu\nu}}{2}\left\{(k^2-D)\ln\frac{\mu^2}{k^2-D} + \frac{5k^2-3D}{6}\right\} + k_\nu k_\mu\left(\ln\frac{k^2-D}{\mu^2} + \frac{11}{6}\right),\end{aligned} \tag{29}$$

$$\mathrm{reg}_\Lambda \frac{i}{\pi^2}\int \frac{p_\nu p_\mu\,dp}{(p^2-2pk+D)^3} = -\frac{g_{\mu\nu}}{4}\ln\frac{\Lambda^2}{\mu^2} + \frac{g_{\mu\nu}}{4}\left(\ln\frac{k^2-D}{\mu^2} + \frac{3}{2}\right) + \frac{1}{2}\frac{k_\mu k_\nu}{k^2-D}. \tag{30}$$

Comparison of these formulae with (22)–(26) reveals that the finite parts of the integrals considered coincide with each other up to finite constants in the case of logarithmically divergent integrals and up to finite polynomials of the first and second degree in the case of linearly and quadratically divergent integrals.

5. Simplest one-loop quadratures. Let us now write out some more results of calculations for a one-loop diagram consisting of two internal scalar lines with different masses:

$$f \sim \frac{i}{\pi^2} \int \frac{dk}{(m^2 - k^2 - i\varepsilon)\,[\mu^2 - (k+p)^2 - i\varepsilon]}\,. \tag{31}$$

The finite part of the function f may be written as follows:

$$f(p^2;\ m^2,\ \mu^2) = \int_0^1 dx \ln \frac{xm^2 + (1-x)\,\mu^2 - x\,(1-x)\,p^2 - i\varepsilon}{xm^2 + (1-x)\,\mu^2}\,.$$

This expression is real for $p^2 < (m + \mu)^2$ and is normalized at zero, i.e.,

$$f(0;\ m^2,\ \mu^2) = 0.$$

Calculating the integral over x, we obtain

$$f = \left(\frac{\Delta}{2p^2} - \frac{\Sigma}{2\Delta}\right) \ln \frac{m^2}{\mu^2} + \frac{K(p)}{2} \operatorname{Ln} \frac{\Sigma - p^2 - p^2K(p)}{\Sigma - p^2 + p^2K(p)} - 1, \tag{32}$$

where $\Sigma = m^2 + \mu^2$, $\Delta = m^2 - \mu^2$, and

$$K(p) = \left[\left(1 - \frac{(m+\mu)^2}{p^2}\right)\left(1 - \frac{(m-\mu)^2}{p^2}\right)\right]^{1/2}.$$

The second term of (32), containing the logarithm, represents three different expressions corresponding to different real values of p^2:

$$\begin{aligned} \frac{K(p)}{2} \operatorname{Ln} \frac{\Sigma - p^2 - p^2K}{\Sigma - p^2 + p^2K} &= \frac{K(p)}{2} \ln \frac{\Sigma - p^2 - p^2K}{\Sigma - p^2 + p^2K} \quad \text{for} \quad p^2 < (m-\mu)^2, \\ &= \frac{K(p)}{2} \ln \frac{\Sigma - p^2 - p^2K}{\Sigma - p^2 + p^2K} - i\pi K(p) \ \text{for}\ p^2 \geqslant (m+\mu)^2, \\ &= Q(p) \operatorname{arctg} \frac{p^2 Q(p)}{\Sigma - p^2} \quad \text{for} \quad (m-\mu)^2 \leqslant p^2 \leqslant (m+\mu)^2, \end{aligned}$$

and

$$p^2 Q(p) = [((m+\mu)^2 - p^2)\,(p^2 - (m-\mu)^2)]^{1/2}.$$

For identical masses ($\mu = m$) we obtain

$$f(p^2;\ m^2,\ m^2) = J\left(\frac{p^2}{4m^2}\right) - 2; \quad J(z) = \left(\frac{z-1}{z}\right)^{1/2} \operatorname{Ln} \frac{\sqrt{z} + \sqrt{z-1}}{\sqrt{z-1} - \sqrt{z}}.$$

Note that the function J was introduced in Section 24 and may be represented in the form

$$J(z) = \int_0^1 dx \ln \frac{1 - 4x\,(1-x)\,z - i\varepsilon}{1 - 4x\,(1-x)} = (1-z) \int_1^\infty \frac{d\sigma}{\sqrt{\sigma\,(\sigma-1)}\,(\sigma - z + i\varepsilon)}\,.$$

If, on the other hand, one of the masses equals zero ($\mu = 0$), then

$$f(p^2;\ m^2,\ 0) = \frac{z-1}{z}\ln(1-z) - 1, \qquad z = p^2/m^2.$$

Such expressions are encountered in the course of evaluating the electron self-energy (see (24.17)) and the vertex part. Note, in this connection, that the logarithmic function arises from the parametric integral

$$\int_0^1 \frac{dx}{x\dfrac{p^2}{m^2} - 1} = \frac{m^2}{p^2}\ln\frac{m^2 - p^2}{m^2},$$

which turns out to be singular on the mass shell, i.e., for $p^2 = m^2$. This singularity is a reflection of the so-called *infrared* divergence and is related to the singularity of the integrand of the initial Feynman integral at *small* values of the virtual momentum k.

Therefore the transition to the limit $\mu = 0$ must in certain cases be executed more carefully. As is not difficult to verify, at small but finite μ the integral on the left-hand side has to be replaced (see Section 29.4) by

$$A(p^2) = \int_0^1 \frac{m^2(x-1)\,dx}{(1-x)(m^2 - xp^2) + x\mu^2} = \begin{cases} \dfrac{m^2}{p^2}\ln\dfrac{m^2 - p^2}{m^2} & \text{for } |p^2 - m^2| \gg \mu^2, \\[2ex] \dfrac{1}{2}\ln\dfrac{\mu^2}{m^2} & \text{for } p^2 = m^2. \end{cases} \tag{29.17}$$

The representation of the function f in terms of A,

$$\lim_{\mu^2 \to 0} f(p^2;\ m^2,\ \mu^2) = \frac{p^2 - m^2}{m^2}A(p^2) - 1 \tag{33}$$

turns out to be useful in the course of the subtraction of the electron mass operator $\Sigma(p)$, and also for the evaluation of the vertex function (see Sections 24.3, 29.4, 29.5).

APPENDIX VII.

KINEMATIC RELATIONS FOR THE VERTICES

1. Three-legged vertex. (See Figure AVII.1.) From three four-vectors p_1, p_2, p_3, the sum of which equals zero,

$$p_1+p_2+p_3=0,$$

three invariant expressions can be made up:

$$p_1^2, \quad p_2^2, \text{ and } p_3^2=(p_1+p_2)^2. \tag{1}$$

These quantities are independent. However, not all triplets of values may be realized for real values of the components of the four-vectors p_1, p_2, and p_3.

In quantum electrodynamics the case of interest is the one when two squares are positive and equal to the square of the electron mass, while the third square is equal to zero:

$$p_1^2=p_2^2=m^2, \quad k^2=0 \qquad (k=p_3). \tag{2}$$

If the condition of reality of the components of all the four-vectors is satisfied, then the relations (2) lead to the requirement that $k_\mu=0, p_1=-p_2$. Thus all three legs of the electron–photon vertex can be on the mass shell only for photons of infinitesimal frequency.

The region

$$p_1^2=p_2^2=m^2, \qquad k^2<0, \tag{3}$$

corresponding to the emission or absorption by the electron of a spacelike photon, as well as the region

$$p_1^2=p_2^2=m^2, \quad k^2\geqslant 4m^2, \tag{4}$$

corresponding to the annihilation of a pair, is kinematically allowed.

In the case of the meson–nucleon three-vertex describing the Yukawa interaction (the meson meomentum $p_3=k$) it is impossible to simultaneously be on the mass shell for all three momenta. The point

$$p_1^2=p_2^2=M^2, \quad k^2=\mu^2<M^2 \tag{5}$$

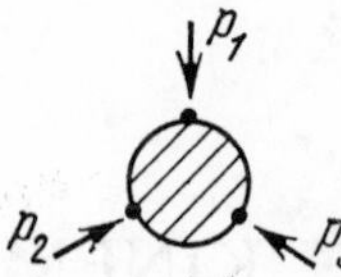

Fig. AVII.1. Kinematic notation for a three-legged vertex.

is not kinematically allowed. It is necessary to take this fact into account in formulating the prescription for the subtraction of divergences in the corresponding models of quantum field theory.

2. Four-legged vertex. The scalar four-legged vertex (Figure AVII.2), by virtue of the Lorentz invariance, is a function of the scalar products of its four-momentum arguments p_ip_j. Because of the law of conservation of the total four-momentum,

$$p_1+p_2+p_3+p_4=0$$

only six of the scalar products are independent. It is natural to choose four of them in the form of squares of the external four-momenta p_i, while representing the two remaining ones by expressions of the type $(p_i+p_j)^2$. Altogether three such different expressions may be composed:

$$s=(p_1+p_2)^2=(p_3+p_4)^2, \quad t=(p_1+p_3)^2, \quad u=(p_1+p_4)^2, \tag{6}$$

which by virtue of the law of conservation of the total four-momentum satisfy the relation

$$s+t+u=p_1^2+p_2^2+p_3^2+p_4^2. \tag{7}$$

To understand the physical meaning of the variables s, t, u, we note that the four-vertex under consideration can describe six different processes of the type 2 particles → 2 particles:

$$\begin{array}{lll} 1+2\to 3+4, & 1+3\to 2+4, & 1+4\to 2+3, \\ 3+4\to 1+2, & 2+4\to 1+3, & 2+3\to 1+4. \end{array} \tag{8}$$

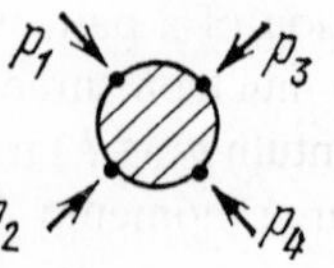

Fig. AVII.2. Kinematic notation for a four-legged vertex.

Here the numbers stand for the particle indices. The last three are connected with the first three by a change of sign of all the four-momenta and correspond to the same values of the quadratic momentum variables. Therefore for our purposes it is sufficient to distinguish among three possibilities:

$$\begin{array}{ll}
\text{I} & \text{II} \\
1+2 \leftrightarrow 3+4 & 1+3 \leftrightarrow 2+4 \\
p_1+p_2 \rightarrow (-p_3)+(-p_4) & p_1+p_3 \rightarrow (-p_2)+(-p_4) \\
p_3+p_4 \rightarrow (-p_1)+(-p_2) & p_2+p_4 \rightarrow (-p_1)+(-p_3)
\end{array}$$

$$\begin{array}{l}
\text{III} \\
1+4 \leftrightarrow 2+3 \\
p_1+p_4 \rightarrow (-p_2)+(-p_3) \\
p_2+p_3 \rightarrow (-p_1)+(-p_4),
\end{array} \tag{9}$$

each one of which unites a certain reaction (the direct one) with its inverse and is said to correspond to a definite *channel.*

In the first channel, the variable $s = +(p_1 + p_2)^2$ plays the role of the relativistically invariant square of the total energy, and the variable $t = (p_2 + p_3)^2$ plays the role of the invariant square of the four-momentum transferred from particle 1 to particle 3. The variable $u = (p_1 + p_4)^2$ describes the four-momentum transfer from particle 1 to particle 4. Therefore within the physical region of channel I

$$s \geqslant (m_1+m_2)^2, \quad s \geqslant (m_3+m_4)^2, \quad u,\ t < 0.$$

Henceforth we shall assume the masses of all four particles to be equal to each other. Then

$$s \geqslant 4m^2, \quad u \leqslant 0, \quad t \leqslant 0. \tag{10-I}$$

Similarly, the physical regions of channels II and III are

$$t \geqslant 4m^2, \quad s \leqslant 0, \quad u \leqslant 0, \tag{10-II}$$

$$u \geqslant 4m^2, \quad s \leqslant 0, \quad t \leqslant 0. \tag{10-III}$$

Then, in accordance with (7), on the mass shell the variables s, t, u satisfy the sum rule

$$s+u+t=4m^2. \tag{11}$$

This picture may be represented on a plane where the axes of the variables s, t, u differ in their directions by 120° (Figure AVII.3).

In the physical regions (10), the scattering amplitude, proportional to the matrix element of the scattering matrix, is a complex quantity. Therefore the subtraction point of the four-vertex cannot be chosen to be within these

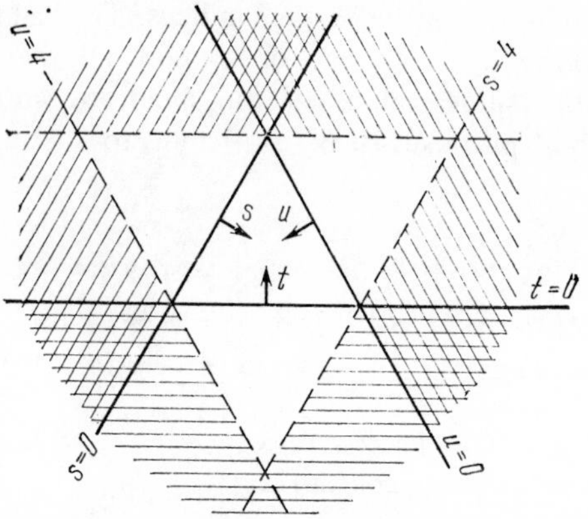

Fig. AVII.3. The plane of the kinematic variables s, u, t of a four-legged vertex. The double cross-hatching indicates the physical regions. The central triangle, defined by the conditions $s, t, u \leq 4m^2$, corresponds to the region in which the four-vertex is real.

regions. An exception to this is given by points corresponding to reaction thresholds:

$$s = 4m^2, \quad t = u = 0, \tag{12-I}$$

$$t = 4m^2, \quad s = u = 0, \tag{12-II}$$

$$u = 4m^2, \quad s = t = 0. \tag{12-III}$$

The vertex turns out to be real inside a triangle bounded by the straight lines $s = 4m^2$, $u = 4m^2$, $t = 4m^2$ corresponding to the two-particle reaction thresholds I, II, and III. Already in the second perturbation order in the φ^4 model, contributions to the four-vertex come from the "fish" diagrams, which, as was shown by explicit calculation in Section 24.1, have an imaginary part differing from zero above the two-particle threshold.

APPENDIX VIII

FEYNMAN RULES FOR YANG–MILLS FIELDS

In this appendix the rules are given for constructing the scattering matrix in perturbation theory for quantum-field models containing non-Abelian gauge fields. Since the quantization procedure is complicated, we cannot discuss how these rules are derived. We may merely point out that the Yang–Mills fields were quantized only during the second half of the sixties and that to this end such complicated methods were utilized as the representation of the scattering matrix by a path integral (see, for example, the monograph by Faddeev and Slavnov (1978)). In addition, models in quantum field theory based on Yang–Mills fields are now becoming more and more important both for electromagnetic and weak interactions (unified models like the Weinberg–Salam model; see Section 32) and for strong interactions (parton–gluon models).

1. The free field of the group *SU*(2). We start with the simplest case of the free Yang-Mills field B^a corresponding to the gauge group $SU(2)$. We write the classical Lagrangian in the form

$$\mathscr{L}(B) = -\frac{1}{4} F_{\mu\nu}F^{\mu\nu} - \frac{1}{2\alpha}(\partial B)(\partial B), \tag{1}$$

which differs from the expression for $\mathscr{L}_{YM}$ used in Section 11.1 by the inclusion of the second term fixing the gauge of the field B_ν. The introduction of such a term into the Lagrangian of an Abelian gauge field was discussed in Section 4.3.

It is convenient to break up the Lagrangian $\mathscr{L}$ into two terms:

$$\mathscr{L} = \mathscr{L}_0(B) + \mathscr{L}_1(B, g), \tag{2}$$

to the first of which we assign the terms quadratic in the components of B and their derivatives,

$$\mathscr{L}_0(B) = -\frac{1}{4} H_{\mu\nu}H^{\mu\nu} - \frac{1}{2\alpha}(\partial B)^2, \tag{3}$$

where

$$H_{\mu\nu} = \partial_\mu B_\nu - \partial_\nu B_\mu,$$

while the second contains the higher-order terms

$$\mathscr{L}_1(B,\ g) = -\frac{g}{2} H_{\mu\nu}[B_\mu \times B_\nu] - \frac{g^2}{4}([B_\mu \times B_\nu])^2. \tag{4}$$

We recall (compare Section 4.3) that due to the existence of the second term the differential operator of the Lagrangian

$$\mathscr{L}_0(B) = -\frac{1}{2} B_\nu^a K^{\nu\mu} B_\mu^a, \tag{5}$$

$$K^{\nu\mu} = g^{\nu\mu}\partial^2 + \frac{1-\alpha}{\alpha}\partial^\nu\partial^\mu \qquad (\partial^2 = -\Box)$$

turns out to be nonsingular, i.e., its inverse

$$K_{\nu\mu}^{-1} = \frac{1}{\Box^2}\{g_{\nu\mu}\partial^2 - (1-\alpha)\,\partial_\mu\partial_\nu\}. \tag{6}$$

exists.

The quadratic Lagrangian (3), (5) is often referred to as the "free" Lagrangian. In this case, field described by a Lagrangian of the form of (11.10) and (1) are called "Yang–Mills fields in vacuum". To avoid confusion (and in accordance with the terminology of Section 19), we shall call the field described by linear equations following from the quadratic Lagrangians (3), (5) the linear-approximation field or the "linear Yang–Mills field". The Yang–Mills field described by the Lagrangians (11.10) and (1) and satisfying nonlinear equations of the form of (11.11) and not containing interaction terms with other fields (i.e., with fields of matter) will here be referred to as the "free Yang–Mills field," in correspondence with the physics under consideration.

If we now consider the constant g to be a small quantity ($g \ll 1$), then it is possible to formulate the problem of constructing a perturbation theory in orders of g with the linear approximation as a starting point. Such a perturbation theory was schematically considered in Section 19.4. In this case the diagramatic technique involves a massless propagator in an arbitrary gauge

$$\langle TB_\nu^a(x)\,B_\mu^b(y)\rangle = i\delta^{ab}\Delta_{\nu\mu}^c(x-y),$$

where

$$\Delta_{\nu\mu}^c(x;\ \alpha) = \frac{1}{(2\pi)^4}\int \frac{e^{-ikx}\,dk}{-k^2-i\varepsilon}\left\{g_{\nu\mu} - (1-\alpha)\frac{k_\nu k_\mu}{k^2+i\varepsilon}\right\}, \tag{7}$$

as well as vertices corresponding to third-order and fourth-order components from (4). As was explained in Section 19, this technique leads to difficulties concerning unitarity, which may be overcome by introducing a fictitious massless ghost field (Faddeev–Popov) $\xi^a(x)$ which transforms according to the adjoint representation of the gauge group (see (19.22)):

$$\mathscr{L} \to \mathscr{L}_{\text{eff}} = \mathscr{L} + \mathscr{L}_0(\xi) + \mathscr{L}_1(\xi, B), \tag{8}$$

where

$$\mathscr{L}_0 \xi = -\partial^\mu \bar{\xi}(x)\, \partial_\mu \xi(x), \tag{9}$$

$$\mathscr{L}_1(\xi, B) = g\bar{\xi}(x)\,[\mathbf{B}_\nu(x) \times \partial^\nu \xi(x)]. \tag{10}$$

The ghost field ξ must then formally be assumed to be quantized according to the Fermi—Dirac scheme, with the result that to each closed ghost loop in the Feynman diagram there corresponds in the matrix element a factor -1.

Thus, the Feynman diagrams for the free Yang–Mills field contain external and internal lines of the Yang–Mills field as well as internal lines of the ghost field. The corresponding Feynman rules in the configuration representation were formulated in Section 19.4.

In the momentum representation, the Feynman rules of the free Yang–Mills field involve the following elements:

(a) the propagator of massless vector particles,

$$\Delta^{ab}_{\mu\nu}(k) = \frac{1}{i}\frac{\delta^{ab}}{k^2 + i\varepsilon}\left[g_{\mu\nu} - (1-\alpha)\frac{k_\mu k_\nu}{k^2 + i\varepsilon}\right] \tag{11}$$

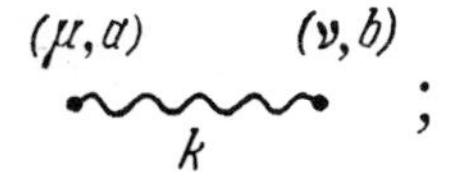

(b) the vertex of the third-order self-interaction of vector particles,

$$V_3 = igt^{abc}\,[(p-k)_\rho\, g_{\mu\nu} + (k-q)_\mu\, g_{\nu\rho} + (q-p)_\nu\, g_{\mu\rho}]\,\delta(p+k+q) \tag{12}$$

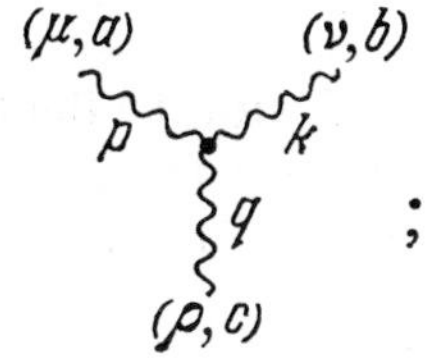

(in this diagram, as well as in all further vertex diagrams, for definiteness all four-momenta are considered to be incoming);

(c) the fourth-order interaction vertex

$$V_4 = g^2 \left\{ t^{abe} t^{cde} (g_{\mu\rho} g_{\nu\sigma} - g_{\mu\sigma} g_{\nu\rho}) + \begin{pmatrix} b \leftrightarrow c \\ \nu \leftrightarrow \rho \end{pmatrix} + \begin{pmatrix} b \leftrightarrow d \\ \nu \leftrightarrow \sigma \end{pmatrix} \right\} \delta(\Sigma k) \tag{13}$$

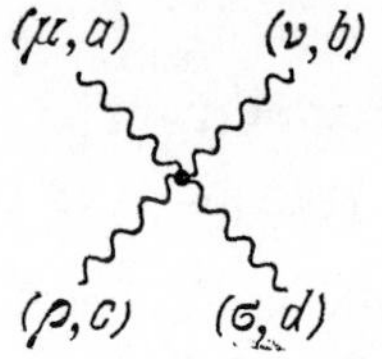

;

(d) the ghost propagator

$$\Delta_{\xi}^{ab}(p) = i \frac{\delta^{ab}}{p^2 + i\varepsilon} \tag{14}$$

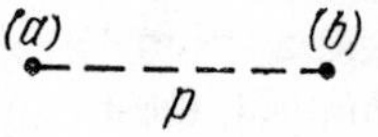

;

(e) the interaction vertex of ghosts with the Yang–Mills field,

$$V_{\xi\xi B} = -\frac{ig}{2} t^{abc} (k - q)_{\mu} \delta(k + q + p) \tag{15}$$

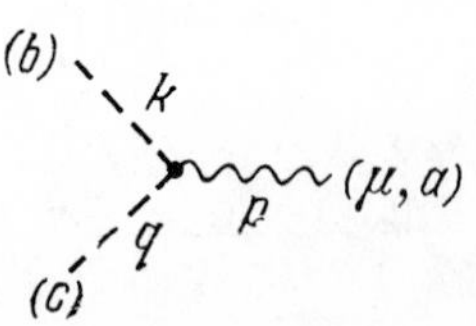

;

(f) a sign factor

$$\eta = (-1)^g, \tag{16}$$

where g is the number of closed loops of the fictitious field ξ.

2. Interaction of the gauge field with the fields of matter. As was pointed out in Section 11.2, the interaction of gauge fields with matter fields is introduced in a unified manner by means of the covariantization of the derivatives. Therefore, to each spinor matter field ψ^a of mass m there corresponds according to the Feynman rules:

(a) a spinor propagator

$$S^{ab}(p) = \frac{\delta^{ab}}{i} \frac{m + \hat{p}}{m^2 - p^2 - i\varepsilon} \tag{17}$$

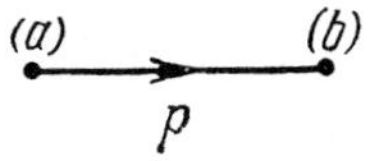

;

(b) the interaction vertex with the gauge field,

$$V_{\psi\psi B} = -g\gamma_\nu (T^c)_{ab}\, \delta(p_1 + p_2 + q) \tag{18}$$

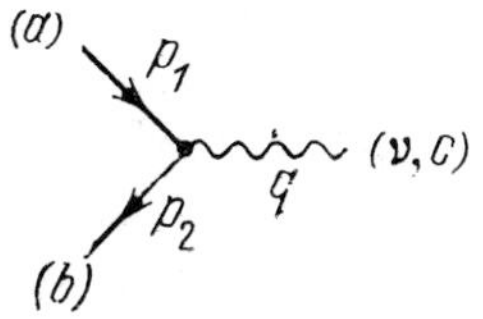

;

(c) a sign factor

$$\eta_f = (-1)^f,$$

where f is the number of closed fermion loops.

To each (pseudo)scalar matter field u^a of mass μ there corresponds:

(a) a scalar propagator

$$D^{ab}(k) = \frac{-i\delta^{ab}}{\mu^2 - k^2 - i\varepsilon} \tag{19}$$

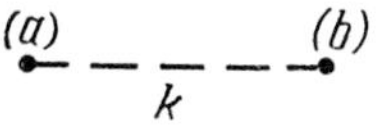

;

(b) a third-order interaction vertex with the gauge field,

$$V_{uuB} = -g(T^c)_{ab}(k - p)_\mu\, \delta(k + p + q) \tag{20}$$

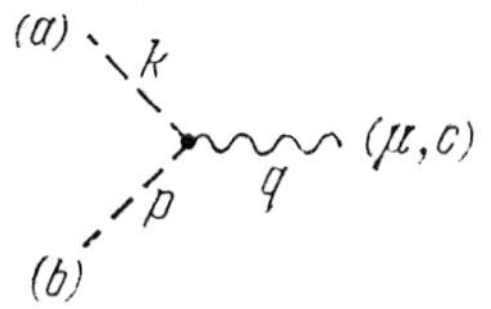

;

(c) a fourth-order interaction vertex with the gauge field,

$$V_{uuBB} = g^2 g_{\mu\nu} (\delta^{ab}\delta^{cd} + \delta^{ac}\delta^{bd} + \delta^{ad}\delta^{bc})\, \delta(k_1 + k_2 + q_1 + q_2) \tag{21}$$

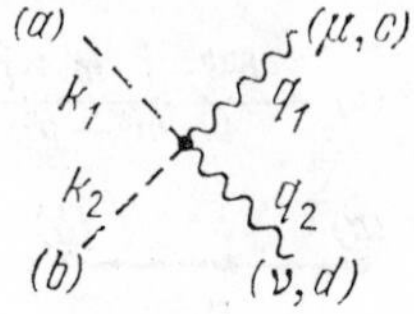

,

(d) other possible vertices of the self-interaction of the scalar field (of the form u^4) or of its interaction with other fields of matter.

3. The massive Yang–Mills field. Let us now formulate Feynman's rules for a gauge field with mass, arising from the Higgs mechanism as a result of spontaneous symmetry breaking by an auxiliary sclar field.

In order to retain conformity with the Weinberg–Salam model, we shall consider the case of a Yang–Mills field of the group $SU(2)$ interacting with the complex scalar field φ^a that transforms in accordance with the spinor representation ($a = 1, 2$) of the gauge group. The field φ thus contains only four real scalar fields. This case is the simplest extension of the model of spontaneous gauge symmetry breaking, described in Section 11.4.

As a result of a shift of the second component of the doublet φ by a real constant η and the transition to four new real scalar fields σ, $C = (C^1, C^2, C^3)$,

$$\varphi^1 = \frac{iC^1 + C^2}{\sqrt{2}}, \quad \varphi^2 = \eta + \frac{\sigma - iC^3}{\sqrt{2}}, \tag{22}$$

the initial Lagrangian of the form (11.20) takes on the following form:

$$\begin{aligned} \mathcal{L}(x) = &-\frac{1}{4} F_{\mu\nu}F^{\mu\nu} - \frac{1}{2\alpha}(\partial B)^2 + \frac{M^2}{2}(B_\mu B^\mu) + \\ &+ \frac{1}{2}[\partial_\mu\sigma\, \partial^\mu\sigma - m_\sigma^2\sigma^2] + \frac{1}{2}(\partial_\mu C)^2 + M B_\mu\, \partial^\mu C - \\ &- \frac{g}{2} B_\mu \{\sigma\partial_\mu C - C\partial_\mu\sigma - [C \times \partial_\mu C]\} - \frac{Mg}{2}\sigma(B_\mu B^\mu) + \\ &+ \frac{g^2}{8}(\sigma^2 + C^2)(B_\mu)^2 + \frac{g m_\sigma^2}{4M}\sigma(\sigma^2 + C^2) - \frac{\lambda^2}{16}(\sigma^2 + C^2)^2. \end{aligned} \tag{23}$$

Here the following notation has been introduced:

$$(B_\mu)^2 = B_\mu^a B^{\mu, a}, \quad (\partial_\mu C)^2 = \partial_\mu C^a \partial^\mu C^a, \quad C^2 = C^a C^a,$$

and also

$$M = \frac{\eta g}{\sqrt{2}}, \quad m_\sigma = \eta\lambda. \tag{24}$$

The Lagrangian (23) contains three massive vector fields B^a (mass M), one scalar Higgs field σ of mass m_σ, and three massless Goldstone fields C^a.

Because of the existence of the nondiagonal term

$$MB_\mu \partial^\mu C \tag{25}$$

the Goldstone field $C(x)$ and the longitudinal part of the vector field are not independent dynamic variables. Therefore by an appropriate choice of the gauge one may make either C or ∂B vanish. The gauge $C = 0$ is called unitary. In this gauge the Lagrangian contains only physical fields, a triplet of massive vector mesons, and a single massive Higgs meson.

For calculations the so-called transverse gauge $\partial B = 0$ turns out to be more convenient, in which the vector propagator takes on the following form:

$$\Delta^{ab}_{\mu\nu}(k) = i\delta^{ab} \frac{g_{\mu\nu} - \frac{k_\mu k_\nu}{k^2}}{M^2 - k^2 - i\varepsilon} \tag{26}$$

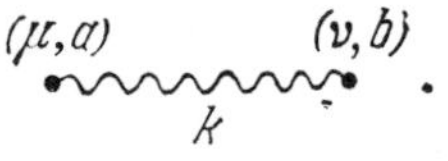

In this case, because of the condition of transversality of the Yang–Mills field, the term $B_\mu \partial^\mu C \sim - C\partial_\mu B^\mu$ turns out to be ineffective, as the time-ordered pairings of the divergence $\partial^\mu B_\mu$ reduce to zero (sometimes it is said that "the field ∂B does not propagate").

Besides the transverse propagator of the vector field (26), Feynman's rules contain the propagation functions of the Faddeev–Popov ghosts and the vertices of their interaction (15) with the Yang–Mills field, the propagation functions of the Goldstone particles

$$\overline{C^a(k)\, C^b}(k) \sim i \frac{\delta^{ab}}{k^2 + i\varepsilon} \tag{27}$$

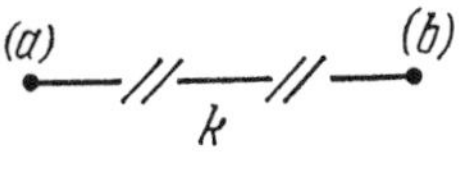

and of the Higgs boson

$$\overline{\sigma(k)\, \sigma}(k) \sim - \frac{i}{m_\sigma^2 - k^2 - i\varepsilon} \tag{28}$$

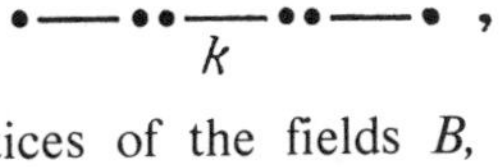

numerous interaction vertices of the fields B, C, σ, which are readily reconstructed by means of the third and fourth rows of the Lagrangian (23), and also, perhaps, propagators and vertices, such as (17)–(21), of various fields of matter.

We recall that the fields corresponding to the Faddeev–Popov ghosts as well as the fields $C(x)$ that describe the Goldstone bosons are encountered only on the internal lines of the Feynman diagrams.

APPENDIX IX

THE RENORMALIZATION GROUP

1. Introduction. The renormalization group method in quantum field theory is the main method for obtaining information on the asymptotic ultraviolet (and infrared, in certain cases) behavior of the Green's functions and matrix elements. It acquired special significance during the seventies in connection with applications to quantum chromodynamics.

The renormalization group (RG) in the general case is the group of finite transformations of the Green's functions and of parameters (coupling constants, masses, gauge parameters) having an effect equivalent to the introduction of finite counterterms that do not violate the renormalizability of the given quantum field model. The finite multiplicative Dyson transformations, mentioned in Sections 28.1 and 29.5, represent examples of such transformations. On the basis of these transformations it turns out to be possible to obtain functional equations for the Green's functions and vertices, as well as group differential equations equivalent to them. The latter can be utilized for an effective improvement of the approximation methods of conventional perturbation theory in the ultraviolet and infrared regions.

In the present appendix we shall give the minimum basic information, including the background and equations of the renormalization group method required for an understanding of the physical consequences mentioned in Section 33.

Let us point out, also, that the RG in quantum field theory was discovered by Stueckelberg and Peterman in 1953. The functional equations were formulated by Gell-Mann and Low in 1954, while the differential group equations were first obtained (and utilized for the summation of logarithms in the ultraviolet and infrared regions) by Bogoliubov and Shirkov in 1955. The reader may find a more detailed exposition of this subject in Chapter IX of the *Introduction*.

2. Functional equations of the renormalization group. Functional equations of the renormalization group are usually written for the Green's functions and vertices considered in the momentum representation. This is connected with the fact that the already mentioned finite arbitrariness which arises in the process of renormalization and which is associated with finite counterterms, is parametrized naturally with the aid of the so-called normalization momentum (or subtraction point), called μ in the momentum representation (see Section 28).

As a typical object we shall consider a scalar function s depending on the single relativistically invariant momentum argument p^2. Such an object could be the transverse part of the photon propagator $d(p^2, \alpha)$ from (29.1); the mass operator of a scalar particle, $M(p^2, \alpha)$, from (27.3); a vertex function of the type (28.18) when all independent arguments are equal to each other; the scalar components $S_i(p^2)$ of the fermion mass operator; and certain other quantities as well.

It will be convenient to deal with a dimensionless quantity normalized to unity in the lowest order of perturbation theory. For definiteness we shall take the transverse part of the photon propagator in spinor electrodynamics, Besides the expansion parameter α and the momentum argument k^2, it depends also on the electron mass m and the square of the normalization momentum (or the dimensional parameter of the method of dimensional regularization) μ:

$$d\left(\frac{k^2}{\mu^2}, \frac{m^2}{\mu^2}, \alpha\right). \tag{1}$$

Then, by definition, $d(\alpha = 0) = 1$. In the case when the ultraviolet divergences are removed with the aid of the R-procedure by means of subtraction at the point $k^2 = \mu^2 \leq 0$—for example, in the one-loop approximation in the notation of Section 29.3

$$d\left(\frac{k^2}{\mu^2}, \frac{m^2}{\mu^2}, \alpha\right) = 1 + \frac{\alpha}{\pi}\left[I\left(\frac{k^2}{4m^2}\right) - I\left(\frac{\mu^2}{4m^2}\right)\right] + O(\alpha^2)$$

—we have also

$$d\left(1, \frac{m^2}{\mu^2}, \alpha\right) = 1. \tag{2}$$

Here we have deliberately diverted our attention from the fact that usually subtraction of the photon propagator is carried out at the point $k^2 = 0$ that physically corresponds, according to the discussion Section 29.3, to the condition that radiative corrections are absent in processes involving photons of long wavelengths and that, under this condition of renormalization, the expansion parameter α is equal to the fine-structure constant, to which we have, in the present context, assigned the lower index 0:

$$\alpha_0^{-1} = 137.036$$

Therefore the function (1) corresponds to the normalization condition at an arbitrary point $\mu^2 < 0$. We shall denote its particular value for $\mu^2 = 0$ by $d_0(k^2/m^2, \alpha_0)$. The finite Dyson transformation from d_0 to an arbitrary d may now be written in the form

$$d_0\left(\frac{k^2}{m^2}, \alpha_0\right) = Z_3 d\left(\frac{k^2}{\mu^2}, \frac{m^2}{\mu^2}, \alpha\right), \qquad \alpha = Z_3\alpha_0 \tag{3}$$

By combining two different transformations of this type we obtain

$$d\left(\frac{k^2}{\mu^2}, \frac{m^2}{\mu^2}, \alpha\right) = \widetilde{Z}_3 d\left(\frac{k^2}{\widetilde{\mu}^2}, \frac{m^2}{\widetilde{\mu}^2}, \widetilde{\alpha}\right), \qquad \widetilde{\alpha} = \widetilde{Z}_3\alpha \tag{4}$$

The formulas (3), (4) explicitly express the fact that a change of the normalization point ($\mu \to \widetilde{\mu}$) of the photon propagator may be "compensated for" by an appropriate change of the coupling constant ($\alpha \to \widetilde{\alpha}$) together with a simultaneous change of the normalization of the propagator ($d \to \widetilde{Z}_3 d$). The transformation (4) obviously exhibits the group property.

Here it must be pointed out that in the general case the renormalization factors of an arbitrary Green's function and the coupling constants are not equal to each other. In the example considered the equality of the multiplicative factors Z_3, $\widetilde{Z}_3$ occurring in both parts of the formulas (3) and (4) represents a special consequence of gauge invariance. In the more general case, for instance, for the scalar component S_p of the electron mass operator

$$G^{-1}(p) = m - \hat{p} - \Sigma(p) = mS_m(p^2) - \hat{p}S_p(p^2)$$

considered in the transverse gauge, the corresponding transformation will be

$$S_p\left(\frac{p^2}{\mu^2}, \frac{m^2}{\mu^2}, \alpha\right) = \widetilde{Z}_2 S_p\left(\frac{p^2}{\widetilde{\mu}^2}, \frac{m^2}{\widetilde{\mu}^2}, \widetilde{Z}_3\alpha\right), \tag{5}$$

where $\widetilde{Z}_3$ is the same as in (4). The situation (5) is more typical of an individual Green's function than (4).

At the same time, (4) represents an example of a transformation of multiplicative combinations of Green's functions that transform similarly to the coupling constant. In this case it turns out to be convenient to introduce the product

$$\bar{\alpha}(x, y, \alpha) \equiv \alpha d(x, y, \alpha), \qquad x = \frac{p^2}{\mu^2}, \quad y = \frac{m^2}{\mu^2}, \tag{6}$$

which is called the invariant charge (or the running coupling), and for which the functional equation takes on the more elegant form

$$\bar{\alpha}(x, y, \alpha) = \bar{\alpha}\left(\frac{x}{t}, \frac{y}{t}, \bar{\alpha}(t, y, \alpha)\right) \tag{7}$$

(In the transition from (4) to (7) the normaliztion condition (2) was utilized, in addition to (6)). An analogous equation of the type of (5) may be written in the form

$$S(x, y, \alpha) = S(t, y, \alpha) S\left(\frac{x}{t}, \frac{y}{t}, \bar{\alpha}(t, y, \alpha)\right). \tag{8}$$

The obtained functional equations (7), (8) are typical for the class of quantum field models involving a single coupling constant. In particular, one encounters them in quantum chromodynamics. Such equations signify the fact that equations in quantum field theory and their solutions possess the feature of being, in a way, model-independent and do not involve the dynamics of any concrete quantum field model in question. The general solutions of equations (7), (8) derived by Ovsyannikov in 1956 contain arbitrary functions of two variables. Thus, taking into account the renormalization-group properties decreases by one the number of independent arguments.

3. Differential equations. To obtain more concrete information concerning the case under consideration one must introduce dynamical information from perturbation theory. To this end the most conveniently used are the group differential equations, since they correspond to an infinitesimal group transformation in (7), (8).

Differentiating (7) and (8) with respect to x, while t and y remain constant, and assuming, upon differentiation, $t = x$, we obtain the Lie differential equations

$$x \frac{\partial \bar{\alpha}(x, y, \alpha)}{\partial x} = \beta\left(\frac{y}{x}, \bar{\alpha}(x, y, \alpha)\right), \tag{9}$$

$$\beta(y, \alpha) = \left.\frac{\partial \bar{\alpha}(\xi, y, \alpha)}{\partial \xi}\right|_{\xi=1}, \tag{10}$$

and

$$x \frac{\partial \ln S(x, y, \alpha)}{\partial x} = \gamma\left(\frac{y}{x}, \bar{\alpha}(x, y, \alpha)\right), \tag{11}$$

$$\gamma(y, \alpha) = \frac{\partial \ln S(\xi, y, \alpha)}{\partial \xi} \bigg|_{\xi=1} \tag{12}$$

On the other hand, if one differentiates (7) and (8) with respect to t, while keeping x and y constant, and then puts $t = 1$, one obtains the following partial differential equations:

$$\left[x \frac{\partial}{\partial x} + y \frac{\partial}{\partial y} - \beta(y, \alpha) \frac{\partial}{\partial \alpha}\right] \bar{\alpha}(x, y, \alpha) = 0, \tag{13}$$

$$\left[x \frac{\partial}{\partial x} + y \frac{\partial}{\partial y} - \beta(y, \alpha) \frac{\partial}{\partial \alpha}\right] \ln S(x, y, \alpha) = \gamma(y, \alpha). \tag{14}$$

The latter equations were first introduced by Ovsyannikov in the course of his derivation of the general solution of the functional equations. Naturally these linear partial differential equations are completely equivalent to the nonlinear equations (9), (11) that are somewhat more convenient for finding solutions corresponding to the usual perturbation theory.

In the current literature; equations (13), (14) are often associated with the names of Callan and Symazik, who discovered an analogous equation by a method different from the renormalization group method one and a half decades later (see p. 509 of the *Introduction*).

We shall obtain some solutions of equations (9), (11) for the ultraviolet asymptotic case—in particular, the formula for the invariant coupling discussed in Section 33.3. To pass to the ultraviolet case we assume $|p^2| \gtrsim |\mu^2| \gg m^2$ and drop the second argument $y = m^2/\mu^2$ occurring in the functional integral equations. Instead of (9) we then obtain

$$\frac{\partial \bar{\alpha}(x, \alpha)}{\partial \ln x} = \beta(\bar{\alpha}) \tag{15}$$

This equation describes the "evolution" of the invariant coupling $\bar{\alpha}$ with increase of the squared momentum.

As was mentioned in Section 33.3, the quantity $\bar{\alpha}$ describes the change of the interaction intensity as the square of the momentum variable increases. As may be demonstrated (see, for example, (41,20) of the *Introduction*), its Fourier transform

$$\rho(r, \alpha) = \frac{2}{\pi} \int_0^\infty \frac{dz}{z} \sin z\, d\left(-\frac{z^2}{r^2}, \alpha\right) \tag{16}$$

represents the effective electron charge, which has changed at short distances r, owing to vacuum polarization effects, and which is connected with the effective potential $v(r)$ of a pointlike electron by the relationship

$$v(r) = -\frac{e\rho(r, \alpha)}{4\pi r}.$$

It is clear, therefore, that the behavior of the function $\overline{\alpha}(x, \alpha) = \alpha d(x, \alpha)$ at large (negative) x is directly connected with the behavior of the effective charge $e\rho(r, \alpha)$ at short distances r.

4. Connection with perturbation theory. Ultraviolet asymptotic behavior. Equation (15) contains an unknown function of one argument, $\beta(\alpha)$. One may obtain its expansion in powers of α by utilizing the usual perturbation theory. As it is not difficult to verify, to this end one must, starting from the definition (10), find the coefficients of the contributions linear in $\ln x$ to the perturbation terms of increasing powers. In other words, if, when $x \to \infty$,

$$d(x, \alpha) = 1 + \alpha\beta_1 \ln x + \alpha^2(\gamma_2 \ln^2 x + \beta_2 \ln x) + \alpha^3(\delta_3 \ln^3 x + \gamma_3 \ln^2 x + \beta_3 \ln x) + \ldots,$$

then

$$\beta(\alpha) = \beta_1 \alpha^2 + \beta_2 \alpha^3 + \beta_3 \alpha^4 + \ldots. \tag{17}$$

The coefficients β_l represent the contributions of l-loop diagrams. In quantum electrodynamics three terms of the expansion of the beta function are known:

$$\beta(\alpha) = \frac{\alpha^2}{3\pi} + \frac{\alpha^3}{4\pi^2} + \frac{\alpha^4}{8\pi^3}\left(\tfrac{8}{3}\zeta(3) - \tfrac{101}{36}\right), \qquad \zeta(3) \simeq 1.2. \tag{18}$$

The most significant fact is that all the terms on the right-hand side of (18) are positive. Therefore $\overline{\alpha}(x,\alpha)$ increases with x. In accordance with (16), the effective charge of the electron increases as the distance decreases, and we thus come to the picture of normal screening qualitatively depicted in Figure 33.1.

To obtain the explicit expression for $\bar{\alpha}$ we restrict ourselves to the first term on the right-hand side of (18) (i.e., to the single-loop contribution). Substituting it into the right-hand side of (15), we obtain, upon integrating (and taking account of the boundary condition $\bar{\alpha}\,(1, \alpha) = \alpha$ following from the normalization condition), the following:

$$\bar{\alpha}_{\mathrm{I}}(x, \alpha) = \frac{\alpha}{1 - (\alpha/3\pi)\ln x}, \tag{19}$$

which coincides with (33.5).

The formula obtained possesses a number of remarkable properties. Firstly, the terms of its expansion in a series in α,

$$\bar{\alpha}_{\mathrm{I}}(x, \alpha) = \alpha + \frac{\alpha^2}{3\pi}\ln x + \frac{\alpha^3}{9\pi^2}\ln^2 x + \ldots,$$

represent an infinite sequence of terms of the form $\alpha(\alpha \ln x)^n$, i.e., of the higher logarithmic contributions from each order of perturbation theory. Thus, the numerical coefficients of the higher logarithmic terms are not independent and can be expressed in terms of the coefficient of the single-loop logarithm.

Secondly, (19), obtained as a result fo the solution of the single-loop (i.e., approximate) Lie differential equation

$$\frac{\partial \bar{\alpha}_{\mathrm{I}}(x, \alpha)}{\partial \ln x} = \frac{1}{3\pi}\bar{\alpha}^2(x, y, \alpha),$$

represents a solution of the exact functional equation (7) in the ultraviolet limit, i.e., it satisfies the relation

$$\bar{\alpha}_{\mathrm{I}}(x, \alpha) = \bar{\alpha}_{\mathrm{I}}\left(\frac{x}{t}, \bar{\alpha}_{\mathrm{I}}(t, \alpha)\right),$$

as may be verified by means of elementary operations. At the same time any finite sum of the terms of the power-series expansion of $\bar{\alpha}_{\mathrm{I}}(x, \alpha)$ will not satisfy this equation. This clearly demonstrates that the synthesis of the equations of the renormalization group together with information from the lower orders of perturbation theory allows one to perform partial summation of the infinite sequence of the "principal" contributions from the higher orders of perturbation theory and thus to significantly improve the approximation.

In the case considered, before the utilization of the renormalization group, the region of applicability of the usual single-loop perturbation theory was determined by the relationship

$$\frac{\alpha}{3\pi}\ln x \ll 1,$$

whereas the region of applicability of (19) was limited by the smallness of the two-loop contribution to the beta function $\beta(\bar{\alpha})$ compared to the single-loop one, which on the basis of (18) gives $\bar{\alpha} \ll 4\pi/3$, i.e.,

$$1-\frac{\alpha}{3\pi}\ln x \gg \frac{3\alpha}{4\pi} \simeq \frac{1}{575}.$$

From the latter estimate, in particular, it follows that one should not attach any special significance to the fact that the right-hand side of (19) formally tends to infinity when $x_* = \exp(3\pi/\alpha)$; for an analysis of the behavior of $\bar{\alpha}(x, \alpha)$ at x-values close to x_* it is necessary to take into account the contributions of the higher-order loops.

5. Some applications to quantum chromodynamics. In quantum chromodynamics, as well, three terms of the expansion of the beta function are known:

$$\beta(\alpha_s) = -\frac{33-2n}{12\pi}\alpha_s^2 - \frac{306-38n}{48\pi^2}\alpha_s^3 - \left(2857 - \tfrac{5033}{9}n + \tfrac{325}{27}n^2\right)\frac{\alpha_s^4}{128\pi^3}. \tag{20}$$

Contrary to QED, here when the number n of flavors (i.e., the number of quarks) is small, all the terms are ngeative. Therefore the effective coupling in quantum chromodynamics, which may be defined by the relation

$$\bar{\alpha}_s(x, \alpha_s) = \alpha_s\Gamma^2(x, \alpha_s)\Delta^3(x, \alpha_s), \qquad \alpha_s = \frac{g_{YM}^2}{4\pi} \tag{21}$$

(where Γ is the three-gluon vertex, and Δ is the gluon propagator), decreases with increasing $|p^2|$; i.e., for small distances we arrive at the qualitative picture in Figure 33.2.

Limiting ourselves to the single-loop approximation for $n = 3$, we obtain, upon solving the differential equation (15),

$$\overline{\alpha}_s(x, \alpha_s) = \frac{\alpha_s}{1 + (9/4\pi)\alpha_s \ln x} \tag{22}$$

—the well-known formula of asymptotic freedom (compare with (33.4)). It is often represented in the form

$$\overline{\alpha}_s(p^2) = \frac{4\pi}{9 \ln(p^2/\Lambda^2)}, \tag{23}$$

where Λ is the so-called scaling parameter, connected with α_s and μ^2 by the relationship

$$\Lambda^2 = \mu^2 \exp\left(-\frac{4\pi}{9\alpha_s}\right). \tag{24}$$

Existing experimental data permit one to determine the numerical values of the parameters in (20) and (21) only approximately. Thus, basing oneself on the data on deep-inelastic scattering of electrons on deuterons as well as on the decay of the J/ψ particle, one may tentatively assume as a rough approximation

$$\overline{\alpha}_s(100\ \mathrm{GeV}^2) \sim 0.2.$$

According to (24) this is equivalent to the value $\Lambda \sim 300$ MeV. The tentative nature of this numerical estimation of Λ must be emphasized, and it must be pointed out, as well, that for $\alpha_s \sim 0.2$ the two-loop contribution to the right-hand side of (20) amounts to 10% of the single-loop contibution, so formulas (22)–(24) need to be made more accurate. We shall not do this here.

We now turn our attention to the group equation of the second type, the equation "typical" of the Green's function (see (8), (11), (12), (14)). The differential equation (11) in the ultraviolet limit takes on the form

$$\frac{\partial \ln S(x, \alpha)}{\partial \ln x} = \gamma_s(\overline{\alpha}(x, \alpha)). \tag{25}$$

The function $\gamma_s(\alpha)$ defined by the relationship

$$\gamma_s(\alpha) = \frac{\partial \ln S(x, \alpha)}{\partial \ln x}\bigg|_{x=1}$$

is called the anomalous dimension of the function $S(x, \alpha)$. Similarly to the beta function, one may define it by means of perturbation theory in the form

$$\gamma_s(\alpha) = \alpha\gamma_1 + \alpha^2\gamma_2 + \dots. \tag{26}$$

Here, as in (17), the notation for the coefficients of the expansion is chosen in such a manner that the γ_l correspond to the contributions of the l-loop diagrams.

By combining (25) and (15) it is possible to show that the solution of (25) may be represented in the form

$$S(x, \alpha) = \exp \int_{\alpha}^{\bar{\alpha}(x, \alpha)} \frac{\gamma_s(a)}{\beta(a)}\, da. \tag{27}$$

Hence we obtain, by expanding the integrand with the aid of (17) and (26) in a series in α and integrating all terms sequentially, the following:

$$S(x, \alpha) = \left[\frac{\bar{\alpha}(x, \alpha)}{\alpha}\right]^{\gamma_1/\beta_1} \exp\left\{\frac{\gamma_2\beta_1 - \gamma_1\beta_2}{\beta_1}[\bar{\alpha}(x, \alpha) - \alpha] + \dots\right\}. \tag{28}$$

The exponential coefficient represents a purely single-loop effect, while the numerical term calculated in the exponent is the two-loop effect. In spinor electrodynamics the electron mass operator has an anomalous dimension independent of the gauge. With the aid of the formulae from Section 29.4 one can obtain in the single-loop approximation the following:

$$\gamma_m(\alpha) = -\frac{3\alpha}{4\pi}.$$

Accordingly

$$\frac{\bar{m}(x, \alpha)}{m} = \left[\frac{\bar{\alpha}(x, \alpha)}{\alpha}\right]^{-9/4}. \tag{29}$$

The mass operator introduced here enters into the asymptotic form of the Green's function of the electron in the following manner:

$$G^{-1}(p, \alpha) = S_p(p^2, \alpha, d_l)\{m^2(p^2, \alpha) - \hat{p}\}.$$

Similar calculations in QCD give for the mass operator of a fermion (i.e. a quark) the following:

$$\gamma_m(\alpha_s) = -\frac{\alpha_s}{\pi},$$

from which, in turn, follows the formula for the effective invariant mass of a quark:

$$m(x, \alpha_s) = m\left[\frac{\bar{\alpha}_s(x, \alpha_s)}{\alpha_s}\right]^{12/(33-2n)}, \tag{30}$$

widely used of late in the literature on the hierarchies of interactions.

It is not difficult to verify that (27), (28), (29), (30), which were obtained as a result of solving approximate differential equations, satisfy the exact functional equation

$$S(x, \alpha) = s(t, \alpha)s\left(\frac{x}{t}, \bar{\alpha}(t, \alpha)\right).$$

"SEPTEMBER" ASSIGNMENT (FOR CHAPTER I)

S1. Proceeding from the definition (2.7) and making use of the equation of motion (2.2), show by explicit calculation that the energy-momentum tensor $T_{\mu\nu}$ satisfies the equation of continuity (2.8).

S2. Derive the formula (2.22) for the angular-momentum tensor of a vector field. To this end, start from (1.6) and obtain an explicit expression for the matrix A in (2.20); then define the matrix Ψ of an infinitesimal increment of the components of the field functions and make use of (2.13).

S3. Using the definition (2.22), show by explicit calculation that the angular-momentum tensor satisfies the continuity equation.

S4. Show that the symmetry of the energy-momentum tensor $T_{\mu\nu}$ leads to the conservation of spin.

S5. Proceeding from the Lagrangian (3.1), obtain with the aid of (2.2) and (2.7) the equation of motion (3.2) and the energy-momentum tensor (3.3) for a real scalar field.

S6. Proceeding from the Lagrangian (3.16), obtain with the aid of (2.2), (2.7), (2.22), and (2.28) the equation of motion and explicit expressions for the energy-momentum and spin-angular-momentum tensors as well as for the current vector of a complex vector field.

S7. Consider the field of pions which transforms as the triplet representation of the isotopic group $SU(2)$. In the real representation $\pi = \{\pi_1, \pi_2, \pi_3\}$, $\overset{*}{\pi}_a = \pi_a$, the isotopic transformations are written down as follows:

$$\pi_a \rightarrow \pi'_a = \Lambda_{ab}(\alpha)\,\pi_b, \quad \Lambda(\alpha) = \exp(-i\omega_l \alpha_l).$$

Here $\alpha_l (l = 1, 2, 3)$ are numerical parameters (angles) of isotopic rotations, and ω_l are generators of rotations which in the real triplet representation may be chosen in the form

$$\omega_1 = \begin{pmatrix} 0 & 0 & 0 \\ 0 & 0 & -i \\ 0 & i & 0 \end{pmatrix}, \quad \omega_2 = \begin{pmatrix} 0 & 0 & i \\ 0 & 0 & 0 \\ -i & 0 & 0 \end{pmatrix}, \quad \omega_3 = \begin{pmatrix} 0 & -i & 0 \\ i & 0 & 0 \\ 0 & 0 & 0 \end{pmatrix},$$

i.e., $(\omega_l)_{jk} = -i\varepsilon_{ljk}$, where ε_{ljk} is a completely antisymmetric tensor of rank three.

Find the matrix O of the unitary transformation $T_l = O\omega_l O^{-1}$ leading to the representation in which

$$T_3 = \begin{pmatrix} 1 & 0 & 0 \\ 0 & 0 & 0 \\ 0 & 0 & -1 \end{pmatrix}.$$

S8. Proceeding from the Lagrangian

$$\mathcal{L}_0(\pi) = \frac{1}{2}\pi_{;\nu}\pi^{;\nu} - \frac{m^2}{2}\pi\pi,$$

construct with the aid of Noether's theorem the corresponding "isotopic" currents and "isotopic charges" which are conserved in time. Check that the Lagrangian is invariant under the transformations considered in Problem S7.

S9. Repeat the preceding problem for the isotopic doublet of nucleons

$$\Psi = \begin{pmatrix} \psi_p \\ \psi_n \end{pmatrix}, \quad \Psi'_\alpha = \Lambda_{\alpha\beta}(\alpha)\,\Psi_\beta, \quad \Lambda(\alpha) = \exp\left(-\frac{i}{2}\tau_l \alpha_l\right),$$

where τ are the Pauli matrices,

$$\mathscr{L}_0=\overline{\Psi}\,(i\hat{\partial}-M)\,\Psi.$$

S10. Demonstrate that the inverses of the formulae (3.8) have the form

$$\tilde{\varphi}^{\pm}(k)=\sqrt{2k_0}\,\varphi^{\pm}(\boldsymbol{k})=(2\pi)^{-3/2}\int e^{\mp ikx}\,[k_0\varphi(x)\mp i\dot{\varphi}(x)]\,d\boldsymbol{x},\quad k_0=\sqrt{\boldsymbol{k}^2+m^2}.$$

S11. Passing to the momentum representation in accordance with (3.8), show by explicit calculation that "contributions with the same frequencies" to the energy and momentum turn out to be zero, i.e.,

$$\int d\boldsymbol{x}\left[\varphi^{\pm}_{;\nu}(x)\,\varphi^{\pm}_{;\nu}(x)+m^2\varphi^{\pm}(x)\,\varphi^{\pm}(x)\right]=\int d\boldsymbol{x}\dot{\varphi}^{\pm}(x)\,\varphi^{\pm}_{;k}(x)=0.$$

S12. By substituting the expansions in momentum (3.18) into the zero components of the formulas (3.17) and integrating over the configuration space, obtain the expressions (3.19)–(3.21) for the vectors of energy-momentum, momentum, and spin and for the electric charge of a complex vector field.

S13.* Find the solution of the Klein–Gordon–Fock equation

$$(\Box-m^2)\,\varphi(x)=0,$$

assuming φ and $\dot{\varphi}=\partial\varphi/\partial t$ to be given on the plane $t=y^0$. Express in terms of $\varphi(t,\,y)$ and $\dot{\varphi}(t,\,y)$ and the frequency parts of the Pauli–Jordan functions

$$D^{\pm}(x)=\frac{1}{(2\pi)^3\,i}\int e^{ikx}\theta(\pm k^0)\,\delta(k^2-m^2)\,dk=\frac{\mp i}{(2\pi)^3}\int\frac{d\boldsymbol{k}}{2k^0}\,e^{\pm ikx}\Big|_{k^0=+\sqrt{\boldsymbol{k}^2+m^2}}$$

the positive- and negative-frequency parts of the field function $\varphi(x)$.

S14. Verify that in the case of a vector field the Lagrangian (3.16) automatically leads to the Klein–Gordon equations and to the subsidiary Lorentz conditions.

S15. Proceeding from the Lagrangian (3.11) for a vector field, obtain expressions for $T_{\mu\nu}$, P_ν, and discuss how they differ from (3.17a) and (3.19).

S16. Utilizing the passage to the local reference frame (3.24), carry out the diagonalization of the quadratic form (3.23). Obtain (3.25).

---

*Problem numbers given in italics denote problems of greater difficulty.

S17. By introducing, with the aid of (3.27), the amplitudes b_n corresponding to circular polarizations, diagonalize the expression (3.26) for the component of the spin of a vector field along the direction of motion.

S18. Proceeding from the Lagrangian (4.13) for the electromagnetic field, obtain the spatial densities of the energy-momentum vector $T^{0\nu}$ and of the spin vector $\boldsymbol{S}$.

S19. By passing to the momentum amplitudes according to the formulae (4.22) and making use of the expansion (4.23) in the local frame, obtain, with the aid of the Lorentz condition, expressions for the energy-momentum tensor and for the spin vector of the electromagnetic field.

S20. Proceeding from the Lagrangian (5.5) for a spinor field, obtain the energy-momentum tensor and the current vector in the form (5.6) and (5.7).

S21. Let

$$\gamma^5 = -i\gamma^0\gamma^1\gamma^2\gamma^3.$$

Show that

$$\gamma^\nu\gamma^5 + \gamma^5\gamma^\nu = 0, \quad \gamma^5\gamma^5 = I \qquad (\nu = 0,\ 1,\ 2,\ 3).$$

S22. Make use of the cyclic property of a trace and of the commutation law for Dirac matrices (5.1) to prove that the traces of all the matrices (AII.2) but the unit matrix are equal to zero.

S23. Taking advantage of the results obtained in the preceding problem prove the linear independence of the 16 matrices (AII.2).

S24. Show that the trace of an odd number of Dirac matrices equals zero, and derive a recurrence formula expressing the trace of the product of $2n$ matrices in terms of the traces of the products of $2n - 2$ matrices.

S25. Evaluate

$$\mathrm{Sp}\,\kappa_1\hat{k}_2, \quad \mathrm{Sp}\,[(\hat{p}_1 + m_1)(\hat{p}_2 + m_2)],$$
$$\mathrm{Sp}\,[\hat{k}_1(\hat{p}_1 + m_1)\hat{k}_2(\hat{p}_2 + m_2)], \quad \mathrm{Sp}\,[k_1k_2(\hat{p}_1 + m_1)(\hat{p}_2 + m_2)].$$

S26. Perform summation over σ in the products

$$\gamma^\sigma\gamma^\nu\gamma_\sigma, \quad \gamma^\sigma\gamma^\mu\gamma^\nu\gamma_\sigma, \quad \gamma^\sigma\gamma^\mu\gamma^\nu\gamma^\rho\gamma_\sigma.$$

S27. Show that

$$\operatorname{Sp}\gamma^{\mu}\gamma^{\nu}\gamma^{5}=0, \quad \operatorname{Sp}\gamma^{\mu}\gamma^{\nu}\gamma^{\rho}\gamma^{\sigma}\gamma^{5}=4i\varepsilon^{\mu\nu\rho\sigma},$$

where $\varepsilon^{\mu\nu\rho\sigma}$ is a completely antisymmetric unit tensor of rank 4.

S28. By making use of the results of Problem S25 and considering the traces of six factors of the type

$$\operatorname{Sp}[\hat{k}_1(\hat{p}_1+m_1)\hat{k}_2(\hat{p}_2+m_2)\hat{k}_3(\hat{p}_3+m_3)], \quad \operatorname{Sp}[\hat{k}_1\hat{k}_2(\hat{p}_1+m_1)(\hat{p}_2+m_2)\hat{k}_3\hat{k}_4], \ldots$$

find the general formula which relates

$$\operatorname{Sp}[\hat{k}_1\hat{p}_1\hat{p}_2\hat{k}_2\hat{p}_3\hat{k}_3 \ldots \hat{p}_m \ldots \hat{k}_n] \text{ and } \operatorname{Sp}[\hat{k}_1\hat{P}_1\hat{P}_2\hat{k}_2\hat{P}_3\hat{k}_3 \ldots \hat{P}_m \ldots \hat{k}_n],$$

where $\hat{P}=\hat{p}_i+m_i$.

S29. Proceeding from formulae (AII.16), (AII.18), (AII.20) for the Lorentz transformations for spinors, obtain explicit expressions for the generators $A^{\psi(\mu\nu)}$ and $A^{\bar{\psi}(\mu\nu)}$ of rotations of the spinors ψ and $\bar{\psi}$ in the plane $x^{\mu}x^{\nu}$, introduced in (2.20), in terms of the spin tensor $\sigma^{\mu\nu}$ defined by the formula (AII.2), and obtain with the aid of (2.24) the expression (5.8) for the spin tensor of a spinor field.

S30. With the aid of the formulas (AII.18) for transformations of a spinor under Lorentz rotations, obtain from (5.12) the formulas (5.26).

S31. Find in the representation (AII.5), (AII.7) the normalized solutions $u^{s,\pm}(k)$, $s=1, 2$, of the Dirac equation which are eigenfunctions of the helicity operator

$$\lambda=\frac{\boldsymbol{k}\boldsymbol{\Sigma}}{|\boldsymbol{k}|}, \quad \boldsymbol{\Sigma}=\frac{1}{2}\gamma^{5}\gamma^{0}\boldsymbol{\gamma}$$

and which satisfy the condition of being orthonormal

$$\overset{*}{u}{}^{r,\mp}(\boldsymbol{k})\,u^{s,\pm}(\boldsymbol{k})=\frac{k_0}{m}\delta^{rs}.$$

S32. Construct a projection operator which singles out states of definite helicity.

"OCTOBER" ASSIGNMENT (FOR CHAPTER II)

O1. Obtain a realization of the eigenfunctions ψ_n and of the operators a and

$\overset{+}{a}$ for a harmonic oscillator both in the configuration and in the momentum representation. Verify that the following relations are satisfied:

$$\overset{+}{a}a=\tilde{n}, \quad a\overset{+}{a}=\tilde{n}+1, \quad \tilde{n}\psi_n=n\psi_n, \quad a\psi_0=0,$$
$$[a, \overset{+}{a}]=a\overset{+}{a}-\overset{+}{a}a=1, \quad [a, a]=[\overset{+}{a}, \overset{+}{a}]=0.$$

O2. Proceeding from the commutation relations of the preceeding problem and the normalization condition for the lowest state,

$$\overset{*}{\Psi}_0\Psi_0=1,$$

evaluate the norm of the state $\Psi_n=(\overset{+}{a})^n\Psi_0$.

O3. Evaluate the norm of the single-particle state

$$\Phi_1=\int d\boldsymbol{q}\, w(\boldsymbol{q})\overset{+}{a}(\boldsymbol{q})\,\Phi_0, \quad \overset{*}{\Phi}_0\Phi_0=1,$$

where the operator $\overset{+}{a}(\boldsymbol{q})$ satisfies the commutation relations (6.8) and $w(\boldsymbol{q})$ is a c-number. Consider the limit

$$w(\boldsymbol{q})\to\delta(\boldsymbol{q}-\boldsymbol{P}),$$

where $\boldsymbol{P}$ is a fixed vector.

O4. Proceeding from the commutation relations (7.4) containing the charge operator Q, show that for a complex field the operators $u^{\pm}$ lower the charge, while $\overset{*}{u}{}^{\pm}$ raise it, i.e., for example, if $Q\Phi_q=q\Phi_q$, then $Q\overset{*}{u}{}^{\pm}\Phi_q=(q+1)\overset{*}{u}{}^{\pm}\Phi_q$.

O5. Obtain the analog of equation (7.4) for isotopic transformations. Consider the particular cases: (a) the nucleon doublet ($I=\frac{1}{2}$), (b) the pion triplet ($I=1$).

O6. Prove the relations (7.14). Obtain the analog of these relations for the pion triplet, making use of the solution of Problem O5.

O7. Derive the angular-momentum operator for a scalar field in the configuration and momentum representations. Obtain commutation relations for its components. Show that

$$[M_{\mu\nu}, \varphi(x)]=i(x_\mu\partial_\nu-x_\nu\partial_\mu)\,\varphi(x)$$

O8. Proceeding from the condition of Hermitian conjugation of operators

$$(a^{\pm}(k))^{+}=\overset{*}{a}{}^{\mp}(k)$$

and the formula (7.19), and making use of the property of the metric being positive in the Hilbert space ($\overset{*}{\Phi}|A|^2\Phi \geq 0$), determine the form of the commutation relations for particles with integer and half-integer spins.

O9. Assume that scalar fields $\varphi(x)$ are subject to the Fermi–Dirac quantization, while spinor fields $\psi(x)$ are subject to the Bose–Einstein quantization, and discuss the behavior of the anticommutator $\{\varphi(x), \varphi(y)\}$ for $x = y$ as well as of the commutator $[\psi(x), \bar{\psi}(y)]$ when $(x - y)^2 < 0$.

O10. Proceeding from the commutators (7.19), obtain, with the aid of the transformation of transition to the local frame (3.24), commutation relations for the vector field (8.14).

O11. Assuming $\mu = \nu = 0$ in the commutation relation (8.17), prove that the commutator for timelike pseudophotons contradicts the property of positivity of the metric in Hilbert space. *Hint*: Multiply the commutation relation for $a_0^+(\boldsymbol{q})$ and $a_0^-(\boldsymbol{k})$ by $f(\boldsymbol{q})\overset{*}{f}(\boldsymbol{k})$ and integrate over $\boldsymbol{q}$ and $\boldsymbol{k}$.

O12. Making use of the anticommutation relations (9.4), evaluate the commutators

$$[\bar{\psi}(x)\gamma^\nu\psi(x),\ \bar{\psi}(y)\gamma^\mu\psi(y)],\quad [\bar{\psi}(x)\gamma^5\psi(x),\ \bar{\psi}(y)\gamma^5\psi(y)]$$

and examine their behavior outside of the light cone.

O13. Determine the behavior outside of the light cone (when $x \sim y$) of the commutators $[\mathcal{L}(x), \mathcal{L}(y)]$ when:

(a) $\mathcal{L}(x) = e{:}\bar{\psi}(x)\hat{A}(x)\psi(x){:}$,
(b) $\mathcal{L}(x) = g{:}\bar{\psi}(x)\gamma^5\psi(x)\varphi(x){:}$.

O14. Evaluate the Fourier transforms $\tilde{F}(q) = \int e^{iqx}F(x)\,dx$ of the following products of the functions D and S of the same frequencies:

(a) $F(x) = D^-(x)D^-(x)$,
(b) $F(x) = \mathrm{Sp}[\gamma^5 S^-(x)\gamma^5 S^+(-x)]$,
(c) $F(x) = \mathrm{Sp}[\gamma^\mu S^-(x)\gamma_\mu S^+(-x)]$.

O15. Evaluate the Fourier transform of the vacuum expectation value of the commutator of free spinor currents $\int e^{ip(x-y)}\langle[J^\mu(x), J^\nu(y)]\rangle_0\,dx$ for

(a) $J^\nu(x) = \bar{\psi}(x)\gamma^\nu\psi(x)$,
(b) $J^\nu(x) = \bar{\psi}(x)\gamma^\nu\gamma^5\psi(x)$.

O16. Proceeding from the two-component Weyl equations (9.15) for a massless spinor field, show that the values of the helicity operator

$$\lambda(\boldsymbol{p}) = \frac{(\boldsymbol{\sigma p})}{p^0}$$

averaged over the state $\tilde{\varphi}_{\pm}$ equal ± 1, i.e., that the spin of the neutrino in the state $\tilde{\varphi}_{+}(\boldsymbol{p})$ is directed along its motion, while in the state $\varphi_{-}(\boldsymbol{p})$ it points in the opposite direction.

O17. Proceeding from (9.24), (9.28), and (9.29) and the unitarity condition of the matrix γ^0, derive the formula (9.30).

O18. With the aid of (9.29), obtain the explicit form of the matrix of charge conjugation C in the standard (AII.5) and the helicity (9.21) representations of the Dirac matrics. Comment on the fact that the matrices C in both of these representations have the same form.

"NOVEMBER" ASSIGNMENT (FOR CHAPTER III)

N1. Construct interaction Lagrangians, containing minimal powers of the fields and not more than a single derivative, between:

(a) a scalar field φ and a spinor field ψ;
(b) a pseudoscalar field φ and a spinor field ψ;
(c) a real scalar field φ and an axial-vector field B_ν;
(d) a spinor field ψ and a vector field A_ν.

N2. Construct isotopically invariant interaction Lagrangians, containing minimal powers of the fields, between:

(a) a doublet of spinor fields and a triplet of scalar fields;
(b) a doublet of spinor fields and a triplet of pseudoscalar fields;
(c) a doublet of spinor fields and a triplet of vector fields;
(d) a doublet of spinor fields and a triplet of axial-vector fields;
(e) a doublet of spinor fields and a triplet of scalar fields with vector coupling;
(f) a doublet of spinor fields and a doublet of pseudoscalar fields.

N3. Construct the minimal locally gauge-invariant Lagrangian for interacting (a) spinor and electromagnetic fields, (b) scalar and electromagnetic fields.

N4. Proceeding from the formulae (11.3) and the definition of a covariant derivative (11.4), show that it transforms like a field function, i.e., prove (11.6).

N5. Proceeding from the definition (11.7) and the transformation formulae (11.5) for the Yang–Mills field, find the transformtion law for the tensor of the Yang–Mills field, find the transformation law for the tensor of the Yang–Mills field $F_{\mu\nu}$.

N6. Construct a gauge model of the Yang–Mills field of the group $SU(2)$, interacting with a triplet of scalar mesons φ and a doublet of fermions ψ.

N7. Find the static ($\partial\varphi/\partial t = 0$) solutions of the following problem with spontaneous symmetry breaking:

$$\mathscr{L}(\varphi) = \frac{1}{2}\left\{\left(\frac{\partial\varphi}{\partial t}\right)^2 - \left(\frac{\partial\varphi}{\partial z}\right)^2\right\} - \frac{h^2}{4}\left[\varphi^2(z,\ t) - \varphi_0^2\right]^2,$$

satisfying the boundary conditions $\varphi(z, \pm\infty) = \pm\varphi_0$, where $\varphi(x, t)$ is a field function in the two-dimensional space-time $x = (z, t)$, and h^2 and φ_0^2 are certain positive constants. Discuss the nature of the localization of energy and momentum.

N8. By means of a linear substitution diagonalize the Lagrangian

$$\mathscr{L}(\varphi_1, \varphi_2) = \mathscr{L}_0(\varphi_1;\ m_1) + \mathscr{L}_0(\varphi_2;\ m_2) + g\varphi_1(x)\,\varphi_2(x),$$

where

$$\mathscr{L}_0(\varphi;\ m) = \frac{1}{2}\,\partial_\nu\varphi\partial^\nu\varphi - \frac{m^2}{2}\varphi^2.$$

N9. The same as the preceding problem, for the Lagrangian

$$\mathscr{L} = \mathscr{L}_0(\varphi;\ m) + g\rho(x)\,\varphi(x).$$

N10. Find the operator $U(\alpha)$ which realizes the unitary transformation of the Bose operators a, $\overset{+}{a}$, satisfying the relations

$$\left[a,\ \overset{+}{a}\right] = a\overset{+}{a} - \overset{+}{a}a = 1, \quad [a,\ a] = \left[\overset{+}{a},\ \overset{+}{a}\right] = 0,$$

into the new operators

$$A = U^{-1}(\alpha)\,aU(\alpha) = a + \alpha,$$
$$\overset{+}{A} = U^{-1}(\alpha)\,\overset{+}{a}U(\alpha) = \overset{+}{a} + \alpha,$$

where α is a real c-number.

N11. Find the explicit form of the operator U realizing the unitary transformation of the Bose operators a, $\overset{+}{a}$, b, $\overset{+}{b}$ satisfying the relations

$$\left[a,\ \overset{+}{a}\right] = \left[b,\ \overset{+}{b}\right] = 1, \quad [a,\ b] = \left[a,\ \overset{+}{b}\right] = \left[\overset{+}{a},\ b\right] = \left[\overset{+}{a},\ \overset{+}{b}\right] = 0,$$

into the new operators

$$A=U^{-1}(\varphi)\,aU(\varphi)=a\cos\varphi+b\sin\varphi,\quad B=U^{-1}(\varphi)\,bU(\varphi)=b\cos\varphi-a\sin\varphi.$$

N12. Perform the canonical quantization of the Lagrangian of Problem N8. At the same time obtain the commutation relations.

N13. Diagonalize the Hamiltonian

$$H=H_0+H_1,\ H_0=\omega\overset{+}{a}a+\varepsilon\overset{+}{b}b,\qquad H_1=g(\overset{+}{a}b+a\overset{+}{b}),$$

containing the Bose operators a, b which satisfy the relations

$$\left[a,\overset{+}{a}\right]=\left[b,\overset{+}{b}\right]=1,\ \ [a,b]=\left[a,\overset{+}{b}\right]=\left[\overset{+}{a},b\right]=\left[\overset{+}{a},\overset{+}{b}\right]=0.$$

Construct eigenstates and express them in terms of (a) the field operators of the diagonal representation; (b) the eigenstates of the free Hamiltonian H_0.

N14. The same as in the preceding problem, but for the Hamiltonian

$$H=\int\omega(\boldsymbol{k})\overset{+}{a}(\boldsymbol{k})\,a(\boldsymbol{k})\,d\boldsymbol{k}+\int g(\boldsymbol{k}^2)\left(\overset{+}{a}(-\boldsymbol{k})+a(\boldsymbol{k})\right)d\boldsymbol{k};\quad\left[a(\boldsymbol{k}),\overset{+}{a}(\boldsymbol{q})\right]=\delta(\boldsymbol{k}-\boldsymbol{q}).$$

N15. Consider the passage to the local relativistic limit

$$g(\boldsymbol{k}^2)=\frac{g}{[2\omega(\boldsymbol{k})]^{1/2}},\qquad\omega(\boldsymbol{k})=(\boldsymbol{k}^2+m^2)^{1/2}$$

in the expressions of Problem N14 for the shift of the energy of the vacuum and of the norm of the eigenstates.

N16. Prove that within the one-nucleon sector the solution of the problem of the heavy nucleon reduces to the solution of equation (13.9).

N17. Check by direct substitution that the expression

$$\Psi_p^{(1)}(\boldsymbol{k})=\overset{+}{b}(\boldsymbol{k})\Psi_p^{(0)}+g(\boldsymbol{k})\Psi_{p+k}^{(0)}\tag{13.23}$$

satisfies the equation

$$H\Psi_p^{(1)}(\boldsymbol{k})=(M-\Delta M+\omega(\boldsymbol{k}))\,\Psi_p^{(1)}(\boldsymbol{k}),\tag{13.22}$$

where the state $\Psi_p^{(0)}$ is defined by the relation (13.17), while H is the Hamiltonian (13.1)–(13.3) of the heavy-nucleon problem.

N18. Determine the general form for the kernels $\omega(\boldsymbol{k},q)$ and $J(\boldsymbol{k}_1,\boldsymbol{k}_2;\boldsymbol{q}_1,\boldsymbol{q}_2)$ for which the operator

$$H=\int d\boldsymbol{k}\,d\boldsymbol{q}\,\omega(\boldsymbol{k},\boldsymbol{q})\overset{+}{a}(\boldsymbol{k})\,a(\boldsymbol{q})+$$
$$+\int d\boldsymbol{k}_1\,d\boldsymbol{k}_2\,d\boldsymbol{q}_1\,d\boldsymbol{q}_2\,J(\boldsymbol{k}_1,\boldsymbol{k}_2;\boldsymbol{q}_1,\boldsymbol{q}_2)\overset{+}{a}(\boldsymbol{k}_1)\overset{+}{a}(\boldsymbol{k}_2)\,a(\boldsymbol{q}_1)\,a(\boldsymbol{q}_2),$$

may be regarded as the Hamiltonian of a system in which the number of particles and the momentum three-vector are conserved.

N19. Determine the explicit form of the functions ω and J as well as the quantization rules for the operators a, $\overset{+}{a}$ occurring in the Hamiltonian of the preceding problem, for which

(a) the Hamiltonian H describes the total energy of a system of identical nonrelativistic Bose particles whose charge has the same sign and that interact according to the classical Coulomb-interaction potential

$$V(\boldsymbol{x}-\boldsymbol{y})=\frac{1}{4\pi}\frac{e^2}{|\boldsymbol{x}-\boldsymbol{y}|},$$

(b) the Hamiltonian H describes correctly the total energy of a system of identical relativistic fermions (for instance, nucleons) with the Yukawa-interaction potential

$$V(\boldsymbol{x})=\frac{g^2}{4\pi}\frac{\exp(-m|\boldsymbol{x}|)}{|\boldsymbol{x}|}.$$

"DECEMBER" ASSIGNMENT (FOR CHAPTER IV)

D1. Solve for the S-matrix by iterating the Schrödinger equation,

$$i\frac{dS(t)}{dt}=H(t)\,S(t).$$

Consider the nth term. Reduce the solution to the form of a time-ordered exponential.

D2. Consider the following interaction Lagrangians (here $\varphi(x)$ is a scalar field, $\boldsymbol{\varphi}(x)$ is the pseudoscalar isotriplet of π-mesons, $\Psi\{\psi_p, \psi_n\}$ is the isodoublet of nucleons, $e(x)$ is the electron field, $\mu(x)$ is the muon field, and $\nu(x)$ is the neutrino field):

(a) $\mathscr{L}= g\varphi^3(x)$;

(b) $\mathscr{L}= h\varphi^4(x)$;

(c) $\mathscr{L}= h(\boldsymbol{\varphi}(x)\boldsymbol{\varphi}(x))^2$;

(d) the interaction Lagrangian of the electron, nucleon, and pion fields with the electromagnetic field;

(e) $\mathscr{L}= g\bar{\Psi}(x)\boldsymbol{\tau}\gamma^5\Psi(x)\boldsymbol{\varphi}(x)$;

(f) $\mathscr{L}= (f/\mu)\bar{\Psi}(x)\boldsymbol{\tau}\gamma^5\gamma^\nu\Psi(x)\partial_\nu\boldsymbol{\varphi}(x)$;

(g) $\mathscr{L}= \mathscr{L}_{(2c)} + \mathscr{L}_{(2e)}$;

(h) $\mathscr{L}= \mathscr{L}_{(2c)} + \mathscr{L}_{(2f)}$;

(i) $\mathscr{L}= (G/\sqrt{2})j_\alpha(x)$, where $j_\alpha = l_\alpha + h_\alpha$,

and

$$l_\alpha = \bar{\nu}_e(x)\gamma_\alpha(1-\gamma_5)e(x) + \bar{\nu}_\mu(x)\gamma_\alpha(1-\gamma_5)\mu(x),$$
$$h_\alpha = \bar{\psi}_p(x)\gamma_\alpha(1-\gamma_5)\psi_n(x) + f_\pi\partial_\alpha\pi_-(x).$$

For each of these Lagrangians, with the aid of Wick's second theorem, write down in the normal form the time-ordered product $T(\mathscr{L}(x)\,\mathscr{L}(y))$.

D3. Do the same for the self-interaction Lagrangian of the Yang-Mills gauge field:

$$\mathscr{L}_{YM} = -(1/4)F^a_{\mu\nu}F^{a,\,\mu\nu},$$

where

$$F^a_{\mu\nu} = \partial_\mu B^a_\nu - \partial_\nu B^a_\mu + igf^{abc}B^b_\mu B^c_\nu.$$

Here f are structure constants of the corresponding Lie group. *Method*: Isolate from Λ_{YM} the quadratic terms and regard the remaining part as the self-interaction Lagrangian.

D4. Do the same for the time-ordered products $T(\mathscr{L}(x)\,\mathscr{L}(y)\,\mathscr{L}\mathscr{L}(z))$ from Problem 2(a)–(c).

D5. Construct Feynman's rules in the momentum representation for the Lagrangians of Problems 2 and 3.

D6. Making use of combinations of Lagrangians from Problem 2, draw the simplest diagrams and write out the respective contributions to the S-matrix in the momentum representation for the following processes: (a) $e^+e^- \to p\bar{p}$; (b) $e^+e^- \to \pi^+\pi^-$; (c) π^+-scattering on a proton; (d) $\pi^+\pi^- \to \pi^+\pi^-$, $\pi^+\pi^+ \to \pi^+\pi^+$; (e) Yukawa scattering $np \to np$; (f) Compton scattering $\gamma p \to \gamma p$; (g) Möller scattering $ep \to ep$; (h) $\pi^0 \to 2\gamma$; (i) interaction of neutrons with photons; (j) $\pi^\pm \to \mu^\pm\nu$; (k) $\pi^+p \to \pi^+\pi^0 p$; (l) $\mu \to e\nu\bar{\nu}$.

"FEBRUARY" ASSIGNMENT (FOR CHAPTER V)

F1. Derive a formula for the differential cross section of elastic scattering of two spinor particles off each other.

F2. Derive the formulas for the transition probabilities of the decay of one particle into two and into three particles as well as the lifetime of the initial particle.

F3. Utilize the results of Problem D6 to evaluate the differential cross sections for the annihilation of an electron and positron by the following channels:

(a) $e^+e^- \to \mu^+\mu^-$;
(b) $e^+e^- \to 2\gamma$;
(c) $e^+e^- \to \pi^+\pi^-$.

F4. Obtain the values of the lifetimes of the initial particles in the following decay processes:

(a) $\mu^- \to e^- + \bar{\nu}_e + \nu_\mu$;
(b) $\pi^\pm \to \mu^\pm \nu_\mu$.

Hint: Make use of the results of Problem D2(d), (e), (i). Compare the results obtained with experimental data. The latter may, for example, be taken from the book by Perkins (1972).

F5. Obtain the energy distribution for electrons emitted in the β-decay

$$n \to p + e^- + \bar{\nu}_e$$

Hint: Make use of the interaction Lagrangian given in Problem D2(i). Compare the results obtained with the experimental data presented in Chapter IV of the book by Perkins (1975).

F6. Derive the partial widths of the decays

$$\varphi \to \begin{cases} K^+K^-, \\ K^0\tilde{K}^0 \end{cases}$$

as functions of the spin of the φ-meson. The characteristics of the particles are the following:

	Mass (MeV)	Spin	Isospin	Parity
φ	1019	?	0	$-$
$K^\pm$	494	0	$\frac{1}{2}$	$-$
K^0, $\bar{K}^0$	498	0	$\frac{1}{2}$	$-$

Make use of the experimental result

$$\Gamma(\varphi \to K^+K^-)/\Gamma(\varphi \to K^0\tilde{K}^0) \simeq 1.5,$$

and put forward arguments on the probable value of the spin of the φ-meson.

F7. Evaluate the branching ratio for the decay probabilities of π^+-mesons for

$$\pi^\pm \to \mu^\pm + \nu_\mu \quad \text{and} \quad \pi^\pm \to e^\pm + \nu_e,$$

and compare the result with the experimental branching ratio equal to $(1.24 \pm 0.03) \times 10^{-4}$. *Hint*: Use the decay mechanism depicted in the diagram

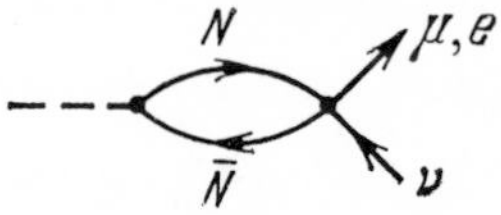

as well as the weak interaction Lagrangian from problem D2.

"MARCH" ASSIGNMENT (FOR CHAPTER VI)

M1. Calculate the difference between the Feynman integrals $I(k) - I(0)$, where

$$I(k) = \frac{i}{\pi^2} \int D^c(q) D^c(k-q)\, dq,$$

corresponding to the second-order self-energy diagram of a meson in the theory with the interaction Lagrangian $\mathscr{L} = (g/3):\varphi^3(x):$. Carry out the calculation in two ways: (a) by passage to the α-representation, (b) by using the Feynman parameters.

M2. By the same two methods evaluate the third-order diagram of the vertex function of a meson in the theory $\mathscr{L} = (g/3):\varphi^3(x):$.

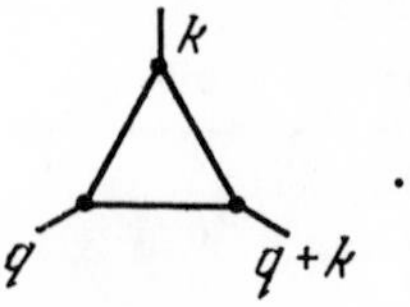

M3. Evaluate the lifetime of the pion due to the decay $\pi^0 \to 2\gamma$ through the virtual $N\bar{N}$state. From the known experimental value of the lifetime determine determine the pion–nucleon interaction coupling constant g. *Hint*: Make use of the results of Problems F2 and D2(d), (e).

M4. Calculate the contribution to the S-matrix of the second-order diagram of the electron self-energy in spinor electrodynamics. Take the photon propagator in the arbitrary gauge

$$D_{\mu\nu}(k)=\frac{1}{k^2}\left\{\left(g_{\mu\nu}-\frac{k_\mu k_\nu}{k^2}\right)+\frac{k_\mu k_\nu}{k^2}\,d_l\right\}.$$

Use three regularization methods:

(a) the Pauli-Villars regularization

$$\frac{1}{m^2-p^2}\to\frac{1}{m^2-p^2}-\frac{1}{M^2-p^2};$$

(b) the cutoff of the momentum of the virtual photon at $p^2=\Lambda^2$;

(c) dimensional regularization.

Investigate the behavior of the obtained expressions when Λ^2, $M^2\to\infty$, $\varepsilon\to 0$. Obtain an analog of the formulas (24.16), (24.17).

Find the value of d_l for which the diagram under consideration does not contain divergences.

M5. Evaluate the Schwinger contribution to the anomalous magnetic moment of the electron corresponding to the vertex diagram of the third order in Figure 24.2b. *Hint*: Make use of the formulas (24.24)–(24.26) and of (30.4).

M6. Evaluate the anomalous magnetic moment of the neutron proceeding from the interaction Lagrangian of Problem D2(d) and (e). *Hint*: Consider vertex diagrams of the form

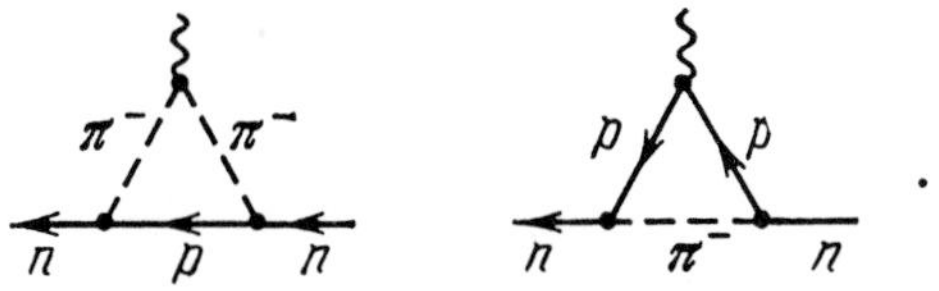

"APRIL" ASSIGNMENT (FOR CHAPTER VII)

A1. Evaluate the contribution to the S-matrix of the meson self-energy second-order diagram in the theory with cubic self-action

$$\mathscr{L}_{\text{int}}(x)=\frac{g}{3}:\varphi^3(x):.$$

Carry out the removal of the divergences.

A2. Find the maximal vertex index for a scalar theory with the exponential self-interaction

$$\mathcal{L}_{\text{int}}(x) = h{:}\varphi^k(x){:}$$

in an N-dimensional space-time. Determine the possible values for pairs of numbers (N, k) for renormalizable models.

A3. Find the $\pi\pi$-scattering amplitudes in the second order of the theory

$$\mathcal{L}_{\text{int}}(x) = -\frac{4\pi^2}{3} h \left(\overset{*}{\varphi}(x)\,\varphi(x)\right)^2.$$

Find the threshold values (at $s = (p_1 + p_2)^2 = 4m_\pi^2$) of the amplitudes and cross sections of elastic $\pi^+\pi^0$ and $\pi^0\pi^0$ scattering processes.

A4. Find the cross section of the process $e^+e^- \to \pi^+\pi^-$ to second order to α and h in the theory with the interaction Lagrangian

$$\mathcal{L}_{\text{int}}(x) = eA^\nu(x)\left\{\bar{\psi}(x)\gamma_\nu\psi(x) + i\overset{*}{\varphi}(x)\,T_3\partial_\nu\varphi(x)\right\} + \frac{4\pi^2}{3} h\left(\overset{*}{\varphi}(x)\varphi(x)\right)^2,$$

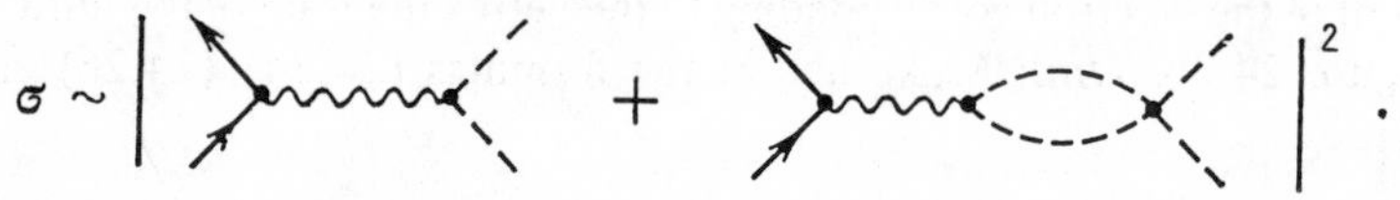

A5. Find the contribution of the vacuum polarization to the correction to the Coulomb law,

$$A^0_{\text{ext}}(q) = \frac{1}{|q|^2},$$

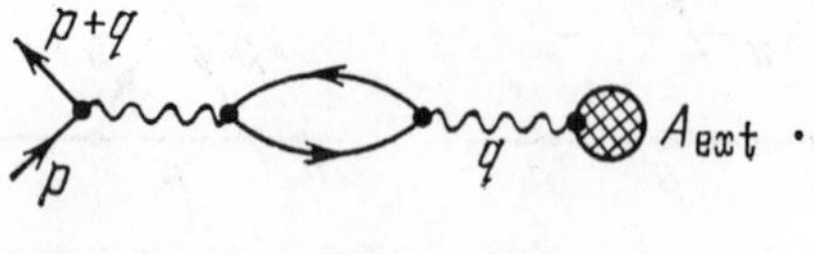

A6. Evaluate the contribution to the S-matrix of the third-order "glass" diagram of the Lagrangian

$$\mathcal{L}_{\text{int}}(x) = -\frac{4\pi^2}{3} h\varphi^4(x),$$

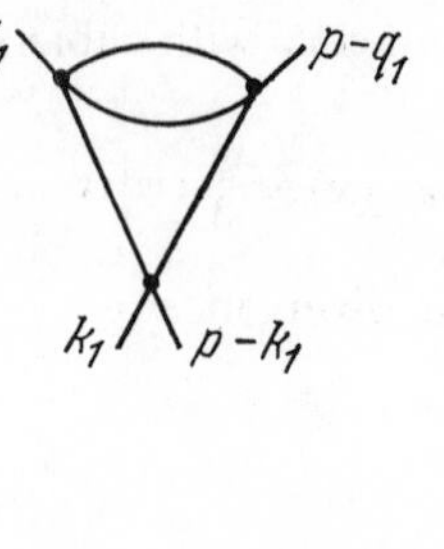

Formulate the R-operation, remove the divergences, and discuss the structure of the finite arbitrariness in the ultimate expression.

RECOMMENDATIONS FOR USE

This book represents a textbook for a year-long course consisting of a little more than thirty lectures and of the same number of seminars. Each section corresponds, approximately, to a single one and a half hour lecture. The exercises, collected in assignments based on specific subjects, are intended for practical work during the seminars. The technical material given in the appendices may also be used for this purpose. It seems reasonable to conclude work on the assignments by holding tutorial sessions twice each semester.

In the table of contents the titles given in italics denote parts which may be omitted in a first reading of the book. The remaining material forms a minimum which is recommended to students insufficiently familiar with the fundamentals of relativistic theory and many-particle quantum theory included in the university course on quantum mechanics, and also in case the university term is incomplete. The said minimum consists of the following:
First semester: Sections 1–5, 6*, 7, 8*, 9*, 10*, 14, 17*, and 18—in all 13 lectures;
Second semester: Sections 19*, 20–22, 23*, 24–26, 27*, 28, 30, 31, and 33—in all 13 lectures. (The asterisks indicate that all the material in the section is meant, except for subsections whose titles appear in italics in the table of contents).

To this minimum one may add, in various combinations (depending on the particular conditions and on the interests of the teacher), the material mentioned in italics. We recommend as a "first reserve" Sections 9.4, 10.3, 12, 23.3 and 29. Of the remaining material we single out the following sets: (a) Sections 8.4, 8.5, 11, 19.4, 32, and Appendix VIII; (b) Sections 15 and 16, which it seems reasonable to include as a whole.

It is desirable for the students who proceed to study this book to have taken, besides the standard course of quantum mechanics which includes the aforementioned two subjects, a short course on the theory of continuous groups and their representations (especially the representations of the Poincaré group), as well as one on the fundamentals of particle physics.

REFERENCES

REFERENCES

Bogoliubov, N.N., and Shirkov, D.V., *Introduction to the Theory of Quantized Fields,* translation from the 1976 third Russian edition, J. Wiley, New York, 1980.

Akhiezer, A.I., and Berestetskii, V.B., *Kvantovaya elektrodinamika* (Quantum Electrodynamics), 3rd edition, Nauka Press, Moscow, 1969.

Blokhintsev, D.I., *Osnovy kvantovoy mekhaniki* (Fundamentals of Quantum Mechanics), 5th edition, Nauka Press, Moscow, 1976.

Bogoliubov, N.N., Lectures on Quantum Statistics, in Vol. 2 of *Selected works* in three volumes (in Russian), Naukova Dumka Press, Kiev, 1970.

Bogoliubov, N.N., *Selected Works in Statistical Physics* (in Russian), Moscow University Press, Moscow, 1979.

Bogoliubov, N.N., Logunov, A.A., and Todorov, I.T., *Osnovy aksiomaticheskogo podkhoda v kvantovoy teorii polya* (Fundamentals of the Axiomatic Approach in Quantum Field Thoery), Nauka Press, Moscow, 1969.

Bjorken, J.D., and Drell, S.D., *Relativistic Quantum Fields,* McGraw-Hill, New York, 1965. Russian translation, Nauka Press, Moscow, 1978.

Davydov, A.S., *Kvantovaya mekhanika* (Quantum Mechanics), 2d edition, Nauka Press, Moscow, 1973.

Dirac, P.A.M., *Lectures on Quantum Mechanics*, Scripta Mathematica, 1964, Russian translation, Mir Press, Moscow, 1968.

Dirac, P.A.M., *Principles of Quantum mechanics*, Oxford University Press, 1958. Russian translation, Nauka Press, Moscow, 1979.

Elyutin, P.V., and Krivchenkov, V.D., *Kvantovaya Mekhanika* (Quantum Mechanics), Nauka Press, Moscow, 1976.

Faddeev, L.D., and Slavnov, A.A. *Gauge Fields. Introduction to Quantum*

Theory, Benjamin/Cummings, Reading, Massachusetts, 1980, translation from the Russian edition, Nauka Press, Moscow, 1978.

Feynman, R.P. *The Theory of Fundamental Processes,* Benjamin, New York, 1961. Russian translation, Nauka Press, Moscow, 1978.

Kirzhnitz, D.A., *Polyeviye metody teorii mnogikh chastits* (Field Methods in the Theory of Many Particles), Atomizdat Press, Moscow, 1963. English translation: Kirzhnits, D.A., *Field Theoretical Methods in Many-Body Systems,* Pergamon, Oxford, 1967.

Klauder, J.R., and Sudarshan, E.C.G., *Fundamentals of Quantum Optics,* W.A. Benjamin, New York, Amsterdam, 1968. Russian translation, Mir Press, Moscow, 1970.

Landau, L.D., and Lifshitz, E.M., *Kvantovaya mekhanika* (Quantum Mechanics), Nauka Press, Moscow, 1974.

Lipkin, H.J., *Quantum Mechanics. New Approaches to Selected Topics,* Elsevier North-Holland, Amsterdam, London, New York, 1973. Russian translation, Mir Press, Moscow, 1977.

Medvedev, B.V., *Nachala teoreticheskoi fiziki* (Fundamentals of Theoretical Physics), Nauka Press, Moscow, 1977.

Messiah, A., *Quantum Mechanics*, North-Holland, Amsterdam, 1962. Russian translation, Nauka Press, Moscow, 1978.

Perkins, D. H., *Introduction to High Energy Physics*, Addison-Wesley, 1972. Russian translation, Mir Press, Moscow, 1975.

Vladimirov, V.S. *Uravneniya matematicheskoi fiziki* (Equations of Mathematical Physics), 3rd edition, Nauka Press, Moscow, 1976.

Wigner, E.P., *Symmetries and Reflections*, Indiana University Press, 1967, Russian translation, Mir Press, Moscow, 1971.

Zav'yalov, O.I., *Perenormirovannye diagrammy Feynmana* (Renormalized Feynman Diagrams), Nauka Press, Moscow, 1979.

INDEX

INDEX